Applications of q-Calculus in Operator Theory

Ali Aral • Vijay Gupta • Ravi P. Agarwal

Applications of q-Calculus in Operator Theory

 Springer

Ali Aral
Department of Mathematics
Kırıkkale University
Yahşihan, Kirikkale, Turkey

Vijay Gupta
School of Applied Sciences
Netaji Subhas Institute of Technology
New Delhi, India

Ravi P. Agarwal
Department of Mathematics
Texas A&M University-Kingsville
Kingsville, Texas, USA

ISBN 978-1-4899-9625-1   ISBN 978-1-4614-6946-9 (eBook)
DOI 10.1007/978-1-4614-6946-9
Springer New York Heidelberg Dordrecht London

Mathematics Subject Classification (2010): 41A36-41A25-41A17-30E10

Printed on acid-free paper

Springer is part of Springer Science+Business Media (www.springer.com)

Preface

Simply, quantum calculus is ordinary classical calculus without the notion of limits. It defines *q-calculus* and *h-calculus*. Here *h* ostensibly stands for Planck's constant, while *q* stands for quantum. A pioneer of *q*-calculus in approximation theory is the former Professor Alexandru Lupas [117], who first introduced the *q*-analogue of Bernstein polynomials. Ten years later Phillips [133] introduced another generalization on Bernstein polynomials [113] based on *q*-integers. Ostrovska [125, 127] studied *q*-Bernstein polynomials. After that several researchers have estimated the approximation properties of several operators. This book is an attempt to compile and present some papers on *q*-calculus in approximation theory.

We divide the book into seven chapters. In Chap. 1, we mention some notations and basic definitions of *q*-calculus, which will be used throughout the book. We also present the generating functions of some of the important *q*-basis functions. In Chap. 2, we present some discrete *q*-operators, which include the *q*-Bernstein polynomials, *q*-Baskakov operators, *q*-Szász operators, *q*-Blemian–Butzer–Hahn operators, and *q*-Meyer–König and Zeller operators. We present the approximation properties of such operators.

In Chap. 3, we present the *q*-analogue of integral operators which include *q*-Picard and *q*-Weierstrass-type singular integral operators and study their rate of convergence and weight approximation. We also discuss error estimation and global smoothness preservation property of such operators. In the last section of this chapter, we study generalized Picard operators and pointwise convergence, order of pointwise convergence, and norm convergence of the generalized operators. In the last section, we study the *q*-Meyer–König–Zeller–Durrmeyer operators and estimate the moments and some direct results.

In Chap. 4, we study the integral modifications of Bernstein operators using the *q*-beta functions of the first kind. We present the approximation properties of the *q*-Bernstein–Kantorovich operators, *q*-Bernstein–Durrmeyer polynomials, discretely defined *q*-Durrmeyer-type operators, and genuine *q*-Bernstein–Durrmeyer operators. We mention the moment estimation, direct results, and the limiting convergence of such operators. We have also included a section on fuzzy approximation and applications.

v

In Chap. 5, we discuss some other recently introduced q-integral operators on the positive real axis. To tackle such operators, we generally use q-beta functions of the second kind. This chapter includes q-Baskakov–Durrmeyer operators, q-Szász-beta operators, q-Szász–Durrmeyer operators, and q-Phillips operators. We present moments, recurrence relations for moments, asymptotic formula, and weighted approximations for such operators.

In Chap. 6, we study the statistical convergence of the q-operators. We mention results for a general class of positive linear operators and present statistical approximation properties in weighted space. We also present the results for q-Szász–King-type operators and q-Baskakov–Kantorovich operators and the study rate of convergence.

In the last chapter, we present the quantitative Voronovskaja-type estimate for certain q-Durrmeyer polynomials. In this way, we put in evidence the overconvergence phenomenon for these q-Durrmeyer polynomials, namely, the extensions of approximation properties (with quantitative estimates) from the real interval $[0, 1]$ to compact disks in the complex plane. Also, we study the complex q-Gauss–Weierstrass integral operators. We show that these operators are an approximation process in some subclasses of analytic functions giving Jackson-type estimates in approximation. Furthermore, we give q-calculus analogues of some shape-preserving properties for these operators satisfied by the classical complex Gauss–Weierstrass integral operators.

Kirikkale, Turkey   Ali Aral
New Delhi, India   Vijay Gupta
Kingsville, TX   Ravi P. Agarwal

Contents

Introduction

Nowadays there is a significant increase of activities in the area of q-calculus due to its applications in various fields such as mathematics, mechanics, and physics. In 1910, Jackson [103] defined and studied the q-integral. He was the first to develop q-integral in a systematic way. Later the integral representations of q-gamma and q-beta functions were proposed by De Sole and Kac [49].

The applications of q-calculus in the area of approximation theory were initiated by Lupas [117], who first introduced q-Bernstein polynomials. Also in the last decade, Phillips [133] proposed other q-Bernstein polynomials, which became popular. Later several researchers obtained the interesting properties of q-Bernstein polynomials and their Durrmeyer variants; we mention some of the papers in this direction as [45, 86, 94, 129, 130]. The q-Bernstein–Durrmeyer-type operators are based on q-beta function of the first kind. The approximation of vector-valued functions by q-Durrmeyer operators with applications to random and fuzzy approximation was discussed by Gal and Gupta [77]. Another important operator, namely, q-Bleimann, Butzer, and Hahn operators, was discussed in [27] and also in [10, 60, 120]. The q-Baskakov operators in two different setups were proposed and studied in [30, 32]; also some direct results in terms of Ditzian–Totik modulus of continuity were discussed in [63]. Recently we [31] proposed the q-analogue of Baskakov–Durrmeyer operators, which is based on q-improper integral, namely, q-analogue of beta function of the second kind. Another important q-generalization of the well-known Szász–Mirakyan operators was proposed by Aral [25] and studied in detail by Aral and Gupta [29]. The authors have also proposed the q-analogue of Szász–Mirakyan–Durrmeyer operators in [33]. Several mixed q-analogues of hybrid summation-integral-type operators were proposed, some of them are discussed in this book.

Other q-analogues of integral operators are the q-Picard and q-Weierstrass-type singular integral operators. It can be observed that the q-Picard and the q-Gauss–Weierstrass singular integral operators give better approximation results than the classical ones. Trif [150] studied some approximation properties of the operators $\widehat{M}_{n,q}f(x)$. Also, Dogru and Gupta [55] proposed some other bivariate q-Meyer–König and Zeller operators having different test functions and established some approximation properties. Govil and Gupta [84] considered q-Meyer–König–Zeller–Durrmeyer operators which are discussed here.

In the recent years, many researchers have studied the statistical convergence for linear positive operators. Here, we present statistical convergence results for a general class of operators. We also discuss results for q-Szasz–King-type operators and q-Baskakov–Kantorovich operators and study rate of convergence.

For the quantitative Voronovskaja-type estimates for certain complex operators, we put in evidence the overconvergence phenomenon, namely, the extensions of approximation properties (with quantitative estimates) from the real interval to compact disks in the complex plane. We study complex operators of q-Durrmeyer-type and complex q-Gauss–Weierstrass integral operators. We show that these operators satisfy approximation process in some subclasses of analytic functions giving Jackson-type estimates in approximation. Furthermore, we give q-calculus analogues of some shape-preserving properties for these operators satisfied by the classical complex Gauss–Weierstrass integral operators.

Chapter 1
Introduction of q-Calculus

In the field of approximation theory, the applications of q-calculus are new area in last 25 years. The first q-analogue of the well-known Bernstein polynomials was introduced by Lupas in the year 1987. In 1997 Phillips considered another q-analogue of the classical Bernstein polynomials. Later several other researchers have proposed the q-extension of the well-known exponential-type operators which includes Baskakov operators, Szász–Mirakyan operators, Meyer–König–Zeller operators, Bleiman, Butzer and Hahn operators (abbreviated as BBH), Picard operators, and Weierstrass operators. Also, the q-analogue of some standard integral operators of Kantorovich and Durrmeyer type was introduced, and their approximation properties were discussed. This chapter is introductory in nature; here we mention some important definitions and notations of q-calculus. We give outlines of q-integers, q-factorials, q-binomial coefficients, q-differentiations, q-integrals, q-beta and q-gamma functions. We also mention some important q-basis functions and their generating functions.

1.1 Notations and Definitions in q-Calculus

In this section we mention some basic definitions of q-calculus, which would be used throughout the book.

Definition 1.1. Given value of $q > 0$, we define the q-integer $[n]_q$ by

$$[n]_q = \begin{cases} \frac{1-q^n}{1-q}, & q \neq 1 \\ n, & q = 1 \end{cases},$$

for $n \in \mathbb{N}$.

We can give this definition for any real number λ. In this case we call $[\lambda]_q$ a q-real.

A. Aral et al., *Applications of q-Calculus in Operator Theory*,
DOI 10.1007/978-1-4614-6946-9_1, © Springer Science+Business Media New York 2013

Definition 1.2. Given the value of $q > 0$, we define the q-factorial $[n]_q!$ by

$$[n]_q! = \begin{cases} [n]_q [n-1]_q \cdots [1]_q , & n = 1, 2, \ldots \\ 1 & n = 0. \end{cases},$$

for $n \in \mathbb{N}$.

Definition 1.3. We define the q-binomial coefficients by

$$\begin{bmatrix} n \\ k \end{bmatrix}_q = \frac{[n]_q!}{[k]_q! \, [n-k]_q!}, \quad 0 \leq k \leq n, \tag{1.1}$$

for $n, k \in \mathbb{N}$.

The q-binomial coefficient satisfies the recurrence equations

$$\begin{bmatrix} n \\ k \end{bmatrix}_q = \begin{bmatrix} n-1 \\ k-1 \end{bmatrix}_q + q^k \begin{bmatrix} n-1 \\ k \end{bmatrix}_q \tag{1.2}$$

and

$$\begin{bmatrix} n \\ k \end{bmatrix}_q = q^{n-k} \begin{bmatrix} n-1 \\ k-1 \end{bmatrix}_q + \begin{bmatrix} n-1 \\ k \end{bmatrix}_q . \tag{1.3}$$

Definition 1.4. The q-analogue of $(1+x)_q^n$ is the polynomial

$$(1+x)_q^n := \begin{cases} (1+x)(1+qx) \ldots (1+q^{n-1}x) & n = 1, 2, \ldots \\ 1 & n = 0. \end{cases}$$

A q-analogue of the common Pochhammer symbol also called a q-shifted factorial is defined as

$$(x;q)_0 = 1, (x;q)_n = \prod_{i=0}^{n-1}(1 - q^i x), (x;q)_\infty = \prod_{i=0}^{\infty}(1 - q^i x).$$

Definition 1.5. The Gauss binomial formula:

$$(x+a)_q^n = \sum_{j=0}^{n} \begin{bmatrix} n \\ j \end{bmatrix}_q q^{j(j-1)/2} a^j x^{n-j}.$$

Definition 1.6. The Heine's binomial formula:

$$\frac{1}{(1-x)_q^n} = 1 + \sum_{j=1}^{\infty} \frac{[n]_q [n+1]_q \ldots [n+j-1]_q}{[j]_q!} x^j.$$

Also, we have the following important property:

$$x^n = \sum_{j=0}^{n} \begin{bmatrix} n \\ j \end{bmatrix}_q (x-1)_q^j.$$

1.2   q-Derivative

Definition 1.7. The q-derivative $D_q f$ of a function f is given by

$$(D_q f)(x) = \frac{f(x) - f(qx)}{(1-q)x}, \quad \text{if } x \neq 0, \tag{1.4}$$

and $(D_q f)(0) = f'(0)$ provided $f'(0)$ exists.

Note that

$$\lim_{q \to 1} D_q f(x) = \lim_{q \to 1} \frac{f(qx) - f(x)}{(q-1)x} = \frac{df(x)}{dx}$$

if f is differentiable.

It is obvious that the q-derivative of a function is a linear operator. That is, for any constants a and b, we have

$$D_q\{af(x) + bg(x)\} = aD_q\{f(x)\} + bD_q\{g(x)\}.$$

Now we calculate the q-derivative of a product at $x \neq 0$, using Definition 1.7, as

$$\begin{aligned} D_q\{f(x)g(x)\} &= \frac{f(qx)g(qx) - f(x)g(x)}{(q-1)x} \\ &= \frac{f(qx)g(qx) - f(qx)g(x) + f(qx)g(x) - f(x)g(x)}{(q-1)x} \\ &= \frac{f(qx)(g(qx) - g(x))}{(q-1)x} + \frac{(f(qx) - f(x))g(x)}{(q-1)x} \\ &= f(qx)D_q g(x) + D_q f(x)g(x). \end{aligned}$$

We interchange f and g and obtain

$$D_q\{f(x)g(x)\} = f(x)D_q g(x) + D_q f(x)g(qx). \tag{1.5}$$

The Leibniz rule for the q-derivative operator is defined as

$$D_q^{(n)}(fg)(x) = \sum_{k=0}^{n} \begin{bmatrix} n \\ k \end{bmatrix}_q D_q^{(k)} f(xq^{n-k}) D_q^{(n-k)} g(x).$$

If we apply Definition 1.7 to the quotient $f(x)$ and $g(x)$, we obtain

$$
\begin{aligned}
D_q \left\{ \frac{f(x)}{g(x)} \right\} &= \frac{1}{(q-1)x} \left\{ \frac{f(qx)}{g(qx)} - \frac{f(x)}{g(qx)} + \frac{f(x)}{g(qx)} - \frac{f(x)}{g(x)} \right\} \\
&= \frac{1}{g(qx)} \left\{ \frac{f(qx) - f(x)}{(q-1)x} \right\} + \frac{1}{(q-1)x} \left\{ \frac{f(x)g(x) - f(x)g(qx)}{g(qx)g(x)} \right\} \\
&= \frac{1}{g(qx)} D_q f(x) + \frac{f(x)}{g(qx)g(x)} \left\{ \frac{g(x) - g(qx)}{(q-1)x} \right\} \\
&= \frac{g(x) D_q f(x) - f(x) D_q g(x)}{g(qx)g(x)}.
\end{aligned}
\tag{1.6}
$$

The above formula can also be written as

$$
D_q \left\{ \frac{f(x)}{g(x)} \right\} = \frac{g(qx) D_q f(x) - f(qx) D_q g(x)}{g(qx)g(x)}.
$$

Note that there does not exist a general chain rule for q-derivative. We can give a chain rule for function of the form $f(u(x))$, where $u = u(x) = \alpha x^\beta$ with α, β being constant. For this chain rule, we can write

$$
\begin{aligned}
D_q \{f(u(x))\} &= D_q \left\{ f\left(\alpha x^\beta\right) \right\} \\
&= \frac{f(\alpha q^\beta x^\beta) - f(\alpha x^\beta)}{(q-1)x} \\
&= \frac{f(\alpha q^\beta x^\beta) - f(\alpha x^\beta)}{\alpha q^\beta x^\beta - \alpha x^\beta} \cdot \frac{\alpha q^\beta x^\beta - \alpha x^\beta}{(q-1)x} \\
&= \frac{f(q^\beta u) - f(u)}{q^\beta u - u} \cdot \frac{u(qx) - u(x)}{(q-1)x}
\end{aligned}
$$

and, hence,

$$
D_q \{f(u(x))\} = \left(D_{q^\beta} f \right)(u(x)) D_q(u(x)).
$$

Proposition 1.1. *For $n \geq 1$,*

$$
D_q (1+x)_q^n = [n]_q (1+qx)_q^{n-1}
$$

$$
D_q \left\{ \frac{1}{(1+x)_q^n} \right\} = -\frac{[n]_q}{(1+x)_q^{n+1}}.
$$

Proof. According to the definition of *q*-derivative we have

$$D_q(1+x)_q^n = \frac{(1+qx)_q^n - (1+x)_q^n}{(q-1)x}$$

$$= (1+qx)_q^{n-1}\frac{\{(1+q^n x - (1+x)\}}{(q-1)x}$$

$$= [n]_q(1+qx)_q^{n-1}.$$

According to (1.6), we have

$$D_q\left\{\frac{1}{(1+x)_q^n}\right\} = -\frac{D_q(1+x)_q^n}{(1+qx)_q^n(1+x)_q^n}$$

$$= -\frac{[n]_q}{(1+q^n x)(1+x)_q^n}$$

$$= -\frac{[n]_q}{(1+x)_q^{n+1}}.$$ ∎

Remark 1.1. Suppose $n \geq 1$ and $a,b,r,s \in \mathfrak{R}$, then by simple computation, we immediately have the following:

$$D_q(a+bx)_q^n = [n]_q b(a+bqx)_q^{n-1},$$

$$D_q(ax+b)_q^n = [n]_q a(ax+b)_q^{n-1},$$

and

$$D_q\frac{(1+ax)_q^r}{(1+bx)_q^s} = [r]_q a\frac{(1+aqx)_q^{r-1}}{(1+bqx)_q^s} - b[s]_q\frac{(1+ax)_q^r}{(1+bx)_q^{s+1}}.$$

1.3   *q*-Series Expansions

Theorem 1.1. *For* $|x| < 1$, $|q| < 1$,

$$\sum_{k=0}^{\infty}\frac{(1-a)_q^k}{(1-q)_q^k}x^k = \frac{(1-ax)_q^{\infty}}{(1-x)_q^{\infty}},$$

where $(1-x)_q^{\infty} = \prod_{k=0}^{\infty}(1-q^k x)$.

Proof. Let

$$f_a(x) = \sum_{k=0}^{\infty} \frac{(1-a)_q^k}{(1-q)_q^k} x^k.$$

Clearly

$$\frac{f_a(x) - f_a(qx)}{x} = \sum_{k=0}^{\infty} \frac{(1-a)_q^k}{(1-q)_q^k} \left(1-q^k\right) x^{k-1}$$

$$= (1-a) \sum_{k=1}^{\infty} \frac{(1-aq)_q^{k-1}}{(1-q)_q^{k-1}} x^{k-1}$$

$$= (1-a) \sum_{k=0}^{\infty} \frac{(1-aq)_q^k}{(1-q)_q^k} x^k = (1-a) f_a(qx)$$

or

$$f_a(x) - f_a(qx) = (1-a) x f_a(qx).$$

Also

$$f_a(x) - f_a(qx) = \sum_{k=0}^{\infty} \frac{(1-qa)_q^{k-1}}{(1-q)_q^k} \left(1-a-1+aq^k\right) x^k$$

$$= -ax f_{aq}(x)$$

or

$$f_a(x) = (1-ax) f_{aq}(x).$$

Combining the above two equations, we get

$$f_a(x) = \frac{1-ax}{1-x} f_a(qx).$$

Iterating this relation n times and letting $n \to \infty$ we have

$$f_a(x) = \frac{(1-ax)_q^n}{(1-x)_q^n} f_a(q^n x) = \frac{(1-ax)_q^{\infty}}{(1-x)_q^{\infty}}.$$

Thus we have the desired result.   ∎

Corollary 1.1. *(a) Taking $a=0$ in Theorem 1.1, we have*

$$\sum_{k=0}^{\infty} \frac{x^k}{(1-q)_q^k} = \frac{1}{(1-x)_q^{\infty}}, \qquad |x| < 1, |q| < 1.$$

(b) *Replacing a with $\frac{1}{a}$, and x with ax, and then taking $a = 0$ in Theorem 1.1, we have*

$$\sum_{k=0}^{\infty} \frac{(-1)^n q^{\frac{k(k-1)}{2}} x^k}{(1-q)_q^k} = (1-x)_q^\infty, \qquad |q| < 1.$$

(c) *Taking $a = q^N$ in Theorem 1.1, we have*

$$\sum_{k=0}^{\infty} \begin{bmatrix} N-k-1 \\ k \end{bmatrix}_q x^k = \frac{1}{(1-x)_q^N}, \qquad |x| < 1.$$

We consider Corollary 1.1(a). We can write

$$\sum_{k=0}^{\infty} \frac{x^k}{(1-q)_q^k} = \sum_{k=0}^{\infty} \frac{\left(\frac{x}{1-q}\right)^k}{\left(\frac{1-q^2}{1-q}\right)\left(\frac{1-q^3}{1-q}\right)\cdots\left(\frac{1-q^k}{1-q}\right)}$$

$$= \sum_{k=0}^{\infty} \frac{\left(\frac{x}{1-q}\right)^k}{[k]_q!}$$

which resembles Taylor's expansion of classical exponential function e^x.

Definition 1.8. A *q*-analogue of classical exponential function e^x is

$$e_q(x) = \sum_{k=0}^{\infty} \frac{x^k}{[k]_q!}.$$

Using Corollary 1.1, (*a*) and (*b*), we see that

$$e_q\left(\frac{x}{1-q}\right) = \frac{1}{(1-x)_q^\infty}$$

and

$$e_q(x) = \frac{1}{(1-(1-q)x)_q^\infty}. \tag{1.7}$$

Definition 1.9. Another *q*-analogue of classical exponential function is

$$E_q(x) = \sum_{k=0}^{\infty} q^{\frac{k(k-1)}{2}} \frac{x^k}{[k]_q!} = (1+(1-q)x)_q^\infty. \tag{1.8}$$

The *q*-exponential functions satisfy following properties:

Lemma 1.1. (*a*) $D_q e_q(x) = e_q(x), \qquad D_q E_q(x) = E_q(qx)$.
(*b*) $e_q(x) E_q(-x) = E_q(x) e_q(-x) = 1$.

Note that for $q \in (0,1)$ the series expansion of $e_q(x)$ has radius of convergence $\frac{1}{1-q}$. On the contrary, the series expansion of $E_q(x)$ converges for every real x.

1.4   Generating Functions

In this section we present the generating functions for some of the important q-basis functions, namely, q-Bernstein basis function, q-MKZ basis function, and q-beta basis functions (see [95]).

We can consider the q-exponential function in the following form:

$$\lim_{n\to\infty} \frac{1}{(1-x)_q^n} = \lim_{n\to\infty} \sum_{k=0}^{\infty} \begin{bmatrix} n+k-1 \\ k \end{bmatrix}_q x^k$$

$$= \lim_{n\to\infty} \sum_{k=0}^{\infty} \frac{(1-q^{n+k-1})\ldots(1-q^n)}{(1-q)(1-q^2)\ldots(1-q^k)} x^k$$

$$= \sum_{k=0}^{\infty} \frac{x^k}{(1-q)(1-q^2)\ldots(1-q^k)} = e_q(x).$$

Another form of q-exponential function is given as follows:

$$\lim_{n\to\infty} (1+x)_q^n = \sum_{k=0}^{\infty} \frac{q^{k(k-1)/2} x^k}{(1-q)(1-q^2)\ldots(1-q^k)} = E_q(x).$$

Based on the q-integers Phillips [132] introduced the q-analogue of the well-known Bernstein polynomials. For $f \in C[0,1]$ and $0 < q < 1$, the q-Bernstein polynomials are defined as

$$\mathcal{B}_{n,q}(f,x) = \sum_{k=0}^{n} b_{k,n}^q(x) f\left(\frac{[k]_q}{[n]_q}\right), \tag{1.9}$$

where the q-Bernstein basis function is given by

$$b_{k,n}^q(x) = \begin{bmatrix} n \\ k \end{bmatrix}_q x^k (1-x)_q^{n-k}, x \in [0,1]$$

and $(a-b)_q^n = \prod_{s=0}^{n-1}(a-q^s b), \quad a,b \in \mathbf{R}$.

Also Trif [150] proposed the q-analogue of well-known Meyer–König–Zeller operators. For $f \in C[0,1]$ and $0 < q < 1$, the q-Meyer–König–Zeller operators are defined as

$$\mathcal{M}_{n,q}(f,x) = \sum_{k=0}^{\infty} m_{k,n}^q(x) f\left(\frac{[k]_q}{[n]_q}\right), \tag{1.10}$$

where the q-MKZ basis function is given by

$$m_{k,n}^q(x) = \begin{bmatrix} n+k+1 \\ k \end{bmatrix}_q x^k (1-x)_q^n, x \in [0,1].$$

For $f \in C[0,\infty)$ and $0 < q < 1$, the q-beta operators are defined as

$$V_n(f,x) = \sum_{k=0}^{\infty} v_{k,n}^q(x) f\left(\frac{[k]_q}{q^{k-1}[n]_q}\right), \tag{1.11}$$

where the q-beta basis function is given by

$$v_{k,n}^q(x) = \frac{q^{k(k-1)/2}}{B_q(k+1,n)} \frac{x^k}{(1+x)_q^{n+k+1}}, x \in [0,\infty)$$

and $B_q(m,n)$ is q-beta function.

Now we give the generating functions for q-Bernstein, q-Meyer–König–Zeller, and q-beta basis functions.

1.4.1 Generating Function for q-Bernstein Basis

Theorem 1.2. $b_{k,n}^q(x)$ *is the coefficient of* $\frac{t^n}{[n]_q!}$ *in the expansion of*

$$\frac{x^k t^k}{[k]_q!} e_q((1-q)(1-x)_q t).$$

Proof. First consider

$$\frac{x^k t^k}{[k]_q!} e_q((1-q)(1-x)_q t) = \frac{x^k t^k}{[k]_q!} \sum_{n=0}^{\infty} \frac{(1-x)_q^n t^n}{[n]_q!}$$

$$= \frac{1}{[k]_q!} \sum_{n=0}^{\infty} \frac{x^k (1-x)_q^n t^{n+k}}{[n]_q!}$$

$$= \sum_{n=0}^{\infty} \frac{[n+1]_q [n+2]_q \cdots\cdots [n+k]_q x^k (1-x)_q^n t^{n+k}}{[n+k]_q! [k]_q!}$$

$$= \sum_{n=0}^{\infty} \begin{bmatrix} n+k \\ k \end{bmatrix}_q \frac{x^k (1-x)_q^n t^{n+k}}{[n+k]_q!}$$

$$= \sum_{n=k}^{\infty} \begin{bmatrix} n \\ k \end{bmatrix}_q \frac{x^k (1-x)_q^{n-k} t^n}{[n]_q!} = \sum_{n=0}^{\infty} b_{k,n}^q(x) \frac{t^n}{[n]_q!}.$$

This completes the proof of generating function for $b_{k,n}^q(x)$. ∎

1.4.2   Generating Function for *q*-MKZ

Theorem 1.3. $m_{k,n}^q(x)$ is the coefficient of t^k in the expansion of $\frac{(1-x)_q^n}{(1-xt)_q^{n+2}}$.

Proof. It is easily seen that

$$\frac{(1-x)_q^n}{(1-xt)_q^{n+2}} = \sum_{k=0}^{\infty} \begin{bmatrix} n+k+1 \\ k \end{bmatrix}_q (1-x)_q^n x^k t^k = \sum_{k=0}^{\infty} m_{k,n}^q(x) t^k.$$

This completes the proof.   ∎

1.4.3   Generating Function for *q*-Beta Basis

Theorem 1.4. *It is observed by us that* $v_{k,n}^q(x)$ *is the coefficient of* $\frac{t^k}{[n+k]_q!}$ *in the expansion of* $\frac{1}{(1+x)_q^{n+1}} E_q\left(\frac{(1-q)xt}{(1+q^{n+1}x)_q} \right)$.

Proof. First using the definition of *q*-exponential $E_q(x)$, we have

$$\frac{1}{(1+x)_q^{n+1}} E_q\left(\frac{(1-q)xt}{(1+q^{n+1}x)_q} \right) = \frac{1}{(1+x)_q^{n+1}} \sum_{k=0}^{\infty} q^{k(k-1)/2} \frac{x^k}{(1+q^{n+1}x)_q^k} \frac{t^k}{[k]_q!}$$

$$= \sum_{k=0}^{\infty} q^{k(k-1)/2} \frac{x^k}{(1+x)_q^{n+k+1}} \frac{t^k}{[k]_q!}$$

$$= \sum_{k=0}^{\infty} q^{k(k-1)/2} \frac{x^k t^k}{(1+x)_q^{n+k+1}} \frac{[k+1]_q [k+2]_q \dots [n+k]_q}{[n+k]_q!}$$

$$= \sum_{k=0}^{\infty} q^{k(k-1)/2} \frac{x^k}{(1+x)_q^{n+k+1}} \begin{bmatrix} n+k \\ n \end{bmatrix}_q \frac{[n]_q t^k}{[n+k]_q!}$$

$$= \sum_{k=0}^{\infty} \frac{1}{B_q(k+1,n)} q^{k(k-1)/2} \frac{x^k}{(1+x)_q^{n+k+1}} \frac{t^k}{[n+k]_q!}$$

$$= \sum_{k=0}^{\infty} v_{k,n}^q(x) \frac{t^k}{[n+k]_q!}.$$

This completes the proof of generating function.   ∎

1.5 q-Integral

The Jackson definite integral of the function f is defined by (see [103], [149]):

$$\int_0^a f(x)\,d_qx = (1-q)a\sum_{n=0}^{\infty} f(aq^n)q^n,\ a\in\mathbb{R}. \tag{1.12}$$

Notice that the series on the right-hand side is guaranteed to be convergent as soon as the function f is such that for some $C>0$, $\alpha>-1$, $|f(x)|<Cx^\alpha$ in a right neighborhood of $x=0$.

One defines the Jackson integral in a generic interval $[a,b]$:

$$\int_a^b f(x)\,d_qx = \int_0^b f(x)\,d_qx - \int_0^a f(x)\,d_qx.$$

Now we give the fundamental theorem of quantum calculus.

Theorem 1.5. *(a) If F is any anti q-derivative of the function f, namely, $D_qF=f$, continuous at $x=0$, then*

$$\int_0^a f(x)\,d_qx = F(a)-F(0).$$

(b) For any function f one has

$$D_q\int_0^x f(t)\,d_qt = f(x).$$

Remark 1.2. (a) The q-analogue of the rule of integration by parts is

$$\int_0^a g(x)D_qf(x)\,d_qx = f(x)g(x)\big|_a^b - \int_0^a f(qx)D_qg(x)\,d_qx.$$

(b) If $u(x)=\alpha x^\beta$, change of variable formula is

$$\int_{u(a)}^{u(b)} f(u)\,d_qu = \int_a^b f(u(x))D_{q^{1/\beta}}u(x)\,d_{q^{1/\beta}}x.$$

Definition 1.10. For $m,n>0$ the q-beta function [104] is defined as

$$B_q(m,n) = \int_0^1 t^{m-1}(1-qt)_q^{n-1}\,d_qt.$$

It can be easily seen that for $m>1, n>0$ after integrating by parts:

$$B_q(m,n) = \frac{[m-1]_q}{[n]_q}B_q(m-1,n+1).$$

Also from Definition 1.10, we have

$$B_q(m, n+1) = \int_0^1 t^{m-1}(1-qt)_q^{n-1}(1-q^n t)d_q t$$

$$= \int_0^1 t^{m-1}(1-qt)_q^{n-1}d_q t - q^n \int_0^1 t^m(1-qt)_q^{n-1}d_q t$$

$$= B_q(m,n) - q^n B_q(m+1,n).$$

The improper integral of function f is defined by [49, 107]:

$$\int_0^{\infty/A} f(x)d_q x = (1-q)\sum_{n=-\infty}^{\infty} f\left(\frac{q^n}{A}\right)\frac{q^n}{A}, \quad A \in \mathbb{R}. \tag{1.13}$$

Remark 1.3. If the function f satisfies the conditions $|f(x)| < Cx^\alpha, \forall x \in [0, \varepsilon)$, for some $C > 0, \alpha > -1, \varepsilon > 0$ and $|f(x)| < Dx^\beta, \forall x \in [N, \infty)$, for some $D > 0, \beta < -1$, $N > 0$, then the series on the right hand side is convergent. In general even though when these conditions are satisfied, the value of sum in the right side of (1.13) will be dependent on the constant A. In order to get the integral independent of A, in the anti q-derivative, we have to take the limits as $x \to 0$ and $x \to 1$, respectively.

Definition 1.11. The q-gamma function defined by

$$\Gamma_q(t) = \int_0^{1/1-q} x^{t-1}E_q(-qx)d_q x, \quad t > 0 \tag{1.14}$$

satisfies the following functional equation:

$$\Gamma_q(t+1) = [t]_q \Gamma_q(t),$$

where $[t]_q = \frac{1-q^t}{1-q}$ and $\Gamma_q(1) = 1$.

Remark 1.4. Note that the q-gamma integral given by (1.14) can be rewritten via an improper integral by using definition (1.13). From (1.8) we can easily see that $E_q\left(-\frac{q^n}{1-q}\right) = 0$ for $n \leq 0$. Thus, we can write

$$\Gamma_q(t) = \int_0^{\infty/1-q} x^{t-1}E_q(-qx)d_q x, \quad t > 0.$$

Definition 1.12. The q-beta function is defined as

$$B_q(t,s) = K(A,t)\int_0^{\infty/A} \frac{x^{t-1}}{(1+x)_q^{t+s}}d_q x, \tag{1.15}$$

and the q-gamma function is defined as

$$\Gamma_q(t) = K(A,t) \int_0^{\infty/A(1-q)} x^{t-1} e_q(-x) d_q x, \qquad (1.16)$$

where $K(x,t) = \frac{1}{x+1} x^t \left(1 + \frac{1}{x}\right)_q^t (1+x)_q^{1-t}$.

Remark 1.5. The q-gamma and q-beta functions are related to each other by the following identities:

$$B_q(t,s) = \frac{\Gamma_q(t)\Gamma_q(s)}{\Gamma_q(t+s)} \qquad (1.17)$$

and

$$\Gamma_q(t) = \frac{B_q(t,\infty)}{(1-q)^t}.$$

The function $K(x,t)$ is a q-constant, i.e., $K(qx,t) = K(x,t)$. In particular, for any positive integer n

$$K(x,n) = q^{\frac{n(n-1)}{2}}, \quad K(x,0) = 1.$$

Also

$$\lim_{q \to 1} K(x,t) = 1, \forall x, t \in \Re$$

and

$$\lim_{q \to 0} K(x,t) = x^t + x^{t-1}, \forall t \in (0,1), \qquad x \in \Re.$$

It also satisfies $K(x,t+1) = q^t K(x,t)$ (see [49]).

Chapter 2
q-Discrete Operators and Their Results

This chapter deals with the q-analogue of some discrete operators of exponential type. We study some approximation properties of the q-Bernstein polynomials, q-Szász–Mirakyan operators, q-Baskakov operators, and q-Bleimann, Butzer, and Hahn operators. Here, we present moment estimation, convergence behavior, and shape-preserving properties of these discrete operators.

2.1 q-Bernstein Operators

After the development of quantum calculus, A. Lupaş was the first who gave the q-analogue of the Bernstein polynomials. Let $f \in C[0,1]$. The linear operator $L_{n,q} : C[0,1] \to C[0,1]$, defined by

$$L_{n,q}(f;x) := \sum_{k=0}^{n} f\left(\frac{[k]}{[n]}\right) b_{n,k}^{q}(x), \qquad (2.1)$$

where

$$b_{n,k}^{q}(x) = \frac{\begin{bmatrix} n \\ k \end{bmatrix} q^{k(k-1)/2} x^k (1-x)^{n-k}}{\prod_{j=0}^{n-1}(1-x+q^j x)}$$

is called Lupaş q-analogue of Bernstein polynomials. He established some direct results for the operators $L_{n,q}$, which were later studied in details by Ostrovska [127].

In the year 1997 Phillips [133] introduced another q-analogue of Bernstein polynomials by using the q-binomial coefficients and the q-binomial theorem. Phillips and his colleagues have intensively studied these operators and many applications and generalizations have been investigated (see [134] and references therein). Also Ostrovska (see [125, 126, 129, 130]) established some interesting properties on such operators. In [128], she gave a systematic study on these

A. Aral et al., *Applications of q-Calculus in Operator Theory*,
DOI 10.1007/978-1-4614-6946-9_2, © Springer Science+Business Media New York 2013

operators on the completion of one of the decade on q-Bernstein polynomials. Also the other researcher who worked on q-Bernstein polynomials, we mention the work of Wang Heping and collaborators [96–98, 100]. Recently Il'inski and Ostrovska [102, 125] obtained new results about convergence properties of the q-Bernstein polynomials. This section is based on the q-Bernstein polynomials by Phillips [133].

2.1.1   Introduction

We can verify by induction, using (1.2) or (1.3), that

$$(1+x)(1+qx)\ldots(1+q^{k-1}x) = \sum_{r=0}^{k} q^{\frac{r(r-1)}{2}} \begin{bmatrix} k \\ r \end{bmatrix}_q x^r \qquad (2.2)$$

which generalizes the binomial expansion.

For any real function f we define q-differences recursively from

$$\Delta_q^0 f_i = f_i$$

for $i = 0, 1, \ldots, n$, where n is a fixed positive integer, and

$$\Delta_q^{k+1} f_i = \Delta_q^k f_{i+1} - q^k \Delta_q^k f_i \qquad (2.3)$$

for $k = 0, 1, \ldots, n - i - 1$, where f_i denotes $f(\frac{[i]_q}{[n]_q})$. When $q = 1$, these reduce to ordinary forward differences. It is easily established by induction that the q-differences satisfy

$$\Delta_q^k f_i = \sum_{r=0}^{k} (-1)^r q^{\frac{r(r-1)}{2}} \begin{bmatrix} k \\ r \end{bmatrix}_q f_{i+k-r}. \qquad (2.4)$$

See Schoenberg [140], Lee and Phillips [108] and [134, p. 46].

2.1.2   Bernstein Polynomials

Theorem 2.1. *For each positive integer n, we define*

$$B_n(f;x) = \sum_{r=0}^{n} f_r \begin{bmatrix} n \\ r \end{bmatrix}_q x^r \prod_{r=0}^{n-r-1} (1 - q^s x), \qquad (2.5)$$

where an empty product denotes 1 *and, as above,* $f_r = f(\frac{[r]_q}{[n]_q})$. *When* $q = 1$, *we*

obtain the classical Bernstein polynomial. We observe immediately from (2.5) that, independently of q,

$$B_n(f;0) = f(0), \ B_n(f;1) = f(1) \tag{2.6}$$

for all functions f. *We now state a generalization of the well-known forward difference form (see, e.g., Davis [46]) of the classical Bernstein polynomial.*

Theorem 2.2. *The generalized Bernstein polynomial defined by (2.5) may be expressed in the q-difference form*

$$B_n(f;x) = \sum_{r=0}^{n} \begin{bmatrix} n \\ r \end{bmatrix}_q \Delta_q^r f_0 x^r. \tag{2.7}$$

Proof. The coefficient of x^k in (2.5) is

$$\sum_{s=0}^{\infty} f_{k-s} \begin{bmatrix} n \\ k-s \end{bmatrix}_q (-1)^s q^{\frac{s(s-1)}{2}} \begin{bmatrix} n-k+s \\ s \end{bmatrix}_q = \begin{bmatrix} n \\ k \end{bmatrix}_q \sum_{s=0}^{k} (-1)^s q^{\frac{s(s-1)}{2}} \begin{bmatrix} k \\ s \end{bmatrix}_q f_{k-s}.$$

We see immediately from the expansion of the q-difference (2.4) that the coefficient of x^k in (2.5) simplifies to give $\begin{bmatrix} n \\ k \end{bmatrix}_q \Delta_q^k f_0$, thus verifying (2.7). ∎

We note in passing that (2.7) provides an efficient means of computing $B_n(f;x)$, using (1.2) or (1.3) to evaluate the q-binomial coefficient recursively and (2.3) to compute the q-difference recursively. Let us write the interpolating polynomial for f at the points $x_0, \ldots, x_n$ in the Newton divided difference form

$$p_n(x) = \sum_{r=0}^{n} (\prod_{s=0}^{r-1} (x - x_s)) f[x_0, \ldots, x_r],$$

where the empty product denotes 1. For the choice of points $x_r = \frac{[r]_q}{[n]_q}$, $0 \le r \le n$, we can express the divided differences in the form of q-differences. Specifically, we may verify by induction on k that

$$f[x_i, \ldots, x_{i+k}] = q^{\frac{-k(2i+k-1)}{2}} [n]_q^k \frac{\Delta_q^k f_i}{[k]!}. \tag{2.8}$$

(See Schoenberg [140], Lee and Phillips [108].) From the uniqueness of the interpolating polynomial it is clear that if f is a polynomial of degree m, then $\Delta_q^r f_0 = 0$ for $r > m$ and $\Delta_q^m f_0 \ne 0$. Thus it follows from (2.7) that, if f is a polynomial of degree m, then $B_n(f;x)$ is a polynomial of degree $\min(m,n)$. In particular, we will evaluate $B_n(f;x)$ explicitly for $f(x) = 1, x$ and x^2. First we obtain

$$B_n(1;x) = 1 \tag{2.9}$$

and with $f(x) = x$, we have $f_0 = 0$ and

$$\Delta_q f_0 = f_1 - f_0 = \frac{[1]_q}{[n]_q} - \frac{[0]_q}{[n]_q} = \frac{1}{[n]_q}.$$

Since $\begin{bmatrix} n \\ 1 \end{bmatrix}_q = [n]_q$, we deduce from (2.7) that

$$B_n(x;x) = x. \tag{2.10}$$

For $f(x) = x^2$, we compute $f_0 = 0$,

$$\Delta_q f_0 = \left(\frac{[1]_q}{[n]_q}\right)^2 - \left(\frac{[0]_q}{[n]_q}\right)^2 = \frac{1}{[n]_q^2}$$

and using (2.4),

$$\Delta_q^2 f_0 = \left(\frac{[2]_q}{[n]_q}\right)^2 - \begin{bmatrix} 2 \\ 1 \end{bmatrix}_q \left(\frac{[1]_q}{[n]_q}\right)^2 + q\left(\frac{[0]_q}{[n]_q}\right)^2 = \frac{(1+q)^2 - (1+q)}{[n]_q^2} = \frac{q(1+q)}{[n]_q^2}.$$

Thus

$$B_n(x^2;x) = [n]_q \frac{1}{[n]_q^2}x + \frac{[n]_q[n-1]_q}{[2]_q} \frac{q(1+q)}{[n]_q^2}x^2$$

and, since $[2]_q = 1+q$ and $q[n-1]_q = [n]_q - 1$, we obtain

$$B_n(x^2;x) = x^2 + \frac{x(1-x)}{[n]_q}. \tag{2.11}$$

We note that the relations (2.9) and (2.10) are identical to those obtained for the classical Bernstein polynomials (corresponding to the case $q = 1$), while (2.11) differs only in having $[n]_q$ in place of n.

2.1.3 Convergence

In the classical case, the uniform convergence of $B_n(f;x)$ to $f(x)$ on $[0,1]$ for each $f \in C[0,1]$ is assured by the following two properties:

1. B_n is a positive operator.
2. $B_n(f;x)$ converges uniformly to $f \in C[0,1]$ for $f(x) = 1, x$ and x^2.

Recall that if a linear operator L maps an element $f \in C[0,1]$ to $Lf \in C[0,1]$, then L is said to be monotone if $f(x) \geq 0$ on $[0,1]$ implies that $Lf(x) \geq 0$ on $[0,1]$. We observe that the generalized Bernstein operator defined by (2.5) is monotone for $0 < q < 1$. On the other hand, for a fixed value of q with $0 < q < 1$, we see that

$$[n]_q \to \frac{1}{1-q} \text{ as } n \to \infty$$

and in the case it is clear from (2.11) that $B_n(x^2;x)$ does not converge to x^2. To obtain a sequence of generalized Bernstein polynomials with $q \neq 1$ which converges, we let $q = q_n$ depend on n. We then choose a sequence (q_n) such that

$$1 - \frac{1}{n} \leq q_n < 1.$$

Then we have

$$1 - \frac{r}{n} \leq q_n^r < 1 \text{ for } 1 \leq r \leq n-1$$

and thus

$$[n]_{q_n} = 1 + q_n + q_n^2 + \ldots + q_n^{n-1} \geq n - \frac{1}{2}(n-1) = \frac{1}{2}(n+1),$$

so that $[n]_{q_n} \to \infty$ as $n \to \infty$.

We now state formally our result on convergence.

Theorem 2.3. *Let $q = (q_n)$ satisfy $0 < q_n < 1$ and let $q_n \to 1$ as $n \to \infty$. Then, if $f \in C[0,1]$,*

$$B_n(f;x) = \sum_{r=0}^{n} f_r \begin{bmatrix} n \\ r \end{bmatrix}_{q_n} x^r \prod_{s=0}^{n-r-1} (1 - q_n^s x)$$

converges uniformly to f on $[0,1]$.

Proof. This is a special case of the Bohman–Korovkin theorem. (See, e.g., Cheney [42], Lorentz [114].) Alternatively, we may follow the proof given in Rivlin [138] for the convergence of the classical Bernstein polynomials, except that n must be replaced by $[n]_q$ when estimating how closely $B_n(x^2;x)$ approximates to x^2, as in (2.11) above. ∎

Given a function f defined on $[0,1]$, let

$$w(\delta) = \sup_{|x_1-x_2|<\delta} |f(x_1) - f(x_2)|,$$

the usual modulus of continuity, where the supremum is taken over all $x_1, x_2 \in [0,1]$ such that $|x_1 - x_2| \leq \delta$. Then we have:

Theorem 2.4. *If f is bounded on $[0,1]$ and B_n denotes the generalized Bernstein operator defined by (2.5), then*

$$\| f - B_n f \|_\infty \le \frac{3}{2} w(\frac{1}{[n]_q^{1/2}}). \tag{2.12}$$

Proof. Rivlin [138] states this theorem for the case where $q_n = 1$ for all n, and his proof is easily adapted to justify (2.12). ∎

2.1.4   Voronovskaya's Theorem

In this section we will follow Davis [46], beginning with the sums

$$S_m(x) = \sum_{r=0}^{n} ([r]_q - [n]_q x)^m \begin{bmatrix} n \\ r \end{bmatrix}_q x^r \prod_{s=0}^{n-r-1} (1 - q^s x). \tag{2.13}$$

Let us write

$$([r]_q - [n]_q x)^m = [n]_q^m \sum_{s=0}^{m} \binom{m}{s} \left(\frac{[r]_q}{[n]_q} \right)^s (-x)^{m-s}$$

and so express $S_m(x)$ in the form

$$S_m(x) = [n]_q^m \sum_{s=0}^{m} \binom{m}{s} (-x)^{m-s} B_n(x^s; x). \tag{2.14}$$

We have already noted that $B_n(x^s; x)$ is a polynomial of degree $\min(s, n)$ in x. Thus $S_m(x)$ is a polynomial of degree at most m in x. Since B_n is a linear operator we also obtain from (2.14) that

$$S_m(1) = [n]_q^m B_n((x-1)^m; 1) = 0, \tag{2.15}$$

using the property (2.6) that $B_n(f; x)$ interpolates $f(x)$ at $x = 0$ and $x = 1$. We deduce from (2.15) that $(1 - x)$ is a factor of $S_m(x)$, for $m > 0$. From (2.7) we find by direct calculation (using the symbolic language Maple) that S_6 has the form

$$S_6(x) = (1 - x) \sum_{r=1}^{5} a_r x^r [n]_q^r, \tag{2.16}$$

where a_r is a polynomial in x and q. In the lemma below, we are concerned with the dependence of S_6 on $[n]_q$. The coefficients a_1, a_2 and a_3 are as follows:

$$a_1 = 1 - (4 + 10q + 10q^2 + 5q^3 + q^4)x +$$
$$(6 + 20q + 35q^2 + 39q^3 + 29q^4 + 15q^5 + 5q^6 + q^7)x^2 -$$
$$(4 + 15q + 31q^2 + 46q^3 + 51q^4 + 44q^5 + 29q^6 + 14q^7 + 5q^8 + q^9)x^3 +$$
$$(1 + 4q + 9q^2 + 15q^3 + 20q^4 + 22q^5 + 20q^6 + 15q^7 + 9q^8 + 4q^9 + q^{10})x^4,$$

$$a_2 = (-1 + 10q + 10q^2 + 5q^3 + q^4) +$$
$$(3 - 4q - 36q^2 - 52q^3 - 40q^4 - 19q^5 - 6q^6 - q^7)x$$
$$(-3 - 3q + 12q^2 + 46q^3 + 72q^4 + 67q^5 + 43q^6 + 19q^7 + 6q^8 + q^9)x^2 +$$
$$(1 + 2q - 7q^3 - 20q^4 - 30q^5 - 32q^6 - 25q^7 - 13q^8 - 5q^9 - q^{10})x^3,$$

and

$$a_3 = (1 - 16q + q^2 + 13q^3 + 11q^4 + 4q^5 + q^6) +$$
$$(-2 + 6q + 30q^2 + 5q^3 - 22q^4 - 26q^5 - 15q^6 - 5q^7 - q^8)x$$
$$(1 - 6q^2 - 14q^3 - 6q^4 + 9q^5 + 15q^6 + 11q^7 + 4q^8 + q^9)x^2.$$

We have quoted the values of a_1, a_2 and a_3 for the sake of completeness, although we do not need to know their values. However, we do require the values of

$$a_4 = (1 - q)^2 (q(10 + 10q + 5q^2 + q^3)(1 - qx) - 1 + x) \qquad (2.17)$$

and

$$a_5 = (1 - q)^4. \qquad (2.18)$$

The presence of the factors $(1 - q)^2$ and $(1 - q)^4$ in (2.17) and (2.18), respectively, proves to be significant. For we observe that, with $0 < q < 1$, we have

$$0 < 1 - q < \frac{1}{[n]_q} \qquad (2.19)$$

for any positive integer n. Thus if $[n]_{q_n} \to \infty$ as $n \to \infty$, we see from (2.16) that $S_6(x)$ behaves like $[n]_{q_n}^3$ for large n and not like $[n]_{q_n}^5$. We now give a generalization of Lemma 6.3.5 of Davis [46] as a prelude to a generalization of Voronovskaya's theorem.

Lemma 2.1. *Let* $q = q_n$ *satisfy* $0 < q_n < 1$ *and let* $q_n \to 1$ *as* $n \to \infty$. *Then there exists a constant C independent of n such that, for all* $x \in [0, 1]$,

$$\sum_{\left|\frac{[r]_{q_n}}{[n]_{q_n}} - x\right| \geq [n]_{q_n}^{-\frac{1}{4}}} \begin{bmatrix} n \\ r \end{bmatrix}_{q_n} x^r \prod_{s=0}^{n-r-1} (1 - q^s x) \leq \frac{C}{[n]_{q_n}^{\frac{3}{2}}}.$$

Proof. From the inequality (2.19), together with (2.16)–(2.18), we deduce that there exists a constant C independent of n such that

$$| S_6(x) | \leq C [n]_{q_n}^3$$

for all $x \in [0,1]$. Since

$$| \frac{[r]_{q_n}}{[n]_{q_n}} - x | \geq [n]_{q_n}^{-\frac{1}{4}} \Rightarrow \frac{([r] - [n]_{q_n} x)^6}{[n]_{q_n}^{\frac{9}{2}}} \geq 1,$$

it follows that

$$\sum_{| \frac{[r]_{q_n}}{[n]_{q_n}} -x| \geq [n]^{-\frac{1}{4}}} \begin{bmatrix} n \\ r \end{bmatrix}_{q_n} x^r \prod_{s=0}^{n-r-1} (1 - q_n^s x) \leq \frac{1}{[n]_{q_n}^{\frac{9}{2}}} S_6(x) \leq \frac{C}{[n]_{q_n}^{\frac{3}{2}}}. \qquad \blacksquare$$

Theorem 2.5. *Let f be bounded on $[0,1]$ and let x_0 be a point of $[0,1]$ at which $f''(x_0)$ exists. Further, let $q = q_n$ satisfy $0 < q_n < 1$ and let $q_n \to 1$ as $n \to \infty$. Then the rate of convergence of the sequence of generalized Bernstein polynomials is governed by*

$$\lim_{n \to \infty} [n]_{q_n} (B_n(f;x_0) - f(x_0)) = \frac{1}{2} x_0 (1 - x_0) f''(x_0). \qquad (2.20)$$

Proof. We replace Lemma 6.3.5 of Davis [46] by the lemma stated above and then the proof of Theorem 6.3.6 of Davis is readily extended to justify (2.20). Thus the error $B_n(f;x) - f(x)$ tends to zero like $\frac{1}{[n]_{q_n}}$. At best this is like $\frac{1}{n}$, for the classical Bernstein polynomials. However, through our choice of the sequence (q_n), we can achieve a rate of convergence which is slower than $\frac{1}{n}$ and indeed may be as slow as we please. Such a birthday gift! $\qquad \blacksquare$

2.2 q-Szász Operators

In this section, we give a generalization of Szász–Mirakyan operators based on q-integers that we call q-Szász–Mirakyan operators. Depending on the selection of q, these operators are more flexible than the classical Szász–Mirakyan operators while retaining their approximation properties. For these operators, we give a Voronovskaya-type theorem related to q-derivatives. Furthermore, we obtain convergence properties for functions belonging to particular subspaces of $C[0, \infty)$ and give some representation formulae of q-Szász–Mirakyan operators and their rth q-derivatives. This section is based on [25, 29].

2.2.1 Introduction

In this section, as Phillips has done for Bernstein operators, we introduce a similar modification of the Szász–Mirakyan operators [148] that we call q-Szász–Mirakyan operators and examine the main properties of this new approximation process. Recall that the Bernstein operators were defined with the aid of the functions defined on [0, 1] as opposed to the classical Szász–Mirakyan operators which are defined on $\mathbb{R}_0 := [0, \infty)$ in order to analyze the approximation problems for the functions defined on the same interval. Although, from the structural point of view q-Szász–Mirakyan operators have some resemblances to classical Szász–Mirakyan operators, they have some similarities to Bernstein–Chlodowsky operators from the properties of convergence standpoint. That is, the interval of convergence grows as $n \to \infty$ as in Bernstein–Chlodowsky operators. Our new operators with this construction are sensitive or flexible to the rate of convergence to f. That is, the proposed estimate with rates in terms of modulus of continuity tells us that, depending on our selection of q, the rates of convergence in weighted norm of the new operators are better than the classical Bernstein–Chlodowsky operators.

2.2.2 Construction of Operators

For $0 < q < 1$, we now define new operators that we call the q-Szász–Mirakyan operators as follows:

$$S_n(f; q; x) := S_n^q(f; x)$$

$$:= E_q\left(-[n]_q \frac{x}{b_n}\right) \sum_{k=0}^{\infty} f\left(\frac{[k]_q b_n}{[n]_q}\right) \frac{\left([n]_q x\right)^k}{[k]_q! (b_n)^k}, \qquad (2.21)$$

where $0 \le x < \alpha_q(n)$, $\alpha_q(n) := \frac{b_n}{(1-q)[n]_q}$, $f \in C(\mathbb{R}_0)$, and (b_n) is a sequence of positive numbers such that $\lim_{n \to \infty} b_n = \infty$.

We observe that these operators are positive and linear. Furthermore, in the case of $q = 1$, the operators (2.21) are similar to the classical Szász–Mirakyan operators.

By the properties of the series in (1.7), the interval of domain of operators (2.21) is the interval $0 \le x < \alpha_q(n)$ for $0 < q < 1$; in the mean while the operators are interpolating the function f on $\mathbb{R}_0$. Note that the interval of convergence grows as $n \to \infty$. A similar situation arises for Bernstein–Chlodowsky operators (see [35, 43, 70, 113]).

We denote by e_m the test functions defined by $e_m(t) := t^m$ for every integer $m \ge 0$ and for each $x \ge 0$, $\varphi_x(t) := t - x$ such that $t - x \ge 0$.

2.2.3 Auxiliary Result

In the sequel, we need the following results:

Lemma 2.2. *For* $0 \leq x < \alpha_q(n)$, *when* $0 < q < 1$ *and integer* $m \geq 0$, *we have*

$$S_n^q(e_m; x) = q^{\frac{m(m-1)}{2}} x^m + \sum_{j=1}^{m-1} \left(\frac{b_n}{[n]_q} \right)^{m-j} S_q(m, j) q^{\frac{j(j-1)}{2}} x^j, \qquad (2.22)$$

where $S_q(m, j)$ *are* q-*Stirling polynomials of the second kind.*

Proof. Using (1.8) and the Cauchy rule for multiplication of two series [21, p. 376], from (2.21) we have the representation

$$S_n^q(f; x) = \sum_{j=0}^{\infty} \sum_{i=0}^{j} (-1)^i q^{\frac{i(i-1)}{2}} f\left(\frac{[j-i]_q b_n}{[n]_q} \right) \frac{\left([n]_q x\right)^j}{[i]_q! [j-i]_q! b_n^j}$$

$$= \sum_{j=0}^{\infty} \left(\frac{[n]_q}{b_n} \right)^j \Delta_q^j f_0 \frac{x^j}{[j]_q!}, \qquad (2.23)$$

where $\Delta_q^j f_0$ as in (2.4).

We can easily see from (2.4) for $f(x) = x^m$, $m = 0, 1, 2, \ldots$

$$\Delta_q^j (t^m)(0) = \sum_{i=0}^{j} (-1)^i q^{\frac{i(i-1)}{2}} \begin{bmatrix} j \\ i \end{bmatrix} \left(\frac{b_n}{[n]_q} \right)^m [j-i]_q^m$$

for $j \geq 0$.

Also we know that the connection with q-differences $\Delta_q^j f_0$ and jth derivative $f^{(j)}$ is the following:

$$\frac{\Delta_q^j f_0}{q^{j(j-1)} [j]_q!} = \frac{f^{(j)}(\xi)}{j!},$$

where $\xi \in (0, [j])$ (see [134, p. 268]). From this equality, it is obvious that q-differences of monomial t^m of order greater than m are zero. Thus, we have

$$S_n^q(e_m; x) = \sum_{j=0}^{m} \left(\frac{b_n}{[n]_q} \right)^{m-j} S_q(m, j) q^{\frac{j(j-1)}{2}} x^j$$

$$= q^{\frac{m(m-1)}{2}} x^m + \sum_{j=1}^{m-1} \left(\frac{b_n}{[n]_q} \right)^{m-j} S_q(m, j) q^{\frac{j(j-1)}{2}} x^j,$$

where

$$\mathbb{S}_q(m, j) = \frac{1}{[j]_q! q^{\frac{j(j-1)}{2}}} \sum_{i=0}^{j} (-1)^i q^{\frac{i(i-1)}{2}} \begin{bmatrix} j \\ i \end{bmatrix} [j-i]_q^m$$

are the Stirling polynomials of the second kind satisfying the equality

$$\mathbb{S}_q(m+1, j) = \mathbb{S}_q(m, j-1) + [j]_q \mathbb{S}_q(m, j),$$

for $m \geq 0$ and $j \geq 1$ with $\mathbb{S}_q(0, 0) = 1$, $\mathbb{S}_q(m, 0) = 0$ for $m > 0$. Also $\mathbb{S}_q(m, j) = 0$ for $j > m$. Thus the proof is completed. ■

Lemma 2.2 gives the explicit expression of $S_n^q(e_m; x)$ for $m = 0, 1, 2$:

$$S_n^q(e_0; x) = 1 \tag{2.24}$$

$$S_n^q(e_1; x) = x \tag{2.25}$$

$$S_n^q(e_2; x) = qx^2 + \frac{b_n}{[n]_q} x. \tag{2.26}$$

The equality (2.24) can also be obtained from (1.1, b).

Remark 2.1. Since for a fixed value of q with $0 < q < 1$,

$$\lim_{n \to \infty} [n]_q = \frac{1}{1-q},$$

to ensure the convergence properties of (2.21), we will assume $q = q_n$ as a sequence such that $q_n \to 1$ as $n \to \infty$ for $0 < q_n < 1$ and so that $[n]_{q_n} \to \infty$ as $n \to \infty$.

2.2.4 Convergence of $S_n^{q_n}(f)$

Proposition 2.1 (q-L'Hopital's Rule). *Suppose that throughout some interval containing a, each of f and g are q-differentiable and continuous functions and* $(D_q g)(x) \neq 0$ *for* $q \in (0, 1) \cup (1, \infty)$. *If*

$$\lim_{x \to a} f(x) = \lim_{x \to a} g(x) = 0$$

and there exists $\widehat{q} \in (0, 1)$ *such that for all* $q \in (\widehat{q}, 1) \cup (1, \widehat{q}^{-1})$

$$\lim_{x \to a} \frac{(D_q f)(x)}{(D_q g)(x)} = L,$$

then

$$\lim_{x \to a} \frac{f(x)}{g(x)} = L.$$

Proof. Suppose that x is close enough to a so that throughout the interval between a and x, f and g are q-differentiable with $(D_q g)(x) \neq 0$. Then by q-Lagrange theorem (see [137]), there exists $\widehat{q} \in (0,1)$ such that for all $q \in (\widehat{q}, 1) \cup (1, \widehat{q}^{-1})$

$$\frac{f(x)}{g(x)} = \frac{(D_q f)(\mu_1)}{(D_q g)(\mu_2)},$$

where $\mu_1, \mu_2 \in (a, x)$.

Since μ_1 and μ_2 are between a and x, $x \to a$ implies that $\mu_1 \to a$ and $\mu_2 \to a$. Hence for all $q \in (\widehat{q}, 1) \cup (1, \widehat{q}^{-1})$,

$$\lim_{x \to a} \frac{f(x)}{g(x)} = \lim_{x \to a} \frac{(D_q f)(\mu_1)}{(D_q g)(\mu_2)}$$

$$= \lim_{x \to a} \frac{(D_q f)(x)}{(D_q g)(x)}. \qquad \blacksquare$$

Now we give a Voronovskaya-type relation for the operator (2.21).

Theorem 2.6. *Let $f \in C(\mathbb{R}_0)$ be a bounded function and (q_n) denote a sequence such that $0 < q_n < 1$ and $q_n \to 1$ as $n \to \infty$. Suppose that the second derivative $D_{q_n}^2 f(x)$ exists at a point $x \in [0, \alpha_{q_n}(n))$ for n large enough. If $\lim\limits_{n \to \infty} \frac{b_n}{[n]_{q_n}} = 0$, then*

$$\lim_{n \to \infty} \frac{[n]_{q_n}}{b_n} \left(S_n^{q_n}(f; x) - f(x) \right) = \frac{1}{2} x \lim_{q_n \to 1} D_{q_n}^2 f(x).$$

Proof. By the q-Taylor formula [137] for f, we have

$$f(t) = f(x) + D_q f(x)(t - x) + \frac{1}{[2]_q} D_q^2 f(x)(t - x)_q^2 + \Phi_q(x; t)(t - x)_q^2$$

for $0 < q < 1$ where $(t - x)_q^2 = (t - x)(t - qx)$. By application of q-L'Hopital's Rule, there exists $\widehat{q}_1 \in (0,1)$ such that for all $q \in (\widehat{q}_1, 1)$

$$\lim_{t \to x} \Phi_q(x; t) = \lim_{t \to x} \frac{D_q f(t) - D_q f(x) - D_q^2 f(x)(t - x)}{[2]_q (t - x)}$$

where we use the equality

$$\left(D_q (t - x)_q^n \right)(t) = [n]_q (t - x)_q^{n-1},$$

where

$$(t-x)_q^n = \prod_{k=0}^{n-1} \left(t - q^k x\right)$$

(see [59]).

By applying again of q-L'Hopital's Rule, there exist $\widehat{q}_2 \in (0,1)$ $(\widehat{q}_1 < \widehat{q}_2)$ such that for all $q \in (\widehat{q}_2, 1)$

$$\lim_{t \to x} \Phi_q(x; t) = \lim_{t \to x} \frac{D_q^2 f(t) - D_q^2 f(x)}{[2]_q} = 0. \tag{2.27}$$

By assumption the function $\Phi_q(t) := \Phi_q(t; x)$ is a bounded function for all $q \in (\widehat{q}_2, 1)$. Consequently, we can write

$$\frac{[n]_{q_n}}{b_n} \left(S_n^{q_n}(f; x) - f(x)\right)$$

$$= D_{q_n} f(x) \frac{[n]_{q_n}}{b_n} S_n^{q_n}(\varphi_x; x) + \frac{D_{q_n}^2 f(x)}{[2]_{q_n}} \frac{[n]_{q_n}}{b_n} S_n^{q_n}\left((t-x)_{q_n}^2; x\right)$$

$$+ \frac{[n]_{q_n}}{b_n} S_n^{q_n}\left(\Phi_{q_n}(t)(t-x)_{q_n}^2; x\right)$$

$$= D_{q_n} f(x) \frac{[n]_{q_n}}{b_n} S_n^{q_n}(\varphi_x; x) + \frac{D_{q_n}^2 f(x)}{[2]_{q_n}} \frac{[n]_{q_n}}{b_n} S_n^{q_n}(\varphi_x^2; x)$$

$$+ \frac{[n]_{q_n}}{b_n} \left(\frac{D_{q_n}^2 f(x)}{[2]_{q_n}} x(1-q_n) S_n^{q_n}(\varphi_x; x) + S_n^{q_n}\left(\Phi_{q_n}(t)(t-x)_{q_n}^2; x\right)\right)$$

By (2.24)–(2.26), we get

$$\lim_{n \to \infty} \frac{[n]_{q_n}}{b_n} S_n^{q_n}(\varphi_x; x) = 0 \tag{2.28}$$

$$\lim_{n \to \infty} \frac{[n]_{q_n}}{b_n} S_n^{q_n}(\varphi_x^2; x) = x \tag{2.29}$$

and thus

$$\lim_{n \to \infty} \frac{[n]_{q_n}}{b_n} \left(S_n^{q_n}(f; x) - f(x)\right)$$

$$= \frac{1}{2} x \lim_{q_n \to 1} D_{q_n}^2 f(x) + \lim_{n \to \infty} \frac{[n]_{q_n}}{b_n} S_n^{q_n}\left(\Phi_{q_n}(t)(t-x)_{q_n}^2; x\right)$$

Now, the last term on the right-hand side can be estimated in the following way. Since $\lim_{t \to x} \Phi_{q_n}(t) = 0$, then for all $\varepsilon > 0$, there exists $\delta > 0$ such that $|t - x| < \delta$ implies $|\Phi_{q_n}(t)| < \varepsilon$ for $x \in [0, \alpha_{q_n}(n))$ where n is large enough. While if $|t - x| \geq \delta$, then $|\Phi_{q_n}(t)| \leq \frac{M}{\delta^2} \varphi_x^2(t)$, where $M > 0$ is a constant. Hence we can infer

$$\frac{[n]_{q_n}}{b_n} \left(S_n^{q_n} \left(\Phi_{q_n}(t)(t - x)_{q_n}^2 ; x \right) \right)$$

$$= \frac{[n]_{q_n}}{b_n} \left(S_n^{q_n} \left(\Phi_{q_n}(t) \varphi_x^2; x \right) \right)$$

$$+ \frac{[n]_{q_n}}{b_n} x (1 - q_n) \left(S_n^{q_n} \left(\Phi_{q_n}(t) \varphi_x; x \right) \right)$$

$$\leq \varepsilon \frac{[n]_{q_n}}{b_n} \left(S_n^{q_n} \left(\varphi_x^2; x \right) + x(1 - q_n) S_n^{q_n} \left(\varphi_x; x \right) \right)$$

$$+ \frac{M}{\delta^2} \frac{[n]_{q_n}}{b_n} \left(S_n^{q_n} \left(\varphi_x^4; x \right) + x(1 - q_n) S_n^{q_n} \left(\varphi_x^3; x \right) \right). \tag{2.30}$$

If we calculate $S_n^{q_n} \left(\varphi_x^4; x \right)$ by using Lemma 2.2, we get

$$S_n^{q_n} \left(\varphi_x^4; x \right) = x^4 \left(q_n^6 - 4q_n^3 + 6q_n - 3 \right)$$

$$+ x^3 \frac{\left(q_n^3 \left(1 + [2]_{q_n} + [3]_{q_n} \right) - 4 \left(1 + [2]_{q_n} \right) q_n + 6 \right) b_n}{[n]_{q_n}}$$

$$+ x^2 \frac{\left(q_n \left(1 + [2]_{q_n} + [2]_{q_n}^2 \right) - 4 \right) b_n^2}{[n]_{q_n}^2} + x \frac{b_n^3}{[n]_{q_n}^3}.$$

Since $\lim_{n \to \infty} b_n = \infty$ then we have

$$\lim_{n \to \infty} \frac{[n]_{q_n} \left(q_n^6 - 4q_n^3 + 6q_n - 3 \right)}{b_n} = \lim_{n \to \infty} \frac{(1 - q_n^n) \left(\frac{q_n^6 - 4q_n^3 + 6q_n - 3}{1 - q_n} \right)}{b_n} = 0.$$

We thus obtain

$$\lim_{n \to \infty} \frac{[n]_{q_n}}{b_n} S_n^{q_n} \left(\varphi_x^4; x \right) = 0 \tag{2.31}$$

for fixed $x \in [0, \alpha_{q_n}(n))$ where n is large enough. Using Lemma 2.2 we can easily see that

$$\lim_{n \to \infty} (1 - q_n) \frac{[n]_{q_n}}{b_n} \left(S_n^{q_n} \left(\varphi_x^3; x \right) \right) = 0 \tag{2.32}$$

and therefore by (2.31), (2.32), and (2.30), we conclude that

$$\lim_{n\to\infty} \frac{[n]_{q_n}}{b_n} \left(S_n^{q_n} \left(\Phi_{q_n}(t)(t-x)_{q_n}^2 ; x \right) \right) = 0$$

for fixed $x \in [0, \alpha_{q_n}(n))$ where n is large enough and therefore, we have the desired result.   ■

We know that if f is differentiable n times, then $\lim_{q\to 1} D_q^n f(x) = f^{(n)}(x)$ (see [81, p. 22]). Using this property we have the following corollary.

Corollary 2.1. *Let $f \in C(\mathbb{R}_0)$ be a bounded function and (q_n) denote a sequence such that $0 < q_n < 1$ and $q_n \to 1$ as $n \to \infty$. Suppose that the second derivative $f''(x)$ exists at a point $x \in [0, \alpha_n(q_n))$ for n large enough. If $\lim_{n\to\infty} \frac{b_n}{[n]_{q_n}} = 0$, then*

$$\lim_{n\to\infty} \frac{[n]_{q_n}}{b_n} \left(S_n^{q_n}(f;x) - f(x) \right) = \frac{1}{2} x f''(x).$$

Recall that a continuous function on an interval, which does not include 0, is continuous q-differentiable. According to this, for every x in an interval not including 0, since q-derivatives of f become finite, we deduce the q-differentiable condition in Theorem 2.6. In other words, Voronovskaya-type theorem is valid only for continuous and bounded functions.

Corollary 2.2. *Let $f \in C(\mathbb{R}_0)$ be a bounded function and (q_n) denote a sequence such that $0 < q_n < 1$ and $q_n \to 1$ as $n \to \infty$. If $\lim_{n\to\infty} \frac{b_n}{[n]_{q_n}} = 0$, then*

$$\lim_{n\to\infty} \frac{[n]_{q_n}}{b_n} \left(S_n^{q_n}(f;x) - f(x) \right) = \frac{1}{2} x \lim_{q_n\to 1} D_{q_n}^2 f(x)$$

for every point $x \in (0, \alpha_n(q_n))$, where n is large enough.

Remark 2.2. If the assumption of Theorem 2.6 holds for the function f, then the pointwise convergence rate of the operators (2.21) to f is $\mathcal{O}\left(\frac{b_n}{[n]_{q_n}}\right)$. Also this convergence rate can be made better depending on the chosen q_n and is at least as fast as $\frac{b_n}{n}$ which is the convergence rate of the classical Bernstein–Chlodowsky operators (see [35, 133]).

2.2.5 Convergence Properties in Weighted Space

As we mentioned above, when $q_n \to 1$ as $n \to \infty$, the interval $[0, \alpha_{q_n}(n))$ which is the domain of the operator $S_n^{q_n}(f)$ grows. In this case the uniform norm is not valid to compute the rate of convergence for these operators. So we will consider

weighted function spaces and the weighted norm. In this section, we obtain a direct
approximation theorem in weighted norm and an estimate in terms of the modulus
of continuity. These types of theorems are given in [65, 66]. Now we recall this
theorem.

Let φ be a continuous and monotonically increasing function on the positive real
axis, such that $\lim\limits_{x \to \infty} \varphi(x) = \pm\infty$ and $\rho(x) = 1 + \varphi^2(x)$.

Let $B_\rho(\mathbb{R}_0)$ be the set of all functions f satisfying the condition $|f(x)| \leq
M_f \rho(x), x \in \mathbb{R}_0$ with some constant M_f, depending only on f. We denote by $C_\rho(\mathbb{R}_0)$
the space of all continuous functions belongs $B_\rho(\mathbb{R}_0)$ with the norm

$$\|f\|_\rho := \sup_{x \in \mathbb{R}_0} \frac{|f(x)|}{\rho(x)}$$

and $C_\rho^0(\mathbb{R}_0) = \left\{ f \in C_\rho(\mathbb{R}_0) : \lim\limits_{x \to \infty} \frac{|f(x)|}{\rho(x)} < \infty \right\}$.

Theorem 2.7 ([65]). *Let $\{A_n\}$ be a sequence of positive linear operators acting
from $C_\rho(\mathbb{R}_0)$ to $B_\rho(\mathbb{R}_0)$ satisfying the following three conditions:*

$$\lim_{n \to \infty} \|A_n(\varphi^\nu; x) - \varphi^\nu(x)\|_\rho = 0, \quad \nu = 0, 1, 2.$$

Then

$$\lim_{n \to \infty} \|A_n(f; x) - f(x)\|_\rho = 0$$

for any function $f \in C_\rho^0(\mathbb{R}_0)$.

The definitions of the spaces $C_\rho(\mathbb{R}_0)$ and $C_\rho^0(\mathbb{R}_0)$ are the same as $C_m(\mathbb{R}_0)$ and
$C_m^0(\mathbb{R}_0)$, respectively, if we take $\rho(x) = 1 + x^m$ ($m \geq 2$) instead of $\rho(x) = 1 + \varphi^2(x)$.

Theorem 2.8. *Let (q_n) denote a sequence such that $0 < q_n < 1$ and $q_n \to 1$ as
$n \to \infty$. For any function $f \in C_{2m}^0(\mathbb{R}_0)$, if $\lim\limits_{n \to \infty} \frac{b_n}{[n]_{q_n}} = 0$, then*

$$\lim_{n \to \infty} \sup_{0 \leq x \leq \alpha_{q_n}(n)} \frac{|S_n^{q_n}(f; x) - f(x)|}{1 + x^{2m}} = 0.$$

Moreover, for n large enough

$$\sup_{0 \leq x \leq \alpha_{q_n}(n)} \frac{|S_n^{q_n}(f; x) - f(x)|}{1 + x^{2m}} \leq (2 + \sqrt{2}) \, \omega\left(f; \sqrt{\frac{b_n}{[n]_{q_n}}}\right)$$

where $\omega(f; \cdot)$ is the classical modulus of continuity.

Proof. Applying Theorem 2.7 with $\varphi(t) = e_m(t)$, $m \geq 1$, to the operators

$$A_n(f; x) = \begin{cases} S_n^{q_n}(f; x) & \text{if } 0 \leq x \leq \alpha_{q_n}(n) \\ f(x) & \text{if } x > \alpha_{q_n}(n) \end{cases}$$

to complete the proof, it is sufficient to show that the conditions

$$\lim_{n\to\infty}\sup_{0\le x\le\alpha_{q_n}(n)}\frac{\left|S_n^{q_n}\left(e_m^v\left(t\right);x\right)-x^{mv}\right|}{1+x^{2m}}=0,\quad v=0,1,2 \tag{2.33}$$

are satisfied. As a consequence of Lemma 2.2, since $\left|S_n^{q_n}\left(1+t^{2m};x\right)\right|\le C\left(1+x^{2m}\right)$ for $x\in[0,\alpha_{q_n}(n))$ where n is large enough and C is a positive constant, $\left\{S_n^{q_n}\right\}$ is a sequence of linear positive operators acting from $C_{2m}\left(\mathbb{R}_0\right)$ to $C_{2m}\left(\mathbb{R}_0\right)$.

From (2.24)

$$\lim_{n\to\infty}\sup_{0\le x\le\alpha_{q_n}(n)}\frac{\left|S_n^{q_n}\left(e_0;x\right)-1\right|}{1+x^{2m}}=0$$

holds. Thus the condition (2.33) holds for $v=0$. Since $\lim\limits_{n\to\infty}\frac{b_n}{[n]_{q_n}}=0$, then there exists $n_0\in\mathbb{N}$ such that $\left(\frac{b_n}{[n]_{q_n}}\right)^{m-j}\le\frac{b_n}{[n]_{q_n}}$ for $n>n_0$ and $j=1,2,\ldots,m-1$. By Lemma 2.2 we have, for $n>n_0$,

$$\sup_{x\in[0,\alpha_{q_n}(n))}\frac{\left|S_n^{q_n}\left(e_m;x\right)-x^m\right|}{1+x^{2m}}$$

$$\le\left(1-q_n^{\frac{m(m-1)}{2}}\right)+\sup_{x\in[0,\alpha_{q_n}(n))}\frac{\sum_{j=1}^{m-1}\left(\frac{b_n}{[n]_{q_n}}\right)^{m-j}\mathbb{S}_{q_n}\left(m,j\right)q_n^{\frac{j(j-1)}{2}}x^j}{1+x^{2m}}$$

$$\le\left(1-q_n^{\frac{m(m-1)}{2}}\right)+\frac{b_n}{[n]_{q_n}}\left(\sum_{j=1}^{m-1}\mathbb{S}_{q_n}\left(m,j\right)q_n^{\frac{j(j-1)}{2}}\right).$$

Hence we obtain

$$\lim_{n\to\infty}\sup_{x\in[0,\alpha_{q_n}(n))}\frac{\left|S_n^{q_n}\left(e_m;x\right)-x^m\right|}{1+x^{2m}}=0.$$

Thus the condition (2.33) holds for $v=1$.

Similarly, we have, for $n>n_0$

$$\sup_{x\in[0,\alpha_{q_n}(n))}\frac{\left|S_n^{q_n}\left(e_{2m};x\right)-x^{2m}\right|}{1+x^{2m}}$$

$$\le\left(1-q_n^{m(2m-1)}\right)+\sup_{x\in[0,\alpha_{q_n}(n))}\frac{\sum_{j=1}^{2m-1}\left(\frac{b_n}{[n]_{q_n}}\right)^{2m-j}\mathbb{S}_{q_n}\left(2m,j\right)q_n^{\frac{j(j-1)}{2}}x^j}{1+x^{2m}}$$

$$\le\left(1-q_n^{m(2m-1)}\right)+\frac{b_n}{[n]_{q_n}}\left(\sum_{j=1}^{2m-1}\mathbb{S}_{q_n}\left(2m,j\right)q_n^{\frac{j(j-1)}{2}}\right).$$

That is, for $v = 2$, the condition (2.33) is satisfied. Therefore, the proof is completed from Theorem 2.7.

For the second part of the theorem, using the property of the modulus of continuity $\omega(f, \cdot)$ for every $\delta > 0$, $t \geq 0$ and $x \geq 0$,

$$|f(t) - f(x)| \leq \left(1 + \delta^{-2}(t-x)^2\right)\omega(f, \delta).$$

Using this inequality we can write

$$|S_n^{q_n}(f; x) - f(x)| \leq 2\omega\left(f, \sqrt{S_n^{q_n}(\varphi_x^2; x)}\right)$$

for $f \in C_{2m}^0(\mathbb{R}_0)$. Since

$$\sup_{x \in [0, \alpha_{q_n}(n))} \frac{S_n^{q_n}(\varphi_x^2; x)}{1 + x^{2m}} \leq (1 - q_n) + \frac{b_n}{[n]_{q_n}}$$

$$= (1 - q_n) + (1 - q_n)\alpha_{q_n}(n)$$

$$\leq 2(1 - q_n)\alpha_{q_n}(n)$$

$$= 2\frac{b_n}{[n]_{q_n}}$$

for n large enough, we have the desired result. ∎

Remark 2.3. In [35, Theorem 2.1] it has been shown that for any function f satisfying Theorem 2.8, the weighted rate of convergence of classical Bernstein–Chlodowsky operators is $\mathcal{O}\left(\frac{b_n}{n}\right)$. As a consequence of Theorem 2.8 we say that the rate of convergence of $S_n^{q_n}(f)$ to f in the weighted norm is $\frac{b_n}{[n]_{q_n}}$, which is at least as fast as $\frac{b_n}{n}$.

2.2.6   Other Properties

In this section, we give two representations of the rth q-derivative of q-Szász–Mirakyan operators in terms of the q-differences and the divided difference and then obtain a representation of q-Szász–Mirakyan operators in terms of the divided differences which is the modified form of the representation of the classical Szász–Mirakyan operator given in [145, pp. 1183–1184]. Note that these representations are not obtained using classical derivatives and forward differences.

Proposition 2.2. *For each integer r > 0*

$$D_q^r\left(S_n^q(f;x)\right) = E_q\left(-[n]_q\, q^r \frac{x}{b_n}\right) \sum_{j=0}^{\infty} \left(\frac{[n]_q}{b_n}\right)^r \Delta_q^r f_j \frac{\left([n]_q x\right)^j}{[j]_q!\,(b_n)^j}. \qquad (2.34)$$

Proof. The proof is by induction on r. According to (1.4) we set $D_q\left(E_q\left(-[n]\frac{x}{b_n}\right)\right)$
$= -\frac{[n]}{b_n} E_q\left(-[n]q\frac{x}{b_n}\right)$. Applying the D_q-differential operator to (2.21) and using Lemma 1.1, (1.5), and (2.3) we find

$$D_q\left(S_n^q(f;x)\right) = -\frac{[n]_q}{b_n} E_q\left(-[n]_q\, q\frac{x}{b_n}\right) \sum_{j=0}^{\infty} f\left(\frac{[j]_q b_n}{[n]_q}\right) \frac{\left([n]_q x\right)^j}{[j]_q!\,(b_n)^j}$$

$$+\frac{[n]_q}{b_n} E_q\left(-[n]_q\, q\frac{x}{b_n}\right) \sum_{j=0}^{\infty} f\left(\frac{[j+1]_q b_n}{[n]_q}\right) \frac{\left([n]_q x\right)^j}{[j]_q!\,(b_n)^j}$$

$$= E_q\left(-[n]_q\, q\frac{x}{b_n}\right) \sum_{j=0}^{\infty} \frac{[n]_q}{b_n} \left(f\left(\frac{[j+1]_q b_n}{[n]_q}\right) - f\left(\frac{[j]_q b_n}{[n]_q}\right)\right) \frac{\left([n]_q x\right)^j}{[j]_q!\,(b_n)^j}$$

$$= E_q\left(-[n]_q\, q\frac{x}{b_n}\right) \sum_{j=0}^{\infty} \frac{[n]_q}{b_n} \Delta_q^1 f_j \frac{\left([n]_q x\right)^j}{[j]_q!\,(b_n)^j}.$$

Similarly,

$$D_q^2\left(S_n^q(f;x)\right)$$
$$= D_q\left(D_q\left(S_n^q(f;x)\right)\right)$$

$$= -q\left(\frac{[n]_q}{b_n}\right)^2 E_q\left(-[n]_q\, q^2\frac{x}{b_n}\right) \sum_{j=0}^{\infty} \left(f\left(\frac{[j+1]_q b_n}{[n]_q}\right) - \left(\frac{[j]_q b_n}{[n]_q}\right)\right) \frac{\left([n]_q x\right)^j}{[j]_q!\,(b_n)^j}$$

$$+\left(\frac{[n]_q}{b_n}\right)^2 E_q\left(-[n]_q\, q^2\frac{x}{b_n}\right) \sum_{j=0}^{\infty} \left(f\left(\frac{[j+2]_q b_n}{[n]_q}\right) - \left(\frac{[j+1]_q b_n}{[n]_q}\right)\right) \frac{\left([n]_q x\right)^j}{[j]_q!\,(b_n)^j}$$

$$= E_q\left(-[n]_q\, q^2\frac{x}{b_n}\right) \sum_{j=0}^{\infty} \left(\frac{[n]_q}{b_n}\right)^2 \Delta_q^2 f_j \frac{\left([n]_q x\right)^j}{[j]_q!\,(b_n)^j}.$$

Thus (2.34) holds for $r = 1$ and $r = 2$. Let us assume that it holds for some $r \geq 3$. Applying the q-differential operator to (2.34) we find

$$D_q \left(D_q^r \left(S_n^q (f; x) \right) \right)$$

$$= D_q^{r+1} \left(S_n^q (f; x) \right)$$

$$= -q^r E_q \left(-[n]_q q^{r+1} \frac{x}{b_n} \right) \sum_{j=0}^{\infty} \left(\frac{[n]_q}{b_n} \right)^{r+1} \Delta_q^r f_j \frac{\left([n]_q x \right)^j}{[j]_q! (b_n)^j}$$

$$+ E_q \left(-[n]_q q^{r+1} \frac{x}{b_n} \right) \sum_{j=0}^{\infty} \left(\frac{[n]_q}{b_n} \right)^{r+1} \Delta_q^r f_{j+1} \frac{\left([n]_q x \right)^j}{[j]_q! (b_n)^j}$$

$$= E_q \left(-[n]_q q^{r+1} \frac{x}{b_n} \right) \sum_{j=0}^{\infty} \left(\frac{[n]_q}{b_n} \right)^{r+1} \Delta_q^{r+1} f_j \frac{\left([n]_q x \right)^j}{[j]_q! (b_n)^j}$$

by using (2.3). This shows that (2.34) holds when r is replaced by $r + 1$, which completes the proof. ∎

Using the following connection between the divided differences and the q-differences given in [134, p. 44]

$$\Delta_q^r f_j = \left(\frac{b_n}{[n]} \right)^r [r]! q^{rj} q^{\frac{r(r-1)}{2}} f \left[\frac{b_n [j]}{[n]}, \frac{b_n [j+1]}{[n]}, \cdots, \frac{b_n [j+r]}{[n]} \right], \qquad (2.35)$$

then we have the following representation formula.

Corollary 2.3. *For each integer $r > 0$*

$$D_q^r \left(S_n^q (f; x) \right)$$

$$= q^{\frac{r(r-1)}{2}} [r]! E_q \left(-[n]_q q^r \frac{x}{b_n} \right) \sum_{j=0}^{\infty} q^{rj} f \left[\frac{b_n [j]}{[n]}, \frac{b_n [j+1]}{[n]}, \cdots, \frac{b_n [j+r]}{[n]} \right] \frac{\left([n]_q x \right)^j}{[j]_q! (b_n)^j}.$$

Corollary 2.4. *The q-Szász–Mirakyan operator can be represented as*

$$S_n^q (f; x) = \sum_{j=0}^{\infty} q^{\frac{j(j-1)}{2}} f \left[0, \frac{b_n [1]}{[n]}, \cdots, \frac{b_n [j]}{[n]} \right] x^j.$$

Proof. From the equalities (2.23) and (2.35), the proof is obvious. ∎

2.3 *q*-Baskakov Operators

In this section we propose a generalization of the Baskakov operators, based on *q*-integers. We also estimate the rate of convergence in the weighted norm. We also study some shape-preserving and monotonicity properties of the *q*-Baskakov operators and also different generalizations of classical Baskakov operators based on q-integers defined in [30, 136].

First, we recall classical Baskakov operators [37], which for $f \in C[0, \infty)$ are defined as

$$B_n(f,x) = \sum_{k=0}^{\infty} \binom{n+k-1}{k} x^k (1+x)^{-n-k} f\left(\frac{k}{n}\right)$$

This section is based on [32].

2.3.1 *Construction of Operators and Some Properties of Them*

For $f \in C[0, \infty)$, $q > 0$, and each positive integer n, a new *q*-Baskakov operators can be defined as

$$B_{n,q}(f,x) = \sum_{k=0}^{\infty} \begin{bmatrix} n+k-1 \\ k \end{bmatrix}_q q^{\frac{k(k-1)}{2}} x^k (-x,q)_{n+k}^{-1} f\left(\frac{[k]_q}{q^{k-1}[n]_q}\right)$$

$$= \sum_{k=0}^{\infty} \mathcal{P}_{n,k}^q(x) f\left(\frac{[k]_q}{q^{k-1}[n]_q}\right). \tag{2.36}$$

While for $q = 1$ these polynomials coincide with the classical ones.

Definition 2.1. Let f be a function defined on an interval (a,b) and h be a positive real number. The *q*-forward differences Δ_h^r of f are defined recursively as

$$\Delta_q^0 f(x_j) := f(x_j)$$

$$\Delta_q^{r+1} f(x_j) := q^r \Delta_q^r f(x_{j+1}) - \Delta_q^r f(x_j)$$

for $r \geq 0$.

Note that the above definition is different from definition given in [134, p. 44].

As usual, we show divided differences with $f[x_0, x_1, \ldots, x_n]$ at the abscissas $x_0, x_1, \ldots, x_n$.

We now show the following general relation that connect the divided differences $f[x_0, x_1, \ldots, x_n]$ and *q*-forward differences.

Lemma 2.3. *For all* $j, r \geq 0$, *we have*

$$f\left[x_j, x_{j+1}, \ldots, x_{j+r}\right] = q^{\frac{r(2j+r-1)}{2}} \frac{\Delta_q^r f(x_j)}{[r]_q!}, \qquad (2.37)$$

where $x_j = \frac{[j]_q}{q^{j-1}}$.

Proof. Let us use induction on r. By Definition 2.1, the result is obvious for $r = 0$. Let us assume that the equality (2.37) is true for some $r \geq 0$ and all $j \geq 0$. Since

$$x_{j+r+1} - x_j = \frac{[r+1]_q}{q^{j+r}},$$

we have

$$
\begin{aligned}
f\left[x_j, x_{j+1}, \ldots, x_{j+r+1}\right] &= \frac{f\left[x_{j+1}, \ldots, x_{j+r+1}\right] - f\left[x_j, \ldots, x_{j+r}\right]}{x_{j+r+1} - x_j} \\
&= \frac{q^{j+r}}{[r+1]_q} \left(q^{\frac{r(2j+r+1)}{2}} \frac{\Delta_q^r f(x_{j+1})}{[r]_q!} - q^{\frac{r(2j+r-1)}{2}} \frac{\Delta_q^r f(x_j)}{[r]_q!} \right) \\
&= q^{\frac{r(2j+r-1)}{2}+j+r} \left(\frac{q^r \Delta_q^r f(x_{j+1}) - \Delta_q^r f(x_j)}{[r+1]_q!} \right) \\
&= q^{\frac{(r+1)(2j+r)}{2}} \frac{\Delta_q^{r+1} f(x_j)}{[r+1]_q!}.
\end{aligned}
$$
∎

Lemma 2.4. *For* $n, k \geq 0$, *we have*

$$\mathcal{D}_q\left[x^k(-x,q)_{n+k}^{-1}\right] = [k]_q x^{k-1}(-x,q)_{n+k}^{-1} - q^k x^k [n+k]_q(-x,q)_{n+k+1}^{-1} \qquad (2.38)$$

Proof. First, we prove that $\mathcal{D}_q(-x,q)_n = [n]_q(-qx,q)_{n-1}$. Using q-derivative operator (1.5) we have

$$
\begin{aligned}
\mathcal{D}_q(-x,q)_n &= \frac{1}{(q-1)x} \left(\prod_{j=0}^{n-1}(1+q^{j+1}x) - \prod_{j=0}^{n-1}(1+q^j x) \right) \\
&= \frac{1}{(q-1)x} \prod_{j=0}^{n-2}(1+q^{j+1}x)\left((1+q^n x) - (1+x)\right) \\
&= \frac{q^n-1}{q-1} \prod_{j=0}^{n-2}(1+q^{j+1}x) \\
&= [n]_q(-qx,q)_{n-1}.
\end{aligned}
$$

The q-derivative formula for a quotient (1.6) imply that

$$D_q(-x,q)_{n+k}^{-1} = \frac{-[n+k]_q(-qx,q)_{n+k-1}}{(-x,q)_{n+k}(-qx,q)_{n+k}}$$

$$= -[n+k]_q(-x,q)_{n+k+1}^{-1}. \tag{2.39}$$

Also it is obvious that

$$D_q x^k = [k]_q x^{k-1}. \tag{2.40}$$

Then using (2.40) and (2.39), the result follows by (1.5)

$$D_q\left[x^k(-x,q)_{n+k}^{-1}\right] = [k]_q x^{k-1}(-x,q)_{n+k}^{-1} - q^k x^k[n+k]_q(-x,q)_{n+k+1}^{-1}. \qquad \blacksquare$$

We wish to calculate the moments. For this purpose we give q-derivative of $B_{n,q}$. Next theorem gives a representation of the rth derivative of $B_{n,q}$ in terms of q-forward differences.

Theorem 2.9. *Let $r \geq 0$. Then the rth derivative of q-Baskakov operator has the representation*

$$D_q^r B_{n,q}(f,x) = \frac{[n+r-1]_q!}{[n-1]_q!}\sum_{k=0}^{\infty}q^{rk}\mathcal{P}_{n+r,k}^q(x)\Delta_q^r f\left(\frac{[k]_q}{q^{k-1}[n]_q}\right) \tag{2.41}$$

Proof. We use induction on r. Equality (2.38),

$$\begin{bmatrix} n+k \\ k+1 \end{bmatrix}_q [k+1]_q = [n]_q \begin{bmatrix} n+k \\ k \end{bmatrix}_q$$

and

$$\begin{bmatrix} n+k-1 \\ k \end{bmatrix}_q [n+k]_q = [n]_q \begin{bmatrix} n+k \\ k \end{bmatrix}_q,$$

imply that

$$D_q B_{n,q}(f,x) = \sum_{k=1}^{\infty}\begin{bmatrix} n+k-1 \\ k \end{bmatrix}_q q^{\frac{k(k-1)}{2}}[k]_q x^{k-1}(-x,q)_{n+k}^{-1}f\left(\frac{[k]_q}{q^{k-1}[n]_q}\right)$$

$$- \sum_{k=0}^{\infty}\begin{bmatrix} n+k-1 \\ k \end{bmatrix}_q q^{\frac{k(k-1)}{2}}q^k x^k[n+k]_q(-x,q)_{n+k+1}^{-1}f\left(\frac{[k]_q}{q^{k-1}[n]_q}\right)$$

$$= [n]_q\sum_{k=0}^{\infty}\begin{bmatrix} n+k \\ k \end{bmatrix}_q q^{\frac{k(k-1)}{2}+k}x^k(-x,q)_{n+k+1}^{-1}\left(f\left(\frac{[k+1]_q}{q^k[n]_q}\right) - f\left(\frac{[k]_q}{q^{k-1}[n]_q}\right)\right)$$

$$= [n]_q\sum_{k=0}^{\infty}q^k\mathcal{P}_{n+1,k}^q(x)\Delta_q^1 f\left(\frac{[k]_q}{q^{k-1}[n]_q}\right).$$

It is clear that (2.41) holds for $r = 1$. Let us assume that (2.41) holds for some $r \geq 2$. Applying q-derivative operator to (2.41), we have

$$
\mathcal{D}_q^{r+1}\left(B_{n,q}(f,x)\right) = \frac{[n+r-1]_q!}{[n-1]_q!} \sum_{k=1}^{\infty} \begin{bmatrix} n+k+r-1 \\ k \end{bmatrix}_q q^{\frac{k(k-1)}{2}+rk}[k]_q
$$

$$
\times x^{k-1}(-x,q)_{n+k+r}^{-1} \Delta_q^r f\left(\frac{[k]_q}{q^{k-1}[n]_q}\right)
$$

$$
- \frac{[n+r-1]_q!}{[n-1]_q!} \sum_{k=0}^{\infty} \begin{bmatrix} n+k+r-1 \\ k \end{bmatrix}_q [n+k+r]_q q^k q^{\frac{k(k-1)}{2}+rk}
$$

$$
\times x^k(-x,q)_{n+k+r+1}^{-1} \Delta_q^r f\left(\frac{[k]_q}{q^{k-1}[n]_q}\right)
$$

$$
= \frac{[n+r]_q!}{[n-1]_q!} \sum_{k=0}^{\infty} \begin{bmatrix} n+k+r \\ k \end{bmatrix}_q q^{\frac{k(k-1)}{2}+(r+1)k} x^k(-x,q)_{n+k+r+1}^{-1}
$$

$$
\times \left(q^r \Delta_q^r f\left(\frac{[k+1]_q}{q^k[n]_q}\right) - \Delta_q^r f\left(\frac{[k]_q}{q^{k-1}[n]_q}\right)\right)
$$

$$
= \frac{[n+r]_q!}{[n-1]_q!} \sum_{k=0}^{\infty} q^{(r+1)k} \mathcal{P}_{n+r+1,k}^q(x) \Delta_q^{r+1} f\left(\frac{[k]_q}{q^{k-1}[n]_q}\right)
$$

This completes the proof of the theorem. ∎

Corollary 2.5. *q-Baskakov operators can be represented as*

$$
B_{n,q}(f,x) = \sum_{r=0}^{\infty} \frac{[n+r-1]_q!}{[n-1]_q!} \Delta_q^r f(0) \frac{x^r}{[r]_q!}.
$$

Proof. By Theorem 2.9, we have

$$
\mathcal{D}_q^r\left(B_{n,q}(f,x)\right)\big|_{x=0} = \frac{[n+r-1]_q!}{[n-1]_q!} \mathcal{P}_{n+r,0}^q(0) \Delta_q^r f(0)
$$

$$
= \frac{[n+r-1]_q!}{[n-1]_q!} \Delta_q^r f(0)
$$

for $r \geq 1$. By using the above equality in q-Taylor formula given in [137], we get

$$
B_{n,q}(f,x) = \sum_{r=0}^{\infty} \frac{[n+r-1]_q!}{[n-1]_q!} \Delta_q^r f(0) \frac{x^r}{[r]_q!}. \tag{2.42}
$$

∎

From Lemma 2.3 and Corollary 2.5, we have the following corollary.

Corollary 2.6. *The q-Baskakov operators can be represented as*

$$B_{n,q}(f,x) = \sum_{r=0}^{\infty} \frac{[n+r-1]_q!}{[n-1]_q!} q^{-\frac{r(r-1)}{2}} f\left[0, \frac{1}{[n]_q}, \frac{[2]_q}{q[n]_q}, \ldots, \frac{[r]_q}{q^{r-1}[n]_q}\right] x^r.$$

We are now in a position to give the moments of the first and second orders of the operators $B_{n,q}$.

Lemma 2.5. *For $B_{n,q}(t^m,x)$, $m = 0,1,2$, one has*

$$B_{n,q}(1,x) = 1.$$

$$B_{n,q}(t,x) = x,$$

$$B_{n,q}(t^2,x) = x^2 + \frac{x}{[n]_q}\left(1 + \frac{1}{q}x\right).$$

Proof. It is well known [134, p. 10] that

$$f[x_0,x_1,\ldots,x_r] = \frac{f^{(r)}(\xi)}{r!}, \tag{2.43}$$

where $\xi \in (x_0, x_r)$. We also see from Lemma 2.3 and (2.43)

$$q^{\frac{r(r-1)}{2}} \frac{\Delta_q^r f(x_0)}{[n]_q^r [r]_q!} = \frac{f^{(r)}(\xi)}{r!}.$$

Thus it is observed that rth q-forward differences of x^m, $m > r$ are zero. From (2.42), we have

$$B_{n,q}(1,x) = 1. \tag{2.44}$$

For $f(x) = x$ we have $\Delta_q^0 f(0) = f(0) = 0$ and $\Delta_q^1 f(0) = f\left(\frac{1}{[n]_q}\right) - f(0) = \frac{1}{[n]_q}$ and it follows from (2.42)

$$B_{n,q}(t,x) = x \tag{2.45}$$

For $f(x) = x^2$ we have $\Delta_q^0 f(0) = f(0) = 0$ and $\Delta_q^1 f(0) = f\left(\frac{1}{[n]_q}\right) - f(0) = \frac{1}{[n]_q^2}$ and $\Delta_q^2 f(0) = qf\left(\frac{[2]_q}{q[n]_q}\right) - (1+q)f\left(\frac{1}{[n]_q}\right) - f(0)$

$$B_{n,q}(t^2,x) = \frac{[n+1]_q}{[n]_q}\left(\frac{1}{q}[2]_q - 1\right)x^2 + \frac{x}{[n]_q}$$

$$= \frac{q[n]_q + 1}{[n]_q}\left(\frac{1}{q}(1+q) - 1\right)x^2 + \frac{x}{[n]_q}$$

$$= x^2 + \frac{1}{q[n]_q}x^2 + \frac{x}{[n]_q}$$

$$= x^2 + \frac{x}{[n]_q}\left(1 + \frac{1}{q}x\right). \tag{2.46}$$

∎

The following proposition is another application of *q*-derivatives, which enables us to give the estimation of moments:

Proposition 2.3. *If we define*

$$U_{n,m}^q(x) := B_{n,q}(t^m, x) = \sum_{k=0}^{\infty} \mathcal{P}_{n,k}^q(x)\left(\frac{[k]_q}{q^{k-1}[n]_q}\right)^m,$$

then $U_{n,0}^q(x) = 1, U_{n,1}^q(x) = x$ *and there holds the following recurrence relation:*

$$[n]_q U_{n,m+1}^q(qx) = qx(1+x)D_q U_{n,m}^q(x) + qx[n]_q U_{n,m}^q(qx), m > 1.$$

Proof. Obviously $\sum_{k=0}^{\infty} \mathcal{P}_{n,k}^q(x) = 1$; thus, by this identity and (2.1), the values of $U_{n,0}^q(x)$ and $U_{n,1}^q(x)$ easily follow. From Lemma 2.4, it is obvious that $x(1 + q^{n+k}x)D_q\mathcal{P}_{n,k}^q(x) = ([k]_q - q^k[n]_q x)\mathcal{P}_{n,k}^q(x)$, which implies that

$$x(1+x)D_q\mathcal{P}_{n,k}^q(x) = \left(\frac{[k]_q}{q^{k-1}[n]_q} - qx\right)\frac{[n]_q}{q}\mathcal{P}_{n,k}^q(qx)$$

Thus using this identity, we have

$$qx(1+x)D_q U_{n,m}^q(x) = \sum_{k=0}^{\infty} qx(1+x)D_q\mathcal{P}_{n,k}^q(x)\left(\frac{[k]_q}{q^{k-1}[n]_q}\right)^m$$

$$= [n]_q \sum_{k=0}^{\infty}\left(\frac{[k]_q}{q^{k-1}[n]_q} - qx\right)\mathcal{P}_{n,k}^q(qx)\left(\frac{[k]_q}{q^{k-1}[n]_q}\right)^m.$$

$$= [n]_q U_{n,m+1}^q(qx) - qx[n]_q U_{n,m}^q(qx).$$

This completes the proof of the recurrence relation.   ∎

2.3.2   Approximation Properties

We set

$$E_2(\mathbb{R}_+) := \left\{f \in C(\mathbb{R}_+) : \lim_{x \to \infty}\frac{f(x)}{1+x^2} \text{ exist}\right\}$$

and

$$B_2(\mathbb{R}_+) := \left\{ f : |f(x)| \le B_f (1+x^2) \right\}$$

where B_f is a constant depending on f, endowed with the norm $\|f\|_2 := \sup\limits_{x \ge 0} \frac{|f(x)|}{1+x^2}$.

As a consequence of Lemma 2.5, the operators (2.36) map $E_2(\mathbb{R}_+)$ into $E_2(\mathbb{R}_+)$. Since for a fixed value of q with $q > 0$,

$$\lim_{n \to \infty} [n]_q = \frac{1}{1-q},$$

$B_{n,q}(t^2, x)$ does not converge to x^2 as $n \to \infty$. According to well-known Bohman–Korovkin theorem, relations (2.44), (2.45), and (2.46) don't guarantee that $\lim_{n \to \infty} B_{n,q_n} f = f$ uniformly on compact subset of $\mathbb{R}_+$ for every $f \in E_2(\mathbb{R}_+)$. To ensure this type of convergence properties of (2.36) we replace $q = q_n$ as a sequence such that $q_n \to 1$ as $n \to \infty$ for $q_n > 0$ and so that $[n]_{q_n} \to \infty$ as $n \to \infty$. Also, $B_{n,q_n} f$ are linear and positive operators for $q_n > 0$. In this situation, we can apply Bohman–Korovkin theorem to B_{n,q_n}. That is:

Theorem 2.10. *Let (q_n) be a sequence of real numbers such that $q_n > 0$ and $\lim_{n \to \infty} q_n = 1$. Then for every $f \in E_2(\mathbb{R}_+)$*

$$\lim_{n \to \infty} B_{n,q_n} f = f$$

uniformly on any compact subset of $\mathbb{R}_+$.

Theorem 2.11. *Let $q = q_n$ satisfies $q_n > 0$ and let $q_n \to 1$ as $n \to \infty$. For every $f \in B_2(\mathbb{R}_+)$,*

$$\lim_{n \to \infty} \sup_{x \ge 0} \frac{|B_{n,q_n}(f;x) - f(x)|}{(1+x^2)^3} = 0. \tag{2.47}$$

Proof. Since f is continuous, it is also uniformly continuous; on any closed interval, there exist a number $\delta > 0$, depending on ε and f; for $|t - x| < \delta$ we have

$$|f(t) - f(x)| < \varepsilon.$$

Since $f \in B_2(\mathbb{R}_+)$, we can write for $|t - x| \ge \delta$

$$|f(t) - f(x)| < A_f(\delta) \left\{ (t-x)^2 + (1+x^2) |t-x| \right\},$$

where $A_f(\delta)$ is a positive constant depending on f and δ.

On combining above results, we obtain

$$|f(t) - f(x)| < \varepsilon + A_f(\delta) \left\{ (t-x)^2 + (1+x^2) |t-x| \right\},$$

where $t, x \in \mathbb{R}_+$. Thus, we have

$$\left| B_{n,q_n}(f;x) - f(x) \right| < \varepsilon + A_f(\delta) \left\{ B_{n,q_n}\left((t-x)^2;x\right) + \left(1+x^2\right) B_{n,q_n}\left(|t-x|;x\right) \right\}$$

and from Lemma 2.5

$$\sup_{x \geq 0} \frac{\left| B_{n,q_n}(f;x) - f(x) \right|}{1+x^2} < \varepsilon + A_f(\delta) \left\{ \frac{1}{[n]_{q_n}}\left(1+\frac{1}{q_n}\right) + \sqrt{\frac{1}{[n]_{q_n}}\left(1+\frac{1}{q_n}\right)} \right\},$$

and this completes the proof.   ∎

Remark 2.4. Using the similar method given in [12, p. 301], we have

$$\left| B_{n,q_n}(f;x) - f(x) \right| \leq M\omega_2\left(f; \sqrt{\frac{x}{[n]_q}\left(1+\frac{1}{q}x\right)}\right),$$

where $\omega_2(f;\delta)$ is classical second modulus of smoothness of f and f is bounded uniformly continuous function on $\mathbb{R}_+$. Thus, we say that the rate of convergence of $B_{n,q_n}(f)$ to f in any closed subinterval of $\mathbb{R}_+$ is $\frac{1}{\sqrt{[n]_{q_n}}}$, which is at least as fast as $\frac{1}{\sqrt{n}}$ which is the rate of convergence of classical Baskakov operators.

2.3.3 Shape-Preserving Properties

Definition 2.2 ([115, 116, 131]). Let f be continuous and a nonnegative function such that $f(0) = 0$. A function f is called star-shaped in $[0,a]$; a is a positive real number, if

$$f(\alpha x) \leq \alpha f(x)$$

for each α, $\alpha \in [0, 1]$ and $x \in (0,a]$.

From the definition of q-derivative (1.5), the following lemma is obvious.

Lemma 2.6. *The function f is star-shaped if and only if $xD_q(f)(x) \geq f(x)$ for each $q \in (0,1)$ and $x \in [0,a]$.*

Theorem 2.12. *If f is star-shaped, then $B_{n,q}(f)$ is star-shape.*

Proof. From Theorem 2.9, we can write

$$D_q\left(B_{n,q}(f,x)\right) - \frac{B_{n,q}(f,x)}{x}$$

$$= [n]_q \sum_{k=0}^{\infty} q^k \nabla_q^1 f\left(\frac{[k]_q}{q^{k-1}[n]_q}\right) \left[\begin{matrix} n+k \\ k \end{matrix}\right]_q q^{\frac{k(k-1)}{2}} x^k (-x,q)_{n+k+1}^{-1}$$

$$-\sum_{k=1}^{\infty} f\left(\frac{[k]_q}{q^{k-1}[n]_q}\right)\begin{bmatrix}n+k-1\\k\end{bmatrix}_q q^{\frac{k(k-1)}{2}} x^{k-1}(-x,q)_{n+k}^{-1}$$

$$= [n]\sum_{k=0}^{\infty}\begin{bmatrix}n+k\\k\end{bmatrix}_q q^{\frac{k(k-1)}{2}} q^k x^k (-x,q)_{n+k+1}^{-1}$$

$$\left(f\left(\frac{[k+1]_q}{q^k[n]_q}\right)-f\left(\frac{[k]_q}{q^{k-1}[n]_q}\right)-\frac{1}{[k+1]_q}f\left(\frac{[k+1]_q}{q^k[n]_q}\right)\right).$$

Since

$$1-\frac{1}{[k+1]_q}=\frac{q[k]_q}{[k+1]_q},$$

we have

$$\mathcal{D}_q\left(B_{n,q}(f,x)\right)-\frac{B_{n,q}(f,x)}{x}$$

$$= [n]\sum_{k=0}^{\infty} q^k P_{n+1,k}^q\left(\frac{q[k]_q}{[k+1]_q}f\left(\frac{[k+1]_q}{q^k[n]_q}\right)-f\left(\frac{[k]_q}{q^{k-1}[n]_q}\right)\right). \qquad (2.48)$$

Since f is star-shaped, we have

$$\frac{q[k]_q}{[k+1]_q}f\left(\frac{[k+1]_q}{q^k[n]_q}\right)\geq f\left(\frac{[k]_q}{q^{k-1}[n]_q}\right).$$

From this inequality and (2.48), we have the desired result.   ∎

Now we give a certain monotonicity property of the q-Baskakov operators defined by (2.36). Similar results for the classical Baskakov operators were given in [41].

Theorem 2.13. *Suppose $f(x)$ is defined on $(0,\infty)$ and $f(x)\geq 0$ for $x\in(0,\infty)$. If $\frac{f(x)}{x}$ is decreasing for all $x\in(0,\infty)$, then $\mathcal{D}_q\left(\frac{B_{n,q}(f;x)}{x}\right)\leq 0$ for $x\in(0,\infty)$ and for all $q\in(0,\infty)$.*

Proof. From (2.36) we get

$$\frac{B_{n,q}(f;x)}{x}=\sum_{k=1}^{\infty} f\left(\frac{[k]_q}{q^{k-1}[n]_q}\right)\begin{bmatrix}n+k-1\\k\end{bmatrix}_q q^{\frac{k(k-1)}{2}} x^{k-1}(-x,q)_{n+k}^{-1}+\frac{f(0)}{x}(-x,q)_n^{-1}.$$

If we take q-derivative of above equality and using Lemma 2.4, then we have

$$\mathcal{D}_q\left(\frac{B_{n,q}(f;x)}{x}\right)=\sum_{k=2}^{\infty} f\left(\frac{[k]_q}{q^{k-1}[n]_q}\right)\begin{bmatrix}n+k-1\\k\end{bmatrix}_q q^{\frac{k(k-1)}{2}} [k-1]_q x^{k-2}(-x,q)_{n+k}^{-1}$$

$$-\sum_{k=1}^{\infty} f\left(\frac{[k]_q}{q^{k-1}[n]_q}\right) \begin{bmatrix} n+k-1 \\ k \end{bmatrix}_q q^{\frac{k(k-1)}{2}} [n+k]_q q^{k-1} x^{k-1} (-x,q)_{n+k+1}^{-1}$$

$$+\mathcal{D}_q\left(\frac{f(0)}{x}(-x,q)_n^{-1}\right).$$

Also using (1.5) and (1.6), we get

$$\mathcal{D}_q\left(\frac{f(0)}{x}(-x,q)_n^{-1}\right) = -\frac{f(0)}{qx^2}(-x,q)_n^{-1} - [n]_q\frac{f(0)}{x}(-x,q)_{n+1}^{-1}$$

Therefore,

$$\mathcal{D}_q\left(\frac{B_{n,\,q}(f;x)}{x}\right)$$

$$= \sum_{k=1}^{\infty} f\left(\frac{[k+1]_q}{q^k[n]_q}\right) \begin{bmatrix} n+k \\ k+1 \end{bmatrix}_q q^{\frac{k(k-1)}{2}} q^k [k]_q x^{k-1} (-x,q)_{n+k+1}^{-1}$$

$$- \sum_{k=1}^{\infty} f\left(\frac{[k]_q}{q^{k-1}[n]_q}\right) \begin{bmatrix} n+k-1 \\ k \end{bmatrix}_q q^{\frac{k(k-1)}{2}} [n+k]_q q^{k-1} x^{k-1} (-x,q)_{n+k+1}^{-1}$$

$$- \frac{f(0)}{qx^2}(-x,q)_n^{-1} - [n]_q\frac{f(0)}{x}(-x,q)_{n+1}^{-1}$$

Using the identities

$$\begin{bmatrix} n+k \\ k+1 \end{bmatrix}_q = \begin{bmatrix} n+k \\ k \end{bmatrix}_q \frac{[n]_q}{[k+1]_q}$$

$$\begin{bmatrix} n+k-1 \\ k \end{bmatrix}_q [n+k]_q = \begin{bmatrix} n+k \\ k \end{bmatrix}_q [n]_q,$$

we have

$$\mathcal{D}_q\left(\frac{B_{n,\,q}(f;x)}{x}\right) = \sum_{k=1}^{\infty} \begin{bmatrix} n+k \\ k \end{bmatrix}_q q^{\frac{k(k-1)}{2}} x^{k-1} (-x,q)_{n+k+1}^{-1}$$

$$\left(f\left(\frac{[k+1]_q}{q^k[n]_q}\right)\frac{q^k[n]_q}{[k+1]_q} - f\left(\frac{[k]_q}{q^{k-1}[n]_q}\right)\frac{q^{k-1}[n]_q}{[k]_q}\right)[k]_q$$

$$- \frac{f(0)}{qx^2}(-x,q)_n^{-1} - [n]_q\frac{f(0)}{x}(-x,q)_{n+1}^{-1}.$$

Since $f(x) \geq 0$ and $\frac{f(x)}{x}$ is nonincreasing for $x \in (0, \infty)$,

$$D_q \left(\frac{B_{n,\,q}\,(f;x)}{x} \right) \le 0$$

for all $q \in (0, \infty)$ and $x \in (0, \infty)$.   ∎

2.3.4 Monotonicity Property

Now we give the following relation between two consecutive terms of the sequence $B_{n,q}(f)$. Note that similar result for classical Baskakov operators was given in [122].

Theorem 2.14. *If $f \in C(\mathbb{R}^+)$, then the following formula is valid*

$$B_{n+1,q}(f,x) - B_{n,q}(f,x) = -\frac{q^n}{[n]_q\,[n+1]_q} \sum_{k=0}^{\infty} q^{\frac{k(k+1)}{2}-2k} x^{k+1}\,(-x,q)_{n+k+1}^{-1} \begin{bmatrix} n+k \\ k \end{bmatrix}_q$$

$$\times \frac{[n+k+1]_q}{[n+1]_q} f\left[\frac{[k]_q}{q^{k-1}[n+1]_q}, \frac{[k+1]_q}{q^k[n+1]_q}, \frac{[k+1]_q}{q^k[n]_q} \right]$$

Proof. Using the equality

$$1 = 1 + q^{n+k}x - q^{n+k}x,$$

from (2.36) we can write

$$B_{n+1,q}(f;x) = \sum_{k=0}^{\infty} f\left(\frac{[k]_q}{q^{k-1}[n+1]_q} \right) \begin{bmatrix} n+k \\ k \end{bmatrix}_q q^{\frac{k(k-1)}{2}} x^k (-x,q)_{n+k+1}^{-1}$$

$$= \sum_{k=0}^{\infty} f\left(\frac{[k]_q}{q^{k-1}[n+1]_q} \right) \begin{bmatrix} n+k \\ k \end{bmatrix}_q q^{\frac{k(k-1)}{2}} x^k (-x,q)_{n+k}^{-1}$$

$$- \sum_{k=0}^{\infty} f\left(\frac{[k]_q}{q^{k-1}[n+1]_q} \right) \begin{bmatrix} n+k \\ k \end{bmatrix}_q q^{\frac{k(k-1)}{2}} q^{n+k} x^{k+1} (-x,q)_{n+k+1}^{-1}$$

$$= f(0)(-x,q)_n^{-1} + \sum_{k=1}^{\infty} f\left(\frac{[k]_q}{q^{k-1}[n+1]_q} \right) \begin{bmatrix} n+k \\ k \end{bmatrix}_q q^{\frac{k(k-1)}{2}} x^k (-x,q)_{n+k}^{-1}$$

$$- \sum_{k=0}^{\infty} f\left(\frac{[k]_q}{q^{k-1}[n+1]_q} \right) \begin{bmatrix} n+k \\ k \end{bmatrix}_q q^{\frac{k(k-1)}{2}} q^{n+k} x^{k+1} (-x,q)_{n+k+1}^{-1}$$

Thus, we have

$$B_{n+1,q}(f;x) = f(0)(-x,q)_n^{-1} + \sum_{k=0}^{\infty} f\left(\frac{[k+1]_q}{q^k[n+1]_q}\right)$$

$$\cdot \begin{bmatrix} n+k+1 \\ k+1 \end{bmatrix}_q q^{\frac{k(k-1)}{2}} q^k x^{k+1} (-x,q)_{n+k+1}^{-1}$$

$$- \sum_{k=0}^{\infty} f\left(\frac{[k]_q}{q^{k-1}[n+1]_q}\right) \begin{bmatrix} n+k \\ k \end{bmatrix}_q q^{\frac{k(k-1)}{2}} q^{n+k} x^{k+1} (-x,q)_{n+k+1}^{-1}.$$

Since

$$B_{n,q}(f;x) = f(0)(-x,q)_n^{-1} + \sum_{k=1}^{\infty} f\left(\frac{[k]_q}{q^{k-1}[n]_q}\right) \begin{bmatrix} n+k-1 \\ k \end{bmatrix}_q q^{\frac{k(k-1)}{2}} x^k (-x,q)_{n+k}^{-1}$$

$$= f(0)(-x,q)_n^{-1} + \sum_{k=0}^{\infty} f\left(\frac{[k+1]_q}{q^k[n]_q}\right) \begin{bmatrix} n+k \\ k+1 \end{bmatrix}_q q^{\frac{k(k-1)}{2}} q^k x^{k+1} (-x,q)_{n+k+1}^{-1},$$

we have

$$B_{n+1,q}(f,x) - B_{n,q}(f,x)$$

$$= \sum_{k=0}^{\infty} q^{\frac{k(k-1)}{2}} q^k x^{k+1} (-x,q)_{n+k+1}^{-1}$$

$$\left(f\left(\frac{[k+1]_q}{q^k[n+1]_q}\right) \begin{bmatrix} n+k+1 \\ k+1 \end{bmatrix}_q - q^n f\left(\frac{[k]_q}{q^{k-1}[n+1]_q}\right) \begin{bmatrix} n+k \\ k \end{bmatrix}_q \right.$$

$$\left. - f\left(\frac{[k+1]_q}{q^k[n]_q}\right) \begin{bmatrix} n+k \\ k+1 \end{bmatrix}_q \right)$$

Using the equalities

$$\begin{bmatrix} n+k+1 \\ k+1 \end{bmatrix}_q = \frac{[n+k+1]_q}{[k+1]_q} \begin{bmatrix} n+k \\ k \end{bmatrix}_q$$

and

$$\begin{bmatrix} n+k \\ k+1 \end{bmatrix}_q = \frac{[n]_q}{[k+1]_q} \begin{bmatrix} n+k \\ k \end{bmatrix}_q,$$

we can write

$$B_{n+1,q}(f,x) - B_{n,q}(f,x)$$

$$= -\sum_{k=0}^{\infty} q^{\frac{k(k+1)}{2}} x^{k+1} (-x,q)_{n+k+1}^{-1} \begin{bmatrix} n+k \\ k \end{bmatrix}_q$$

$$\left(q^n f\left(\frac{[k]_q}{q^{k-1}[n+1]_q} \right) - \frac{[n+k+1]_q}{[k+1]_q} f\left(\frac{[k+1]_q}{q^k[n+1]_q} \right) + \frac{[n]_q}{[k+1]_q} f\left(\frac{[k+1]_q}{q^k[n]_q} \right) \right).$$

Using the inequalities

$$\frac{[k+1]_q}{q^k[n]_q} - \frac{[k]_q}{q^{k-1}[n+1]_q} = \frac{[n+k+1]_q}{q^k[n]_q[n+1]_q},$$

$$\frac{[k+1]_q}{q^k[n+1]_q} - \frac{[k]_q}{q^{k-1}[n+1]_q} = \frac{1}{q^k[n+1]_q},$$

and

$$\frac{[k+1]_q}{q^k[n]_q} - \frac{[k+1]_q}{q^k[n+1]_q} = \frac{q^n[k+1]_q}{q^k[n+1]_q[n]_q},$$

we can easily see that

$$f\left[\frac{[k]_q}{q^{k-1}[n+1]_q}, \frac{[k+1]_q}{q^k[n+1]_q}, \frac{[k+1]_q}{q^k[n]_q} \right] = \frac{q^{2k}[n]_q[n+1]_q^2}{q^n[n+k+1]_q} \left(q^n f\left(\frac{[k]_q}{q^{k-1}[n+1]_q} \right) \right.$$

$$\left. - \frac{[n+k+1]_q}{[k+1]_q} f\left(\frac{[k+1]_q}{q^k[n+1]_q} \right) + \frac{[n]_q}{[k+1]_q} f\left(\frac{[k+1]_q}{q^k[n]_q} \right) \right).$$

This proves the theorem. ∎

We know that a function f is convex if and only if all second-order divided differences of f are nonnegative. Using this property and Theorem 2.14, we have the following result:

Corollary 2.7. *If $f(x)$ is a convex function defined on $\mathbb{R}^+$, then the q-Baskakov operator $B_{n,q}(f,x)$ defined by (2.36) is strictly monotonically nondecreasing in n, unless f is the linear function (in which case $B_{n,q}(f,x) = B_{n+1,q}(f,x)$ for all n).*

2.4   Approximation Properties of q-Baskakov Operators

We establish direct estimates for the q-Baskakov operator given by (2.36), using the second-order Ditzian–Totik modulus of smoothness. Furthermore, we define and study the limit q-Baskakov operator.

This section based on [63].

2.4.1   Introduction

We denote by $C_B[0,\infty)$ the space of all real valued, continuous, and bounded functions defined on $[0,\infty)$. This space equipped with the norm $\|f\| = \sup\{|f(x)| : x \in [0,\infty)\}$, $f \in C_B[0,\infty)$ is a Banach space.

We know that from (2.36), q-analogue of Baskakov operators, which for $q \in (0,1)$, $n = 1,2,\ldots$, $f \in C_B[0,\infty)$ and $x \in [0,\infty)$, is defined as

$$\mathcal{B}_{n,q}(f,x) = \sum_{k=0}^{\infty} \begin{bmatrix} n+k-1 \\ k \end{bmatrix}_q q^{\frac{k(k-1)}{2}} x^k (-x,q)_{n+k}^{-1} f\left(\frac{[k]_q}{q^{k-1}[n]_q}\right)$$

$$= \sum_{k=0}^{\infty} \mathcal{P}_{n,k}^q(x) f\left(\frac{[k]_q}{q^{k-1}[n]_q}\right).$$

For $q = 1$, we recover the well-known Baskakov operators [37].

Here, to obtain direct global estimates for the q-Baskakov operators, we use the second-order Ditzian–Totik modulus of smoothness, defined for $f \in C_B[0,\infty)$ by

$$\omega_\varphi^2(f;\delta) = \sup_{0<h\leq\delta} \sup_{x\pm h\varphi(x)\in[0,\infty)} |f(x+h\varphi(x)) - 2f(x) + f(x-h\varphi(x))|, \quad (2.49)$$

where $\varphi(x) = \sqrt{x(1+x)}$, $x \in [0,\infty)$, let us consider the following K-function:

$$\bar{K}_{2,\varphi}(f;\delta) = \inf_{g'\in AC_{loc}[0,\infty)} \{\|f-g\| + \delta\|\varphi^2 g''\| + \delta^2\|g''\|\},$$

where $g' \in AC_{loc}[0,\infty)$ means that g is differentiable and g' is absolutely continuous in every closed finite interval $[a,b] \subset [0,\infty)$. In view of [51, pp. 24–25], we have known that $\omega_2^\varphi(f;\delta)$ and $\bar{K}_{2,\varphi}(f;\delta^2)$ are equivalent, i.e., there exists $C > 0$ such that

$$C^{-1}\omega_\varphi^2(f;\delta) \leq \bar{K}_{2,\varphi}(f;\delta^2) \leq C\omega_\varphi^2(f;\delta). \quad (2.50)$$

Here we mention that C will denote throughout this paper an absolute positive constant which can be different at each occurrence. Analogously, for the K-function

$$K_{2,\varphi}(f;\delta) = \inf_{g' \in AC_{loc}[0,\infty)} \{\|f - g\| + \delta\|\varphi^2 g''\|\},$$

we have the equivalence of $\omega_2^{\varphi}(f;\delta)$ and $K_{2,\varphi}(f;\delta^2)$ (see [51, p. 11, Theorem 2.1.1], i.e., there exists $C > 0$ such that

$$C^{-1}\omega_2^{\varphi}(f;\delta) \le K_{2,\varphi}(f;\delta^2) \le C\omega_2^{\varphi}(f;\delta). \tag{2.51}$$

Furthermore, for $f \in C_B[0,\infty)$, $q \in (0,1)$, and $x \in [0,\infty)$, we define the limit q-Baskakov operator as

$$\mathcal{B}_{\infty,q}(f;x) \equiv (\mathcal{B}_{\infty,q}f)(x) = \sum_{k=0}^{\infty} v_{\infty,k}(q;x)f\left(\frac{1-q^k}{q^{k-1}}\right), \tag{2.52}$$

where

$$b_{\infty,k}(q;x) = q^{k(k-1)/2}(1-q)^{-1}(1-q^2)^{-1}\ldots(1-q^k)^{-1}$$

$$\times x^k \prod_{s=0}^{\infty}(1+xq^s)^{-1}. \tag{2.53}$$

By Euler's identity (see [20, Chap. 10, Corollary 10.2.2]), we have

$$\sum_{k=0}^{\infty} q^{k(k-1)/2}x^k(1-q)^{-1}(1-q^2)^{-1}\ldots(1-q^k)^{-1} = \prod_{s=0}^{\infty}(1+xq^s),$$

where $x \in [0,\infty)$ and $q \in (0,1)$. Due to (2.53), the last identity implies that

$$\sum_{k=0}^{\infty} b_{\infty,k}(q;x) = 1 \tag{2.54}$$

for $q \in (0,1)$ and $x \in [0,\infty)$. Hence

$$|\mathcal{B}_{\infty,q}(f;x)| \le \|f\| \sum_{k=0}^{\infty} b_{\infty,k}(q;x) = \|f\|,$$

i.e., $\|\mathcal{B}_{\infty,q}f\| \le \|f\|$ for $f \in C_B[0,\infty)$. This means that the limit q-Baskakov operator is well defined.

In what follows we shall estimate the rate of approximation $\|\mathcal{B}_{n,q}f - \mathcal{B}_{\infty,q}f\|$ by the second-order Ditzian–Totik modulus of smoothness of f (see (2.49)).

2.4.2   Main Results

We introduce the space $\tilde{C}_B[0,\infty) = \{f \in C_B[0,\infty) : \text{there exists } \lim_{x\to\infty} f(x) = 0\}$. Obviously $\tilde{C}_B[0,\infty) \subset C_B[0,\infty)$ and $\tilde{C}_B[0,\infty)$ is also a Banach space. The following theorems were studied in [63].

Theorem 2.15. *Let $C_0 \in (0,1)$ be an absolute constant with the property that $q = q(n) \in (C_0^{1/n}, 1)$ for every $n = 1, 2, \ldots$. Then there exists $C > 0$ such that*

$$\|\mathcal{B}_{n,q}f - f\| \leq C\,\omega_\varphi^2(f; [n]_q^{-1/2}) \tag{2.55}$$

for all $f \in \tilde{C}_B[0,\infty)$ and $n = 3, 4, \ldots$.

In the next theorem we estimate the rate of approximation $\|\mathcal{B}_{n,q}f - \mathcal{B}_{\infty,q}f\|$ for $f \in \tilde{C}_B[0,\infty)$, using the modulus of smoothness (2.49).

Theorem 2.16. *There exists $C > 0$ such that*

$$\|\mathcal{B}_{n,q}f - \mathcal{B}_{\infty,q}f\| \leq C\,\omega_\varphi^2(f; \sqrt{q^{n-1}/(1-q^n)}), \tag{2.56}$$

for all $f \in \tilde{C}_B[0,\infty)$, $n = 1, 2, \ldots$ and $q \in (0,1)$.

2.4.3   Proofs

The q-forward differences lead us to the moments of the first and second orders of $\mathcal{B}_{n,q}$.

Lemma 2.7. *We have*

$$\|\mathcal{B}_{n,q}f\| \leq \|f\|$$

for all $f \in C_B[0,\infty)$, $n = 1, 2, \ldots$ and $q \in (0,1)$.

Proof. For $x \in [0,\infty)$ one has $|\mathcal{B}_{n,q}(f;x)| \leq \|f\|\,\mathcal{B}_{n,q}(1;x) = \|f\|$, taking into account Lemma 2.5. Thus $\|\mathcal{B}_{n,q}f\| \leq \|f\|$, which completes the proof.

Proof of Theorem 2.15. Let $g \in C_B[0,\infty)$ with $g' \in AC_{loc}[0,\infty)$ be arbitrary. From Taylor's expansion

$$g(t) = g(x) + g'(x)(t-x) + \int_x^t g''(u)(t-u)\,du, \quad t \in [0,\infty),$$

we have, by Lemma 2.5, that

$$\mathcal{B}_{n,q}(g;x) - g(x) = \mathcal{B}_{n,q}\left(\int_x^t g''(u)(t-u)\,du; x\right).$$

Hence, in view of [51, p. 140, Lemma 9.6.1], we obtain

$$|\mathcal{B}_{n,q}(g;x) - g(x)| \leq \mathcal{B}_{n,q}\left(\left|\int_x^t |g''(u)||(t-u)|\,du\right|;x\right)$$

$$\leq \mathcal{B}_{n,q}\left(\frac{(t-x)^2}{x}\left(\frac{1}{1+x} + \frac{1}{1+t}\right);x\right)\|\varphi^2 g''\|$$

$$= \frac{1}{x(1+x)}\mathcal{B}_{n,q}((t-x)^2;x)\|\varphi^2 g''\|$$

$$+ \frac{1}{x}\mathcal{B}_{n,q}\left(\frac{(t-x)^2}{1+t};x\right)\|\varphi^2 g''\|. \tag{2.57}$$

But, by Lemma 2.5,

$$\mathcal{B}_{n,q}((t-x)^2;x) = \mathcal{B}_{n,q}(t^2;x) - 2x\mathcal{B}_{n,q}(t;x) + x_{n,q}^2\mathcal{B}(1;x) \tag{2.58}$$

$$= \frac{1}{q[n]_q}x(q+x) \leq \frac{1}{q[n]_q}\varphi^2(x).$$

Furthermore, by Hölder's inequality, we have

$$\mathcal{B}_{n,q}\left(\frac{(t-x)^2}{1+t};x\right) \leq \{\mathcal{B}_{n,q}((1+t)^{-2};x)\}^{1/2}\{\mathcal{B}_{n,q}((t-x)^4;x)\}^{1/2}. \tag{2.59}$$

Using (2.36), we find that

$$\mathcal{B}_{n,q}((1+t)^{-2};x) = \sum_{k=0}^{\infty}\begin{bmatrix}n+k-1\\k\end{bmatrix}_q q^{k(k-1)/2}$$

$$x^k(1+x)^{-1}(1+xq)^{-1}\ldots(1+xq^{n+k-1})^{-1}$$

$$\left(\frac{q^{k-1}[n]_q}{[k]_q + q^{k-1}[n]_q}\right)^2. \tag{2.60}$$

Because

$$(1+xq^{n+k-2})(1+xq^{n+k-1})(1+q^{-2n-2k+3})$$
$$= (1+q^{-2n-2k+3}) + (1+q^{-2n-2k+3})(q^{n+k-2}+q^{n+k-1})x$$
$$+ (1+q^{-2n-2k+3})q^{2n+2k-3}x^2$$
$$\geq 1+2x+x^2 = (1+x)^2$$

for $x \in [0, \infty)$, we get, by (2.60),

$$\mathcal{B}_{n,q}((1+t)^{-2};x)$$

$$\leq \sum_{k=0}^{\infty} \begin{bmatrix} n+k-3 \\ k \end{bmatrix}_q \frac{[n+k-2]_q[n+k-1]_q}{[n-2]_q[n-1]_q} q^{k(k-1)/2}$$

$$x^k(1+x)^{-1}(1+xq)^{-1}\ldots(1+xq^{n+k-3})^{-1}$$

$$(1+q^{-2n-2k+3})(1+x)^{-2}\frac{q^{2k-2}[n]_q^2}{([k]_q+q^{k-1}[n]_q)^2}. \tag{2.61}$$

Using the identities $[n+k-2]_q = [k]_q + q^k[n-2]_q$ and $[n+k-1]_q = [k]_q + q^k[n-1]_q$, we obtain

$$[n+k-2]_q[n+k-1]_q = [k]_q^2 + q^k[k]_q([n-2]_q + [n-1]_q)$$

$$+q^{2k}[n-2]_q[n-1]_q$$

$$\leq [k]_q^2 + 2q^{k-1}[k]_q[n]_q + q^{2k-2}[n]_q^2$$

$$= ([k]_q + q^{k-1}[n]_q)^2.$$

Hence

$$\frac{[n+k-2]_q[n+k-1]_q}{([k]_q+q^{k-1}[n]_q)^2} \leq 1. \tag{2.62}$$

Analogously, the identities $[n]_q = 1 + q[n-1]_q$ and $[n]_q = 1 + q + q^2[n-2]_q$ for $n = 3,4,\ldots$ imply that

$$\frac{[n]_q^2}{[n-2]_q[n-1]_q} = \frac{[n]_q}{[n-2]_q}\frac{[n]_q}{[n-1]_q}$$

$$= \left(\frac{1+q}{[n-2]_q} + q^2\right)\left(\frac{1}{[n-1]_q} + q\right) \leq 6. \tag{2.63}$$

The condition $q = q(n) \in (C_0^{1/n}, 1)$ implies

$$(1+q^{-2n-2k+3})q^{2k-2} \leq \frac{2}{q^{2n}} \leq \frac{2}{C_0^2}. \tag{2.64}$$

Now combining (2.61)–(2.64), we obtain

$$\mathcal{B}_{n,q}((1+t)^{-2};x) \leq \frac{12}{C_0^2}\frac{1}{(1+x)^2}\mathcal{B}_{n-2,q}(1;x)$$

$$= \frac{12}{C_0^2}\frac{1}{(1+x)^2} \tag{2.65}$$

for $x \in [0,\infty)$ and $n = 3,4,\ldots$.

On the other hand

$$\mathcal{B}_{n,q}((t-x)^4;x) = \mathcal{B}_{n,q}(t^4;x) - 4x\mathcal{B}_{n,q}(t^3;x) + 6x_{n,q}^2\mathcal{B}(t^2;x)$$
$$- 4x_{n,q}^3\mathcal{B}(t;x) + x_{n,q}^4\mathcal{B}(1;x). \tag{2.66}$$

To compute $\mathcal{B}_{n,q}(t^m;x)$, $m = 0,1,2,3,4$, we use Lemma 2.5 and the definition of the q-forward differences given above. Then, by direct computations, we get

$$\mathcal{B}_{n,q}(t^3;x) = \frac{1}{[n]_q^2}x + \frac{1+2q+1}{q^2[n]_q^2}x^2 + \frac{1}{q^3}\frac{[n+1]_q[n+2]_q}{[n]_q^2}x^3$$

and

$$\mathcal{B}_{n,q}(t^4;x)$$
$$= \frac{1}{[n]_q^3}x + \frac{1}{q^3}(1+3q+3q^2)\frac{[n+1]_q}{[n]_q^3}x^2$$
$$+ \frac{1}{q^5(1+q)}(1+3q+5q^2+3q^3)\frac{[n+1]_q[n+2]_q}{[n]_q^3}x^3$$
$$+ \frac{1}{q^6(1+q)(1+q+q^2)(1+q+q^2+q^3)}(1+3q+5q^2+6q^3$$
$$+ 5q^4+3q^5+q^6)\frac{[n+1][n+2]_q[n+3]_q}{[n]_q^3}x^4.$$

Hence, by (2.66),

$$\mathcal{B}_{n,q}((t-x)^4;x)$$
$$= \frac{1}{[n]_q^3}x + \frac{1}{q^3[n]_q^3}\left\{q(1+3q-q^2)[n]_q + (1+3q+3q^2)\right\}x^2$$
$$+ \frac{1}{q^5[n]_q^3}\left\{q^3(1-q)^2[n]_q^2 + q(1+4q+3q^2-2q^3)[n]_q\right.$$
$$\left. + (1+3q+5q^2+3q^3)\right\}x^3$$
$$+ \frac{1}{q^6[n]_q^3}\left\{q^3(1-q)^2[n]_q^2 + q(1+3q-q^3)[n]_q\right.$$
$$\left. + (1+2q+2q^2+q^3)\right\}x^4.$$

Taking into account the condition $q \in (C_0^{1/n}, 1)$, we obtain that $q \in (C_0, 1)$. Then, for $x \geq \frac{1}{[n]_q}$, we have

$$\mathcal{B}_{n,q}((t-x)^4;x)$$

$$\leq \frac{1}{[n]_q^2}x^2 + \frac{1}{C_0^3[n]_q^2}\left\{q(1+3q+q^2)+(1+3q+3q^2)\frac{1}{[n]_q}\right\}x^2$$

$$+\frac{1}{C_0^5[n]_q^2}\left\{q^3(1-q^n)^2\frac{1}{[n]_q}+q(1+4q+3q^2+2q^3)\right.$$

$$\left.+(1+3q+5q^2+3q^3)\frac{1}{[n]_q}\right\}x^3$$

$$+\frac{1}{C_0^6[n]_q^2}\left\{q^3(1-q^n)^2\frac{1}{[n]_q}+q(1+3q+q^3)\right.$$

$$\left.+(1+2q+2q^2+q^3)\frac{1}{[n]_q}\right\}x^4$$

$$\leq \frac{1}{C_0^6[n]_q^2}\left\{x^2+12x^2+23x^3+12x^4\right\}$$

$$\leq \frac{C}{[n]_q^2}\varphi^4(x).$$

Hence, in view of (2.57)–(2.60), we find for $x \geq \frac{1}{[n]}$ that

$$|\mathcal{B}_{n,q}(g;x)-g(x)| \leq \frac{1}{x(1+x)}\frac{1}{C_0[n]}\varphi^2(x)\|\varphi^2 g''\|$$

$$+\frac{1}{x}\frac{C}{1+x}\frac{C}{[n]}\varphi^2(x)\|\varphi^2 g''\|$$

$$\leq \frac{C}{[n]}\|\varphi^2 g''\|. \qquad (2.67)$$

For $0 \leq x \leq \frac{1}{[n]}$ we have, by Taylor's expansion,

$$g(t) = g(x)+g'(x)(t-x)+\int_x^t g''(u)(t-u)\,du, \quad t \in [0,\infty),$$

and Lemma 2.5 and (2.58) that

$$|\mathcal{B}_{n,q}(g;x)-g(x)| \leq \mathcal{B}_{n,q}\left(\left|\int_x^t |g''(u)||t-u|\,du\right|;x\right)$$

$$\leq \mathcal{B}_{n,q}((t-x)^2;x)\|g''\| \leq \frac{1}{q[n]_q}x(1+x)\|g''\|$$

$$\leq \frac{C}{[n]_q^2}\|g''\|. \qquad (2.68)$$

Then (2.67) and (2.68) imply

$$|\mathcal{B}_{n,q}(g;x) - g(x)| \leq C\left\{\frac{1}{[n]_q}\|\varphi^2 g''\| + \frac{1}{[n]_q^2}\|g''\|\right\}$$

for all $x \in [0,\infty)$. Hence, by Lemma 2.7, we obtain, for $f \in \tilde{C}_B[0,\infty)$,

$$\|\mathcal{B}_{n,q}f - f\| \leq \|\mathcal{B}_{n,q}f - \mathcal{B}_{n,q}g\| + \|\mathcal{B}_{n,q}g - g\| + \|g - f\|$$

$$\leq \|f - g\| + C\left\{\frac{1}{[n]_q}\|\varphi^2 g''\| + \frac{1}{[n]_q^2}\|g''\|\right\} + \|g - f\|$$

$$\leq C\left\{\|f - g\| + \frac{1}{[n]_q}\|\varphi^2 g''\| + \frac{1}{[n]_q^2}\|g''\|\right\}.$$

Using the definition of the K-functional $\bar{K}_{2,\varphi}(f; 1/[n]_q)$, we have $\|\mathcal{B}_{n,q}f - f\| \leq C\bar{K}_{2,\varphi}(f; 1/[n]_q)$. Then, in view of (2.50), we get (2.55), which was to be proved.

Proof of Theorem 2.16. Let $g \in C_B[0,\infty)$ with $g' \in AC_{loc}[0,\infty)$ be arbitrary. By Taylor's formula, we have

$$g\left(\frac{[k]_q}{q^{k-1}[n+1]_q}\right) = g\left(\frac{[k+1]_q}{q^k[n+1]_q}\right)$$

$$+ \left(\frac{[k]_q}{q^{k-1}[n+1]_q} - \frac{[k+1]_q}{q^k[n+1]_q}\right)g'\left(\frac{[k+1]_q}{q^k[n+1]_q}\right)$$

$$+ \int_{[k+1]_q/q^k[n+1]_q}^{[k]_q/q^{k-1}[n+1]_q}\left(\frac{[k]_q}{q^{k-1}[n+1]_q} - u\right)g''(u)\,du$$

and

$$g\left(\frac{[k+1]_q}{q^k[n]_q}\right) = g\left(\frac{[k+1]_q}{q^k[n+1]_q}\right)$$

$$+ \left(\frac{[k+1]_q}{q^k[n]_q} - \frac{[k+1]_q}{q^k[n+1]_q}\right)g'\left(\frac{[k+1]_q}{q^k[n+1]_q}\right)$$

$$+ \int_{[k+1]_q/q^k[n+1]_q}^{[k+1]_q/q^k[n]_q}\left(\frac{[k+1]_q}{q^k[n]_q} - u\right)g''(u)\,du.$$

Hence, by Theorem 2.14 and $[n+k+1]_q = [n]_q + q^n[k+1]_q$, we obtain

$$\mathcal{B}_{n,q}(g;x) - \mathcal{B}_{n+1,q}(g;x)$$

$$= \sum_{k=0}^{\infty} q^{k(k+1)/2} x^{k+1} (1+x)^{-1}(1+xq)^{-1} \ldots (1+xq^{n+k})^{-1}$$

$$\begin{bmatrix} n+k \\ k \end{bmatrix}_q \left\{ \left\{ q^n \left(\frac{[k]_q}{q^{k-1}[n+1]_q} - \frac{[k+1]_q}{q^k[n+1]_q} \right) \right. \right.$$

$$+ \frac{[n]_q}{[k+1]_q} \left(\frac{[k+1]_q}{q^k[n]_q} - \frac{[k+1]_q}{q^k[n+1]_q} \right) \right\} g' \left(\frac{[k+1]_q}{q^k[n+1]_q} \right)$$

$$+ q^n \int_{[k+1]_q/q^k[n+1]_q}^{[k]_q/q^{k-1}[n+1]_q} \left(\frac{[k]}{q^{k-1}[n+1]} - u \right) g''(u)\,du$$

$$+ \frac{[n]_q}{[k+1]_q} \int_{[k+1]_q/q^k[n+1]_q}^{[k+1]_q/q^k[n]_q} \left(\frac{[k+1]_q}{q^k[n]_q} - u \right) g''(u)\,du \right\}. \qquad (2.69)$$

Because

$$q^n \left(\frac{[k]_q}{q^{k-1}[n+1]_q} - \frac{[k+1]_q}{q^k[n+1]_q} \right) + \frac{[n]_q}{[k+1]_q} \left(\frac{[k+1]_q}{q^k[n]_q} - \frac{[k+1]_q}{q^k[n+1]_q} \right)$$

$$= -\frac{q^{n-k}}{[n+1]_q} + \frac{1}{q^k}\frac{[n+1]_q - [n]_q}{[n+1]_q}$$

$$= -\frac{q^{n-k}}{[n+1]_q} + \frac{q^{n-k}}{[n+1]_q} = 0,$$

we get, by (2.69),

$$|\mathcal{B}_{n,q}(g;x) - \mathcal{B}_{n+1,q}(g;x)|$$

$$\leq \sum_{k=0}^{\infty} q^{k(k+1)/2} x^{k+1}(1+x)^{-1}(1+xq)^{-1} \ldots (1+xq^{n+k})^{-1}$$

$$\begin{bmatrix} n+k \\ k \end{bmatrix}_q \left\{ q^n \left| \int_{[k+1]_q/q^k[n+1]_q}^{[k]_q/q^{k-1}[n+1]_q} \left| \frac{[k]_q}{q^{k-1}[n+1]_q} - u \right| |g''(u)|\,du \right| \right.$$

$$+ \frac{[n]_q}{[k+1]_q} \left| \int_{[k+1]_q/q^k[n+1]_q}^{[k+1]_q/q^k[n]_q} \left| \frac{[k+1]_q}{q^k[n]_q} - u \right| |g''(u)|\,du \right| \right\}. \qquad (2.70)$$

Taking into account the estimate

$$\left| \int_x^t (t-u)g''(u)\,du \right| \le (t-x)^2 \frac{1}{x}\left(\frac{1}{1+x} + \frac{1}{1+t} \right) \|\varphi^2 g''\|,$$

$t,x \in [0,\infty)$ (see [51, p. 140, Lemma 9.6.1]), we find that

$$q^n \left| \int_{[k+1]_q/q^k[n+1]_q}^{[k]_q/q^{k-1}[n+1]_q} \left| \frac{[k]_q}{q^{k-1}[n+1]_q} - u \right| |g''(u)|\,du \right|$$

$$\le q^n \left(\frac{[k]_q}{q^{k-1}[n+1]_q} - \frac{[k+1]_q}{q^k[n+1]_q} \right)^2 \frac{q^k[n+1]_q}{[k+1]_q}$$

$$\left(\frac{q^k[n+1]_q}{[k+1]_q + q^k[n+1]_q} + \frac{q^{k-1}[n+1]_q}{[k]_q + q^{k-1}[n+1]_q} \right) \|\varphi^2 g''\|$$

$$= q^{n-1} \left(\frac{q}{q^k + [n+k+1]_q} + \frac{1}{q^{k-1} + [n+k]_q} \right) \|\varphi^2 g''\| \qquad (2.71)$$

and

$$\frac{[n]_q}{[k+1]_q} \left| \int_{[k+1]_q/q^k[n+1]_q}^{[k+1]_q/q^k[n]_q} \left| \frac{[k+1]_q}{q^k[n]_q} - u \right| |g''(u)|\,du \right|$$

$$\le \frac{[n]_q}{[k+1]_q} \left(\frac{[k+1]_q}{q^k[n]_q} - \frac{[k+1]_q}{q^k[n+1]_q} \right)^2 \frac{q^k[n+1]_q}{[k+1]_q}$$

$$\left(\frac{q^k[n+1]_q}{[k+1]_q + q^k[n+1]_q} + \frac{q^k[n]_q}{[k+1]_q + q^k[n]_q} \right) \|\varphi^2 g''\|$$

$$= \frac{q^{2n}}{[n]_q[n+1]_q} \left(\frac{[n+1]_q}{q^k + [n+k+1]_q} + \frac{[n]_q}{q^k + [n+k]_q} \right) \|\varphi^2 g''\| \qquad (2.72)$$

Then (2.70)–(2.72) imply

$$|\mathcal{B}_{n,q}(g;x) - \mathcal{B}_{n+1,q}(g;x)|$$

$$\le \sum_{k=0}^{\infty} q^{k(k+1)/2} x^{k+1} (1+x)^{-1} (1+xq)^{-1} \ldots (1+xq^{n+k})^{-1}$$

$$\begin{bmatrix} n+k \\ k+1 \end{bmatrix}_q \frac{[k+1]_q}{[n]_q} \left\{ q^{n-1} \left(\frac{q}{q^k + [n+k+1]_q} + \frac{1}{q^{k-1} + [n+k]_q} \right) \right.$$

$$\left. + \frac{q^{2n}}{[n]_q[n+1]_q} \left(\frac{[n+1]_q}{q^k + [n+k+1]_q} + \frac{[n]}{q^k + [n+k]_q} \right) \right\} \|\varphi^2 g''\|$$

$$\leq \sum_{k=0}^{\infty} q^{k(k+1)/2} x^{k+1} (1+x)^{-1} (1+xq)^{-1} \ldots (1+xq^{n+k})^{-1}$$

$$\begin{bmatrix} n+k \\ k+1 \end{bmatrix}_q \frac{1}{[n]_q} \left\{ q^{n-1} \left(\frac{[k+1]_q}{[n+k+1]_q} + \frac{[k+1]_q}{[n+k]_q} \right) \right.$$

$$\left. + q^{2n} \left(\frac{1}{[n]_q} \frac{[k+1]_q}{[n+k+1]_q} + \frac{1}{[n+1]} \frac{[k+1]_q}{[n+k]_q} \right) \right\} \|\varphi^2 g''\|$$

$$\leq \frac{4q^{n-1}}{[n]_q} \sum_{k=0}^{\infty} q^{k(k+1)/2} x^{k+1} (1+x)^{-1} (1+xq)^{-1} \ldots (1+xq^{n+k})^{-1}$$

$$\begin{bmatrix} n+k \\ k+1 \end{bmatrix}_q \|\varphi^2 g''\|$$

$$\leq \frac{4q^{n-1}}{[n]_q} B_{n,q}(1;x) \|\varphi^2 g''\| = \frac{4q^{n-1}}{[n]_q} \|\varphi^2 g''\|$$

(see Lemma 2.5). Hence

$$\|\mathcal{B}_{n,q} g - \mathcal{B}_{n+1,q} g\| \leq \frac{4q^{n-1}}{[n]_q} \|\varphi^2 g''\|.$$

Now for $p = 1, 2, \ldots,$ we obtain

$$\|\mathcal{B}_{n,q} g - \mathcal{B}_{n+p,q} g\| \leq \|\mathcal{B}_{n,q} g - \mathcal{B}_{n+1,q} g\| + \|\mathcal{B}_{n+1,q} g - \mathcal{B}_{n+2,q} g\|$$

$$+ \ldots + \|\mathcal{B}_{n+p-1,q} g - \mathcal{B}_{n+p,q} g\|$$

$$\leq \frac{4q^{n-1}}{[n]_q} (1 + q + \ldots + q^{n+p-2}) \|\varphi^2 g''\|$$

$$\leq \frac{4q^{n-1}}{[n]_q (1-q)} \|\varphi^2 g''\| = \frac{4q^{n-1}}{1-q^n} \|\varphi^2 g''\|.$$

Then, by Lemma 2.7 for $f \in C_B[0, \infty)$, we have

$$\|\mathcal{B}_{n,q} f - \mathcal{B}_{n+p,q} f\| \leq \|\mathcal{B}_{n,q} f - \mathcal{B}_{n,q} g\| + \|\mathcal{B}_{n,q} g - \mathcal{B}_{n+p,q} g\|$$

$$+ \|\mathcal{B}_{n+p,q} g - \mathcal{B}_{n+p,q} f\|$$

$$\leq \|f - g\| + \frac{4q^{n-1}}{1-q^n} \|\varphi^2 g''\| + \|f - g\|$$

$$\leq 4 \left\{ \|f - g\| + \frac{q^{n-1}}{1-q^n} \|\varphi^2 g''\| \right\}.$$

Hence, in view of (2.51), we obtain

$$\|\mathcal{B}_{n,q}f - \mathcal{B}_{n+p,q}f\| \le 4K_{2,\varphi}(f;q^{n-1}/(1-q^n))$$

$$\le C\omega_\varphi^2(f;\sqrt{q^{n-1}/(1-q^n)}) \qquad (2.73)$$

for every $n,p = 1,2,\ldots$.

On the other hand $\lim\limits_{n\to\infty} \omega_\varphi^2(f;\sqrt{q^{n-1}/(1-q^n)}) = 0$, because of $f \in \tilde{C}_B[0,\infty)$ (see [51, pp. 36–37]). Then (2.73) implies that $\{V_{n,q}f\}$ is a Cauchy-sequence in the Banach space $C_B[0,\infty)$. Thus it converges in $C_B[0,\infty)$. In conclusion, by (2.73), there exists an operator $L : \tilde{C}_B[0,\infty) \to C_B[0,\infty)$ such that

$$\|\mathcal{B}_{n,q}f - Lf\| \le C\omega_\varphi^2(f;\sqrt{q^{n-1}/(1-q^n)}) \qquad (2.74)$$

for all $f \in \tilde{C}_B[0,\infty)$ and $n = 1,2,\ldots$.

Finally, we prove that $Lf \equiv \mathcal{B}_{\infty,q}f$ for $f \in \tilde{C}_B[0,\infty)$. Let $x \in [0,\infty)$ be arbitrary. Then

$$|L(f;x) - \mathcal{B}_{\infty,q}(f;x)| \le |L(f;x) - \mathcal{B}_{n,q}(f;x)|$$

$$+ |\mathcal{B}_{n,q}(f;x) - \mathcal{B}_{\infty,q}(f;x)|.$$

By (2.36), (2.52), and (2.53), we have

$$|B_{n,q}(f;x) - B_{\infty,q}(f;x)|$$

$$= \left| \sum_{k=0}^\infty \begin{bmatrix} n+k-1 \\ k \end{bmatrix}_q q^{k(k-1)/2} x^k \prod_{s=0}^{n+k-1}(1+xq^s)^{-1} \right.$$

$$\left(1 - \prod_{s=n+k}^\infty (1+xq^s)^{-1}\right) f\left(\frac{1-q^k}{q^{k-1}(1-q^n)}\right)$$

$$+ \sum_{k=0}^\infty \left\{ \begin{bmatrix} n+k-1 \\ k \end{bmatrix}_q - (1-q)^{-1}(1-q^2)^{-1}\ldots(1-q^k)^{-1} \right\}$$

$$q^{k(k-1)/2} x^k \prod_{s=0}^\infty (1+xq^s)^{-1} f\left(\frac{1-q^k}{q^{k-1}(1-q^n)}\right)$$

$$+ \sum_{k=0}^\infty q^{k(k-1)/2}(1-q)^{-1}(1-q^2)^{-1}\ldots(1-q^k)^{-1}$$

$$\left. x^k \prod_{s=0}^{n+k-1}(1+xq^s)^{-1} \left\{ f\left(\frac{1-q^k}{q^{k-1}(1-q^n)}\right) - f\left(\frac{1-q^k}{q^{k-1}}\right) \right\} \right|$$

$$\leq \|f\| \sum_{k=0}^{\infty} \left\{ \begin{bmatrix} n+k-1 \\ k \end{bmatrix} q^{k(k-1)/2} x^k \prod_{s=0}^{n+k-1} (1+xq^s)^{-1} \right.$$

$$(1 - \prod_{s=n+k}^{\infty} (1+xq^s)^{-1}) + \|f\| \sum_{k=0}^{\infty} \left| \begin{bmatrix} n+k-1 \\ k \end{bmatrix}_q - (1-q)^{-1} \right.$$

$$(1-q^2)^{-1} \ldots (1-q^k)^{-1} \left| q^{k(k-1)/2} x^k \prod_{s=0}^{\infty} (1+xq^s)^{-1} \right.$$

$$+ \sum_{k=0}^{\infty} q^{k(k-1)/2} (1-q)^{-1} (1-q^2)^{-1} \ldots (1-q^k)^{-1}$$

$$x^k \prod_{s=0}^{\infty} (1+xq^s)^{-1} \left| f\left(\frac{1-q^k}{q^{k-1}(1-q^n)} \right) - f\left(\frac{1-q^k}{q^{k-1}} \right) \right|$$

$$=: I_1 + I_2 + I_3. \tag{2.75}$$

Furthermore, the infinite product $\prod_{s=0}^{\infty} (1+xq^s)^{-1}$ is convergent; thus for every $\varepsilon > 0$ there exists n'_ε such that

$$0 < 1 - \prod_{s=n+k}^{\infty} (1+xq^s)^{-1} \leq 1 - \prod_{s=n}^{\infty} (1+xq^s)^{-1} \leq \frac{\varepsilon}{3\|f\|}$$

for $n > n'_\varepsilon$ and $k = 0,1,2,\ldots$. Hence, by Lemma 2.5,

$$I_1 < \frac{\varepsilon}{3} \mathcal{B}_{n,q}(1;x) = \frac{\varepsilon}{3}. \tag{2.76}$$

In view of (2.54), we have $\mathcal{B}_{\infty,q}(1;x) = 1$. By [99, p. 156, (2.8)], we know that the inequality

$$1 - \prod_{s=j}^{\infty} (1-q^s) \leq \frac{q^j}{q(1-q)} \ln \frac{1}{1-q}$$

holds for $0 < q < 1$ and $j = 1,2,\ldots$. Then we obtain

$$I_2 = \|f\| \sum_{k=0}^{\infty} \left| (1-q)(1-q^2)\ldots(1-q^k) \begin{bmatrix} n+k-1 \\ k \end{bmatrix}_q - 1 \right|$$

$$q^{k(k-1)/2} (1-q)^{-1} (1-q^2)^{-1} \ldots (1-q^k)^{-1} x^k \prod_{s=0}^{\infty} (1+xq^s)^{-1}$$

$$= \|f\| \sum_{k=0}^{\infty} |(1-q^n)(1-q^{n+1})\ldots(1-q^{n+k-1}) - 1|$$

$$q^{k(k-1)/2}(1-q)^{-1}(1-q^2)^{-1}\cdots(1-q^k)^{-1}x^k\prod_{s=0}^{\infty}(1+xq^s)^{-1}$$

$$\leq\|f\|\sum_{k=0}^{\infty}\left(1-\prod_{s=n}^{\infty}(1-q^s)\right)$$

$$q^{k(k-1)/2}(1-q)^{-1}(1-q^2)^{-1}\cdots(1-q^k)^{-1}x^k\prod_{s=0}^{\infty}(1+xq^s)^{-1}$$

$$\leq\|f\|\frac{q^n}{q(1-q)}\ln\frac{1}{1-q}V_{\infty,q}(1;x)$$

$$=\frac{q^n}{q(1-q)}\ln\frac{1}{1-q}\|f\|.$$

In conclusion, if $\varepsilon>0$ is arbitrary, then there exists n_ε'' such that $q^n<\varepsilon q(1-q)/(\|f\|\ln(1-q)^{-1})$ for every $n>n_\varepsilon''$. Thus

$$I_2<\frac{\varepsilon}{3}.\tag{2.77}$$

Finally, because $f\in\tilde{C}_B[0,\infty)$ and $\varepsilon>0$ is arbitrary, there exists $y_\varepsilon>0$ such that $|f(y)|<\varepsilon/12$ for $y>y_\varepsilon$. Because $\dfrac{1-q^k}{q^{k-1}}\leq\dfrac{1-q^k}{q^{k-1}(1-q^n)}$ for all $k=0,1,2,\dots$ and $n=1,2,\dots$, there exists k_ε such that $\dfrac{1-q^k}{q^{k-1}}>y_\varepsilon$ for $k>k_\varepsilon$. Then

$$\left|f\left(\frac{1-q^k}{q^{k-1}}\right)\right|<\frac{\varepsilon}{12}\tag{2.78}$$

and

$$\left|f\left(\frac{1-q^k}{q^{k-1}(1-q^n)}\right)\right|<\frac{\varepsilon}{12}\tag{2.79}$$

for $k>k_\varepsilon$.

On the other hand

$$\frac{1-q^k}{q^{k-1}(1-q^n)}-\frac{1-q^k}{q^{k-1}}=\frac{(1-q^k)q^n}{q^{k-1}(1-q^n)};$$

therefore we obtain for $k=0,1,\dots,k_\varepsilon$ that

$$\left|f\left(\frac{1-q^k}{q^{k-1}(1-q^n)}\right)-f\left(\frac{1-q^k}{q^{k-1}}\right)\right|$$

$$\leq \omega\left(f; \frac{(1-q^k)q^n}{q^{k-1}(1-q^n)}\right) \leq \left(1 + \frac{1-q^k}{q^{k-1}}\right)\omega\left(f; \frac{q^n}{1-q^n}\right)$$

$$= (1-q+q^{-k+1})\,\omega\left(f; \frac{q^n}{1-q^n}\right)$$

$$\leq (1-q+q^{-k_\varepsilon+1})\,\omega\left(f; \frac{q^n}{1-q^n}\right), \tag{2.80}$$

where $\omega(f;\delta) = \sup\{|f(x)-f(y)| : x,y \in [0,\infty), |x-y| \leq \delta\}$ is the modulus of continuity of $f \in \tilde{C}_B[0,\infty)$. Then, for every $\varepsilon > 0$ there exists n'''_ε such that

$$(1-q+q^{-k_\varepsilon+1})\,\omega\left(f; \frac{q^n}{1-q^n}\right) < \frac{\varepsilon}{6}$$

for $n > n'''_\varepsilon$. Hence, by (2.78)–(2.80) and (2.54), we get

$$I_3 < \sum_{k=0}^{k_\varepsilon} q^{k(k-1)/2}(1-q)^{-1}(1-q^2)^{-1}\dots(1-q^k)^{-1}$$

$$x^k \prod_{s=0}^{\infty}(1+xq^s)^{-1}(1-q+q^{-k_\varepsilon+1})\,\omega\left(f; \frac{q^n}{1-q^n}\right)$$

$$+ \sum_{k=k_\varepsilon+1}^{\infty} q^{k(k-1)/2}(1-q)^{-1}(1-q^2)^{-1}\dots(1-q^k)^{-1}$$

$$x^k \prod_{s=0}^{\infty}(1+xq^s)^{-1}\left\{\left|f\left(\frac{1-q^k}{q^{k-1}(1-q^n)}\right)\right| + \left|f\left(\frac{1-q^k}{q^{k-1}}\right)\right|\right\}$$

$$< \frac{\varepsilon}{6}V_{\infty,q}(1;x) + \frac{\varepsilon}{6}V_{\infty,q}(1;x) = \frac{\varepsilon}{3}. \tag{2.81}$$

Now combining (2.74)–(2.77) and (2.81), we find that

$$|L(f;x) - \mathcal{B}_{\infty,q}(f;x)| \leq C\omega_\varphi^2(f; \sqrt{q^{n-1}/(1-q^n)}) + \varepsilon$$

for arbitrary $\varepsilon > 0$ and $n > \max\{n'_\varepsilon, n''_\varepsilon, n'''_\varepsilon\}$. Thus $L(f;x) = \mathcal{B}_{\infty,q}(f;x)$, which was to be proved.

2.5   q-Bleimann–Butzer–Hahn Operators

There are several studies related to the approximation properties of the Bleimann, Butzer, and Hahn operators (or, briefly, BBH). There are many approximating operators that their Korovkin-type approximation properties and rates of convergence are investigated. The results involving Korovkin-type approximation properties can be found in [13] with details. In [68], A.D. Gadjiev and Ö. Çakar gave a Korovkin-type theorem using the test function $\left(\frac{t}{1+t}\right)^{v}$ for $v = 0, 1, 2$. Some generalization of the operators (2.82) were given in [6, 7, 52].

2.5.1 Introduction

In [39], Bleimann, Butzer, and Hahn introduced the following operators:

$$B_n(f)(x) = \frac{1}{(1+x)^n} \sum_{k=0}^{n} f\left(\frac{k}{n-k+1}\right) \binom{n}{k} x^k, \quad x > 0, n \in \mathbb{N}. \tag{2.82}$$

Here we derive a q-integers-type modification of BBH operators that we call q-BBH operators and investigate their Korovkin-type approximation properties by using the test function $\left(\frac{t}{1+t}\right)^{v}$ for $v = 0, 1, 2$. Also, we define a space of generalized Lipschitz-type maximal function and give a pointwise estimation. Then a Stancu-type formula of the remainder of q-BBH is given. We shall also give a generalization of these operators and study on the approximation properties of this generalization. We emphasize that while Bernstein and Meyer–König and Zeller operators based on q-integers depend on a function defined on a bounded interval, these operators defined on unbounded intervals. Also, these operators are more flexible than classical BBH operators. That is, depending on the selection of q, the rate of convergence of the q-BBH operators is better than the classical one.

We refer to readers for additional information on q-Bleimann, Butzer, and Hahn operators to [10, 60, 120]. This section is based on [27].

2.5.2 Construction of the Operators

Also, let us recall the following Euler identity (see [134, p. 293])

$$\prod_{k=0}^{n-1}(1+q^k x) = \sum_{k=0}^{n} q^{\frac{k(k-1)}{2}} \begin{bmatrix} n \\ k \end{bmatrix}_q x^k. \tag{2.83}$$

It is clear that when $q = 1$, these q-binomial coefficients reduce to ordinary binomial coefficients.

According to these explanations, similarly in [53], we defined a new Bleimann, Butzer, and Hahn-type operators based on q-integers as follows:

$$L_n(f;x) = \frac{1}{\ell_n(x)} \sum_{k=0}^{n} f\left(\frac{[k]_q}{[n-k+1]_q q^k}\right) q^{\frac{k(k-1)}{2}} \begin{bmatrix} n \\ k \end{bmatrix}_q x^k, \tag{2.84}$$

where

$$\ell_n(x) = \prod_{s=0}^{n-1}(1+q^s x)$$

and f defined on semiaxis $[0, \infty)$.

Note that taking $f\left(\frac{[k]_q}{[n-k+1]_q}\right)$ instead of $f\left(\frac{[k]_q}{[n-k+1]_q q q^k}\right)$ in (2.84), then we obtain usual generalization of Bleimann, Butzer, and Hahn operators based on q-integers. But in this case it is impossible to obtain explicit expressions for the monomials t^v and $(t/(1+t))^v$ for $v = 1, 2$. If we define the Bleimann, Butzer, and Hahn-type operators as in (2.84), then we can obtain explicit formulas for the monomials $(t/(1+t))^v$ for $v = 0, 1, 2$.

By a simple calculation, we have

$$q^k[n-k+1]_q = [n+1]_q - [k]_q, q[k-1]_q = [k]_q - 1. \tag{2.85}$$

From (2.83) to (2.85), we have

$$L_n(1;x) = 1 \tag{2.86}$$

and

$$
\begin{aligned}
L_n(\frac{t}{1+t};x) &= \frac{1}{\ell_n(x)}\sum_{k=1}^{n}\frac{[k]_q}{[n+1]_q}q^{\frac{k(k-1)}{2}}\begin{bmatrix} n \\ k \end{bmatrix}_q x^k \\
&= \frac{1}{\ell_n(x)}\sum_{k=1}^{n}\frac{[n]_q}{[n+1]_q}q^{\frac{k(k-1)}{2}}\begin{bmatrix} n-1 \\ k-1 \end{bmatrix}_q x^k \\
&= \frac{[n]_q}{[n+1]_q}x\frac{1}{\ell_n(x)}\sum_{k=0}^{n-1}q^{\frac{k(k-1)}{2}}\begin{bmatrix} n-1 \\ k \end{bmatrix}_q (qx)^k \\
&= \frac{x}{x+1}\frac{[n]_q}{[n+1]_q}. \tag{2.87}
\end{aligned}
$$

We can also write

$$
\begin{aligned}
L_n(\frac{t^2}{(1+t)^2};x) &= \frac{1}{\ell_n(x)}\sum_{k=1}^{n}\frac{[k]_q^2}{[n+1]_q^2}q^{\frac{k(k-1)}{2}}\begin{bmatrix} n \\ k \end{bmatrix}_q x^k \\
&= \frac{1}{\ell_n(x)}\sum_{k=2}^{n}\frac{q[k]_q[k-1]_q}{[n+1]_q^2}q^{\frac{k(k-1)}{2}}\begin{bmatrix} n \\ k \end{bmatrix}_q x^k \\
&\quad + \frac{1}{\ell_n(x)}\sum_{k=1}^{n}\frac{[k]_q}{[n+1]_q^2}q^{\frac{k(k-1)}{2}}\begin{bmatrix} n \\ k \end{bmatrix}_q x^k
\end{aligned}
$$

$$= \frac{1}{\ell_n(x)} \sum_{k=0}^{n-2} \frac{[n]_q[n-1]_q}{[n+1]_q^2} q^{\frac{k(k-1)}{2}} \begin{bmatrix} n-2 \\ k \end{bmatrix}_q (q^2 x)^k q^2 x^2$$

$$+ \frac{1}{\ell_n(x)} \sum_{k=0}^{n-1} \frac{[n]_q}{[n+1]_q^2} q^{\frac{k(k-1)}{2}} \begin{bmatrix} n-1 \\ k \end{bmatrix}_q (qx)^k x$$

$$= \frac{[n]_q[n-1]_q}{[n+1]_q^2} q^2 \frac{x^2}{(1+x)(1+qx)} + \frac{[n]_q}{[n+1]_q^2} \frac{x}{x+1}. \tag{2.88}$$

Remark 2.5. Note that, if we choose $q = 1$ then L_n operators turn out into the classical Bleimann, Butzer, and Hahn operators given by (2.82). Also using the similar methods as in [53, 54, 133], to ensure the convergence properties of L_n, we will assume $q = q_n$ as a sequence such that $q_n \to 1$ as $n \to \infty$ for $0 < q_n < 1$.

2.5.3 Properties of the Operators

In this section we will give the theorems on uniform convergence and rate of convergence of the operators (2.82). As in [68], for this purpose we give a space of function ω of the type of modulus of continuity which satisfies the following condition:

(a) ω is a nonnegative increasing function on $[0, \infty)$.
(b) $\omega(\delta_1 + \delta_2) \le \omega(\delta_1) + \omega(\delta_2)$.
(c) $\lim_{\delta \to 0} \omega(\delta) = 0$.

And H_ω is the subspace of real-valued function and satisfies the following condition:
For any $x, y \in [0, \infty)$

$$|f(x) - f(y)| \le \omega\left(\left|\frac{x}{1+x} - \frac{y}{1+y}\right|\right). \tag{2.89}$$

Also $H_\omega \subset C_B[0, \infty)$, where $C_B[0, \infty)$ is the space of functions f which is continuous and bounded on $[0, \infty)$ endowed with norm $\|f\|_{C_B} = \sup_{x \ge 0} |f(x)|$.

It is easy to show that from the condition (b), the function ω satisfies the inequality

$$\omega(n\delta) \le n\omega(\delta) \quad n \in \mathbb{N},$$

and from condition (a) for $\lambda > 0$, we have

$$\omega(\lambda\delta) \le \omega((1 + [|\lambda|])\delta)$$
$$\le (1 + \lambda)\omega(\delta) \tag{2.90}$$

where $[|\lambda|]$ is the greatest integer of λ.

Remark 2.6. The operator L_n maps H_ω into $C_B[0, \infty)$ and it is continuous with respect to sup-norm.

The properties of linear positive operators acting from H_ω to $C_B[0, \infty)$ and the Korovkin-type theorems for them have been studied by Gadjiev and Çakar who have established the following theorem (see [68]).

Theorem 2.17. *If A_n is the sequence of positive linear operators from H_ω to $C_B[0, \infty)$ satisfy the following conditions for $U = 0, 1, 2$.*

$$\left\| \left(A_n \left(\frac{t}{1+t} \right)^\upsilon \right)(x) - \left(\frac{x}{1+x} \right)^\upsilon \right\|_{C_B} \to 0 \quad \text{for } n \to \infty$$

then, for any function f in H_ω, one has

$$\|A_n f - f\|_{C_B} \to 0 \quad \text{for } n \to \infty.$$

Theorem 2.18. *Let $q = q_n$ satisfies $0 < q_n < 1$ and let $q_n \to 1$ as $n \to \infty$. If L_n is defined by (2.84), then for any $f \in H_\omega$,*

$$\lim_{n \to \infty} \|L_n f - f\|_{C_B} = 0.$$

Proof. Using Theorem 2.17 we see that it is sufficient to verify the following three conditions:

$$\lim_{n \to \infty} \left\| L_n \left(\left(\frac{t}{1+t} \right)^\upsilon ; x \right) - \left(\frac{x}{1+x} \right)^\upsilon \right\|_{C_B} = 0, \quad \upsilon = 0, 1, 2. \qquad (2.91)$$

From (2.86), the first condition of (2.91) is fulfilled for $\upsilon = 0$. Now it is easy to see that from (2.87)

$$\left\| L_n \left(\left(\frac{t}{1+t} \right) ; x \right) - \frac{x}{1+x} \right\|_{C_B} \leq \left| \frac{[n]_{q_n}}{[n+1]_{q_n}} - 1 \right|$$

$$\leq \left| \frac{1}{q_n} - \frac{1}{q_n [n+1]_{q_n}} - 1 \right|$$

and since $[n+1]_{q_n} \to \infty$, $q_n \to 1$ as $n \to \infty$, the condition (2.91) holds for $\upsilon = 1$. To verify this condition for $\upsilon = 2$, consider (2.88). We see that

$$\left\| L_n \left(\left(\frac{t}{1+t} \right)^2 ; x \right) - \left(\frac{x}{1+x} \right)^2 \right\|_{C_B} = \sup_{x \geq 0} \left(\frac{x^2}{(1+x)^2} \left(\frac{[n]_{q_n}[n-1]_{q_n}}{[n+1]_{q_n}^2} q_n^2 \frac{1+x}{1+q_n x} - 1 \right) \right.$$

$$\left. + \frac{[n]_{q_n}}{[n+1]_{q_n}^2} \frac{x}{1+x} \right).$$

A small calculation shows that

$$\frac{[n]_{q_n}[n-1]_{q_n}}{[n+1]_{q_n}^2} = \frac{1}{q_n^3}\left(1 - \frac{2+q_n}{[n+1]_{q_n}} + \frac{1+q_n}{[n+1]_{q_n}^2}\right).$$

Thus, we have

$$\left\|L_n\left(\left(\frac{t}{1+t}\right)^2;x\right) - \left(\frac{x}{1+x}\right)^2\right\|_{C_B} \le \frac{1}{q_n^2}\left(1 - q_n^2 - \frac{2}{[n+1]_{q_n}} + \frac{1}{[n+1]_{q_n}^2}\right).$$

This means that the condition (2.91) holds also for $v = 2$ and the proof is completed by Theorem 2.17. ∎

Theorem 2.19. *Let $q = q_n$ satisfies $0 < q_n < 1$ with $q_n \to 1$ as $n \to \infty$. If L_n is defined by (2.84), then for each $x \ge 0$ and for any $f \in H_\omega$, the following inequality holds*

$$|L_n(f;x) - f(x)| \le 2\omega\left(\sqrt{\mu_n(x)}\right)$$

where

$$\mu_n(x) = \left(\frac{x}{1+x}\right)^2\left(1 - 2\frac{[n]_{q_n}}{[n+1]_{q_n}} + \frac{[n]_{q_n}[n-1]_{q_n}}{[n+1]_{q_n}^2}q_n^2\frac{(1+x)}{(1+q_nx)}\right) + \frac{[n]_{q_n}}{[n+1]_{q_n}^2}\frac{x}{1+x}. \qquad (2.92)$$

Proof. Since $L_n(1;x) = 1$, we can write

$$|L_n(f;x) - f(x)| \le L_n(|f(t) - f(x)|;x). \qquad (2.93)$$

On the other hand from (2.89) and (2.90)

$$|f(t) - f(x)| \le \omega\left(\left|\frac{t}{1+t} - \frac{x}{1+x}\right|\right)$$

$$\le \left(1 + \frac{\left|\frac{t}{1+t} - \frac{x}{1+x}\right|}{\delta}\right)\omega(\delta),$$

where we choose $\lambda = \delta^{-1}\left|\frac{t}{1+t} - \frac{x}{1+x}\right|$. This inequality and (2.93) imply

$$|L_n(f;x) - f(x)| \le \omega(\delta)\left(1 + \frac{1}{\delta}L_n\left(\left|\frac{t}{1+t} - \frac{x}{1+x}\right|;x\right)\right).$$

According to the Cauchy–Schwarz inequality we have

$$|L_n(f;x) - f(x)| \le \omega(\delta)\left(1 + \frac{1}{\delta}L_n\left(\left|\frac{t}{1+t} - \frac{x}{1+x}\right|^2;x\right)^{\frac{1}{2}}\right).$$

By choosing $\delta = \mu_n(x) = L_n\left(\left|\frac{t}{1+t} - \frac{x}{1+x}\right|^2; x\right)$, we obtain the desired result. ■

Now we will give an estimate concerning the rate of convergence as given in [8, 52, 109]. We define the space of general Lipschitz-type maximal functions on $E \subset [0, \infty)$ by $W_{\alpha, E}^{\sim}$ as

$$W_{\alpha, E}^{\sim} = \left\{ f : \sup(1+x)^\alpha f_\alpha(x, y) \le M \frac{1}{(1+y)^a}, x \ge 0 \text{ and } y \in E \right\},$$

where f is bounded and continuous on $[0, \infty)$, M is a positive constant, $0 < \alpha \le 1$, and f_α is the following function:

$$f_\alpha(x, t) = \frac{|f(t) - f(x)|}{|x - t|^\alpha}.$$

Also, let $d(x, E)$ be the distance between x and E, that is

$$d(x, E) = \inf\{|x - y|; y \in E\}.$$

Theorem 2.20. *For all $f \in W_{\alpha, E}^{\sim}$ we have*

$$|L_n(f; x) - f(x)| \le M\left(\mu_n^{\frac{\alpha}{2}}(x) + 2(d(x, E))^\alpha\right), \tag{2.94}$$

where $\mu_n(x)$ defined in (2.92).

Proof. Let $\overline{E}$ denote the closure of the set E. Then there exists a $x_0 \in \overline{E}$ such that $|x - x_0| = d(x, E)$, where $x \in [0, \infty)$. Thus we can write

$$|f - f(x)| \le |f - f(x_0)| + |f(x_0) - f(x)|.$$

Since L_n is a positive and linear operator and $f \in W_{\alpha, E}^{\sim}$, by using the above inequality we have

$$|L_n(f; x) - f(x)| \le L_n(|f - f(x_0)|; x) + |f(x_0) - f(x)|$$

$$\le M L_n\left(\left|\frac{t}{1+t} - \frac{x_0}{1+x_0}\right|^\alpha; x\right) + M \frac{|x - x_0|^\alpha}{(1+x)^\alpha(1+x_0)^\alpha}. \tag{2.95}$$

If we use the classical inequality $(a+b)^\alpha \le a^\alpha + b^\alpha$ for $a \ge 0$, $b \ge 0$, one can write

$$\left|\frac{t}{1+t} - \frac{x_0}{1+x_0}\right|^\alpha \le \left|\frac{t}{1+t} - \frac{x}{1+x}\right|^\alpha + \left|\frac{x}{1+x} - \frac{x_0}{1+x_0}\right|^\alpha$$

for $0 < \alpha \le 1$ and $t \in [0, \infty)$. Consequently we obtain

$$L_n\left(\left|\tfrac{t}{1+t} - \tfrac{x_0}{1+x_0}\right|^\alpha; x\right) \le L_n\left(\left|\tfrac{t}{1+t} - \tfrac{x}{1+x}\right|^\alpha; x\right) + \frac{|x-x_0|^\alpha}{(1+x)^\alpha(1+x_0)^\alpha}.$$

Since $L_n(1; x) = 1$, applying Hölder inequality with $p = \frac{2}{\alpha}$ and $q = \frac{2}{2-\alpha}$, we have

$$L_n\left(\left|\tfrac{t}{1+t} - \tfrac{x_0}{1+x_0}\right|^\alpha; x\right) \le L_n\left(\left(\tfrac{t}{1+t} - \tfrac{x}{1+x}\right)^2; x\right)^{\frac{\alpha}{2}} + \frac{|x-x_0|^\alpha}{(1+x)^\alpha(1+x_0)^\alpha}.$$

Thus in view of (2.95), we have (2.94).   ∎

As a particular case of Theorem 2.20, when $E = [0, \infty)$, the following is true:

Corollary 2.8. *If $f \in W^{\sim}_{\alpha, [0, \infty)}$ then we have*

$$|L_n(f; x) - f(x)| \le M\mu_n^{\frac{\alpha}{2}}(x),$$

where $\mu_n(x)$ defined in (2.92).

In the following theorem a Stancu-type formula for the remainder of q-BBH operators is obtained which reduces to the formula of the remainder of classical BBH operators (see [2, p. 151]). Similar formula is obtained for q-Szász–Mirakyan operators in [29].

Here, $[x_0, x_1 \ldots, x_n; f] = f[x_0, x_1 \ldots, x_n]$ denotes the divided difference of the function f given in Lemma 2.3.

Theorem 2.21. *If $x \in (0, \infty) \setminus \left\{ \frac{[k]_q}{[n-k+1]_q q^k} \,\middle|\, k = 0, 1, 2, \ldots, n \right\}$, then the following identity holds:*

$$
\begin{aligned}
L_n(f; x) - f\left(\frac{x}{q}\right) & \\
&= -\frac{x^{n+1}}{\ell_n(x)} \left[\frac{x}{q}, \frac{[n]_q}{q^n}; f\right] \\
&\quad + \frac{x}{\ell_n(x)} \sum_{k=0}^{n-1} \left[\frac{x}{q}, \frac{[k]_q}{[n-k+1]_q q^k}, \frac{[k+1]_q}{[n-k]_q q^{k+1}}; f\right] q^{\frac{k(k+1)}{2}-2} \begin{bmatrix} n+1 \\ k \end{bmatrix}_q x^k.
\end{aligned} \tag{2.96}
$$

Proof. By using (2.84), we have

$$
\begin{aligned}
L_n(f; x) - f\left(\frac{x}{q}\right) &= \frac{1}{\ell_n(x)} \sum_{k=0}^{n} \left[f\left(\frac{[k]_q}{[n-k+1]_q q^k}\right) - f\left(\frac{x}{q}\right) \right] q^{\frac{k(k-1)}{2}} \begin{bmatrix} n \\ k \end{bmatrix}_q x^k \\
&= -\frac{1}{\ell_n(x)} \sum_{k=0}^{n} \left(\frac{x}{q} - \frac{[k]_q}{[n-k+1]_q q^k} \right) \\
&\quad \times \left[\frac{x}{q}, \frac{[k]_q}{[n-k+1]_q q^k}; f\right] q^{\frac{k(k-1)}{2}} \begin{bmatrix} n \\ k \end{bmatrix}_q x^k
\end{aligned}
$$

Since

$$\frac{[k]_q}{[n-k+1]_q} \begin{bmatrix} n \\ k \end{bmatrix}_q = \begin{bmatrix} n \\ k-1 \end{bmatrix}_q,$$

we have

$$L_n(f;x) - f\left(\frac{x}{q}\right) = -\frac{1}{\ell_n(x)} \sum_{k=0}^{n} \left[\frac{x}{q}, \frac{[k]_q}{[n-k+1]_q q^k}; f\right] q^{\frac{k(k-1)}{2}-1} \begin{bmatrix} n \\ k \end{bmatrix}_q x^{k+1}$$

$$+ \frac{1}{\ell_n(x)} \sum_{k=1}^{n} \left[\frac{x}{q}, \frac{[k]_q}{[n-k+1]_q q^k}; f\right] q^{\frac{k(k-1)}{2}-k} \begin{bmatrix} n \\ k-1 \end{bmatrix}_q x^{k}.$$

Rearranging the above equality, we can write

$$L_n(f;x) - f\left(\frac{x}{q}\right) = -\frac{x^{n+1}}{\ell_n(x)} \left[\frac{x}{q}, \frac{[n]_q}{q^n}; f\right] q^{\frac{n(n-1)}{2}-1}$$

$$+ \frac{1}{\ell_n(x)} \sum_{k=0}^{n-1} \left(\left[\frac{x}{q}, \frac{[k+1]_q}{[n-k]_q q^{k+1}}; f\right]\right.$$

$$\left. - \left[\frac{x}{q}, \frac{[k]_q}{[n-k+1]_q q^k}; f\right]\right) q^{\frac{k(k-1)}{2}-1} \begin{bmatrix} n \\ k \end{bmatrix}_q x^{k+1}.$$

Using the equality

$$\frac{[k+1]_q}{[n-k]_q q^{k+1}} - \frac{[k]_q}{[n-k+1]_q q^k} = \frac{[n+1]_q}{[n-k]_q [n-k+1]_q q^{k+1}},$$

we have following formula for divided differences:

$$\left[\frac{x}{q}, \frac{[k]_q}{[n-k+1]_q q^k}, \frac{[k+1]_q}{[n-k]_q q^{k+1}}; f\right] \frac{[n+1]_q}{[n-k]_q [n-k+1]_q q^{k+1}}$$

$$= \left[\frac{x}{q}, \frac{[k+1]_q}{[n-k]_q q^{k+1}}; f\right] - \left[\frac{x}{q}, \frac{[k]_q}{[n-k+1]_q q^k}; f\right]$$

and therefore, we obtain that the remainder formula for q-BBH operators, which is expressible in the form (2.96). ∎

We know that a function is convex on an interval if and only if all second-order divided differences of f are nonnegative. From this property and Theorem 2.21 we have the following result:

Corollary 2.9. *If f is convex and nonincreasing, then*

$$f\left(\frac{x}{q}\right) \leq L_n(f;x) \quad (n=0,1,\ldots)$$

2.5.4 Some Generalization of L_n

In this section, similarly as in [52], we shall define some generalization of the operators L_n.

We consider a sequence of linear positive operators as follows:

$$L_n^\gamma (f; x) = \frac{1}{\ell_n(x)} \sum_{k=0}^n f\left(\frac{[k]_q + \gamma}{b_{n,k}}\right) q^{\frac{k(k-1)}{2}} \begin{bmatrix} n \\ k \end{bmatrix}_q x^k, (\gamma \in \mathbb{R}) \qquad (2.97)$$

where $b_{n,k}$ satisfies the following condition:

$$[k]_q + b_{n,k} = c_n \quad \text{and} \quad \frac{[n]_q}{c_n} \to 1 \quad \text{for } n \to \infty.$$

It is easy to check that if $b_{n,k} = [n-k+1]_q q^k + \beta$ for any n, k and $0 < q < 1$, then $c_n = [n+1]_q + \beta$ and these operators turn out into D.D. Stancu-type generalization of Bleimann, Butzer, and Hahn operators based on q-integers (see [145]). If we choose $\gamma = 0$ and $q = 1$, then the operators become the special case of the Balázs-type generalization of the operators (2.82) given in [52].

Theorem 2.22. Let $q = q_n$ satisfies $0 < q_n \le 1$ and let $q_n \to 1$ as $n \to \infty$. If $f \in W^\sim_{\alpha, [0, \infty)}$, then the following inequality holds for a large n

$$\|L_n^\gamma (f; x) - f(x)\|_{C_B}$$

$$\le 3M \max \left\{ \left(\frac{[n]_{q_n}}{c_n + \gamma}\right)^\alpha \left(\frac{\gamma}{[n]_{q_n}}\right)^\alpha, \left|1 - \frac{[n+1]_{q_n}}{c_n + \gamma}\right|^\alpha \right.$$

$$\left. \times \left(\frac{[n]_{q_n}}{[n+1]_{q_n}}\right)^\alpha, 1 - 2\frac{[n]_{q_n}}{[n+1]_{q_n}} + \frac{[n]_{q_n}[n-1]_{q_n}}{[n+1]_{q_n}^2} q_n \right\}. \qquad (2.98)$$

Proof. Using (2.84) and (2.97) we have

$$|L_n^\gamma (f; x) - f(x)| \le \frac{1}{\ell_n(x)} \sum_{k=0}^n \left| f\left(\frac{[k]_{q_n} + \gamma}{b_{n,k}}\right) - f\left(\frac{[k]_{q_n}}{\gamma + b_{n,k}}\right) \right| q_n^{\frac{k(k-1)}{2}} \begin{bmatrix} n \\ k \end{bmatrix}_{q_n} x^k$$

$$+ \frac{1}{\ell_n(x)} \sum_{k=0}^n \left| f\left(\frac{[k]_{q_n}}{\gamma + b_{n,k}}\right) - f\left(\frac{[k]_{q_n}}{[n-k+1]q_n^k}\right) \right| q_n^{\frac{k(k-1)}{2}} \begin{bmatrix} n \\ k \end{bmatrix}_{q_n} x^k$$

$$+ |L_n (f; x) - f(x)|.$$

Since $f \in W^{\sim}_{\alpha, [0, \infty)}$ and by using Corollary 2.8, we can write

$$
\left| L_n^\gamma(f; x) - f(x) \right| \le \frac{M}{\ell_n(x)} \sum_{k=0}^{n} \left| \frac{[k]_{q_n} + \gamma}{[k]_{q_n} + \gamma + b_{n,k}} - \frac{[k]_{q_n}}{\gamma + [k]_{q_n} + b_{n,k}} \right|^\alpha q_n^{\frac{k(k-1)}{2}} \begin{bmatrix} n \\ k \end{bmatrix}_{q_n} x^k
$$

$$
+ \frac{M}{\ell_n(x)} \sum_{k=0}^{n} \left| \frac{[k]_{q_n}}{[k]_{q_n} + \gamma + b_{n,k}} - \frac{[k]_{q_n}}{[n+1]_{q_n}} \right|^\alpha q_n^{\frac{k(k-1)}{2}} \begin{bmatrix} n \\ k \end{bmatrix}_{q_n} x^k + \mu_n^{\frac{\alpha}{2}}(x)
$$

$$
\le \left(\frac{[n]}{c_n + \gamma} \right)^\alpha \left(\frac{\gamma}{[n]_{q_n}} \right)^\alpha + \left| 1 - \frac{[n+1]_{q_n}}{c_n + \gamma} \right|^\alpha
$$

$$
\times \frac{1}{\ell_n(x)} \sum_{k=0}^{n} \left(\frac{[k]_{q_n}}{[n+1]_{q_n}} \right)^\alpha q_n^{\frac{k(k-1)}{2}} \begin{bmatrix} n \\ k \end{bmatrix}_{q_n} x^k + \mu_n^{\frac{\alpha}{2}}(x).
$$

Using the Hölder inequality for $p = \frac{1}{\alpha}, q = \frac{1}{1-\alpha}$ and (2.87), we obtain

$$
\left| L_n^\gamma(f; x) - f(x) \right| \le M \left(\frac{[n]_{q_n}}{c_n + \gamma} \right)^\alpha \left(\frac{\gamma}{[n]_{q_n}} \right)^\alpha + M \left| 1 - \frac{[n+1]_{q_n}}{c_n + \gamma} \right|^\alpha \left(\frac{x}{x+1} \frac{[n]_{q_n}}{[n+1]_{q_n}} \right)^\alpha
$$

$$
+ \mu_n^{\frac{\alpha}{2}}(x).
$$

Thus the inequality (2.98) holds for $x \in [0, \infty)$.   ∎

Chapter 3
q-Integral Operators

3.1 q-Picard and q-Gauss–Weierstrass Singular Integral Operator

For many years scientists have been investigating to develop various aspects of approximation results of above operators. The recent book written by Anastassiou and Gal [18] includes great number of results related to different properties of these type of operators and also includes other references on the subject. For example, in Chapter 16 of [18], Jackson-type generalization of these operators is one among other generalizations, which satisfy the global smoothness preservation property (GSPP). It has been shown in [19] that this type of generalization has a better rate of convergence and provides better estimates with some modulus of smoothness. Beside, in [22, 23], Picard and Gauss–Weierstrass singular integral operators modified by means of nonisotropic distance and their pointwise approximation properties in different normed spaces are analyzed. Furthermore, in [40, 110], Picard and Gauss Weierstrass singular integrals were considered in exponential weighted spaces for functions of one or two variables.

3.1.1 Introduction

Let f be a real-valued function in $\mathbb{R}$. For $\lambda > 0$ and $x \in \mathbb{R}$, the well-known Picard and Gauss–Weierstrass singular integral operators are defined as

$$P_\lambda\left(f;x\right) := \frac{1}{2\lambda} \int_{-\infty}^{\infty} f\left(x+t\right) e^{-\frac{|t|}{\lambda}} dt$$

and

$$W_\lambda (f; x) := \frac{1}{\sqrt{\pi\lambda}} \int\limits_{-\infty}^{\infty} f(x+t) e^{-\frac{t^2}{\lambda}} dt,$$

respectively.

In this section, we introduce a new generalization of Picard singular integral operator and Gauss–Weierstrass singular integral operator which we call the *q*-Picard singular integral operator and the *q*-Gauss–Weierstrass singular integral operator, respectively. As a result, a connection has been constructed between *q*-analysis and approximation theory.

To be able to construct the generalized operators, we need the following *q*-extension of Euler integral representation for the gamma function given in [14, 34] for $0 < q < 1$:

$$c_q(x) \Gamma_q(x) = \frac{1-q}{\ln q^{-1}} q^{\frac{x(x-1)}{2}} \int\limits_{0}^{\infty} \frac{t^{x-1}}{E_q((1-q)t)} dt, \quad \Re x > 0 \qquad (3.1)$$

where $\Gamma_q(x)$ is the *q*-gamma function defined by

$$\Gamma_q(x) = \frac{(1-q)_q^\infty}{(1-q^x)_q^\infty} (1-q)^{1-x}, \quad 0 < q < 1$$

and $c_q(x)$ satisfies the following conditions:

1. $c_q(x+1) = c_q(x)$.
2. $c_q(n) = 1, n = 0, 1, 2, \ldots$.
3. $\lim\limits_{q \to 1^-} c_q(x) = 1$.

When $x = n+1$ with n a nonnegative integer, we obtain

$$\Gamma_q(n+1) = [n]_q!. \qquad (3.2)$$

In [38], Berg evaluated the following integral:

$$\int\limits_{-\infty}^{\infty} \frac{t^{2k}}{E_q(t^2)} dt = \pi \left(q^{1/2}; q\right)_{1/2} q^{-\frac{k^2}{2}} \left(q^{1/2}; q\right)_k, \quad k = 0, 1, 2\ldots \qquad (3.3)$$

where

$$(a; q)_\alpha = \frac{(1-a)_q^\infty}{(1-aq^\alpha)_q^\infty}$$

for any real number α.

The integrals (3.1) and (3.3) are the starting point of our work. Note that these definitions are kinds of q-deformation of usual ones and are reduced to them in the limit $q \to 1$.

Definition 3.1. Let $f : \mathbb{R} \to \mathbb{R}$ be a function. For $\lambda > 0$ and $0 < q < 1$, the q-generalizations of Picard and Gauss–Weierstrass singular integrals of f are

$$P_\lambda (f; q, x) \equiv P_\lambda (f; x) := \frac{(1-q)}{2[\lambda]_q \ln q^{-1}} \int_{-\infty}^{\infty} \frac{f(x+t)}{E_q\left(\frac{(1-q)|t|}{[\lambda]_q}\right)} dt \qquad (3.4)$$

and

$$W_\lambda (f; q, x) \equiv W_\lambda (f; x) := \frac{1}{\pi\sqrt{[\lambda]_q} \, (q^{1/2}; q)_{1/2}} \int_{-\infty}^{\infty} \frac{f(x+t)}{E_q\left(\frac{t^2}{[\lambda]_q}\right)} dt, \qquad (3.5)$$

respectively.

Note that this construction is sensitive to the rate of convergence to f. That is, the proposed estimate with rates in terms of L_p-modulus of continuity tells us that, depending on our selection of q, the rates of convergence in L_p-norm of the q-Picard and the q-Gauss–Weierstrass singular integral operators are better than the classical ones.

3.1.2   Rate of Convergence in $L_p(\mathbb{R})$

For $f \in L_p(\mathbb{R})$, the modulus of continuity of f is defined by

$$\omega_p(f; \delta) = \sup_{|h| \leq \delta} \|f(\cdot + h) - f(\cdot)\|_p,$$

where $\|f\|_p = \left(\int_{-\infty}^{\infty} |f(x)|^p dx \right)^{1/p}$.

Here are some auxiliary lemmas.

Lemma 3.1. *For every* $\lambda > 0$,

(a) $\int_{-\infty}^{\infty} P_\lambda (f; x) \, dx = 1.$

(b) $\int_{-\infty}^{\infty} W_\lambda (f; x) \, dx = 1.$

Proof. The proof is obvious from (3.1) and (3.3).   ∎

By using Lemma 3.1, for every function $f \in L_p(\mathbb{R})$ with $1 \leq p < \infty$, the operators defined by (3.4) and (3.5) are well defined as expressed in the following lemma.

Lemma 3.2. *Let* $f \in L_p(\mathbb{R})$ *for some* $1 \leq p < \infty$. *Then we have*

$$\|P_\lambda(f; \cdot)\|_p \leq \|f\|_p$$

and

$$\|W_\lambda(f; \cdot)\|_p \leq \|f\|_p.$$

Now we give convergence rates for these new operators. A similar approach for classical Picard and Gauss–Weierstrass singular integral operators can be found in [147, Th. 1.18].

Theorem 3.1. *If* $f \in L_p(\mathbb{R})$ *for some* $1 \leq p < \infty$, *then we have*

$$\|P_\lambda(f; \cdot) - f(\cdot)\|_p \leq \omega_p\left(f; [\lambda]_q\right)\left(1 + \frac{1}{q}\right)$$

and

$$\|W_\lambda(f; \cdot) - f(\cdot)\|_p \leq \omega_p\left(f; \sqrt{[\lambda]_q}\right)\left(1 + \sqrt{q^{-1/2}\left(1 - q^{1/2}\right)}\right).$$

Proof. From Lemma 3.1, we get

$$P_\lambda(f; x) - f(x) = \frac{(1-q)}{2[\lambda]_q \ln q^{-1}} \int\limits_{-\infty}^{\infty} \frac{(f(x+t) - f(x))}{E_q\left(\frac{(1-q)|t|}{[\lambda]_q}\right)} dt.$$

Thus

$$\|P_\lambda(f; \cdot) - f(\cdot)\|_p \leq \frac{(1-q)}{2[\lambda]_q \ln q^{-1}}\left(\int\limits_{-\infty}^{\infty}\left|\int\limits_{-\infty}^{\infty}\frac{f(x+t) - f(x)}{E_q\left(\frac{(1-q)|t|}{[\lambda]_q}\right)} dt\right|^p dx\right)^{1/p}$$

(generalized Minkowski inequality, see [146, p. 271])

$$\leq \frac{(1-q)}{2[\lambda]_q \ln q^{-1}} \int\limits_{-\infty}^{\infty} \frac{\omega_p(f; |t|)}{E_q\left(\frac{(1-q)|t|}{[\lambda]_q}\right)} dt$$

$$\leq \omega_p\left(f; [\lambda]_q\right) \frac{(1-q)}{2[\lambda]_q \ln q^{-1}} \int\limits_{-\infty}^{\infty}\left(1 + \frac{|t|}{[\lambda]_q}\right) \frac{dt}{E_q\left(\frac{(1-q)|t|}{[\lambda]_q}\right)}$$

$$= \omega_p\left(f; [\lambda]_q\right)\left(1 + \frac{1}{q}\right),$$

where we use (3.1) and (3.2) and the well-known inequality

$$\omega_p\left(f;C\delta\right)\le\left(1+C\right)\omega_p\left(f;\delta\right)$$

for $C>0$.

Similarly,

$$\|W_\lambda\left(f;\cdot\right)-f\left(\cdot\right)\|_p$$

$$\le \frac{\omega_p\left(f;\sqrt{[\lambda]_q}\right)}{\pi\sqrt{[\lambda]_q}\,(q^{1/2};q)_{1/2}}\int_{-\infty}^{\infty}\left(1+\frac{|t|}{\sqrt{[\lambda]_q}}\right)\frac{dt}{E_q\left(\frac{t^2}{[\lambda]_q}\right)}$$

$$\le \omega_p\left(f;\sqrt{[\lambda]_q}\right)\left(1+\left(\frac{1}{\pi\sqrt{[\lambda]_q}\,(q^{1/2};q)_{1/2}}\int_{-\infty}^{\infty}\frac{t^2}{[\lambda]_q}\frac{dt}{E_q\left(\frac{t^2}{[\lambda]_q}\right)}\right)^{1/2}\right)$$

$$\le \omega_p\left(f;\sqrt{[\lambda]_q}\right)\left(1+\sqrt{q^{-1/2}\left(1-q^{1/2}\right)}\right),$$

where we use (3.3).   ∎

Since for a fixed value of q with $0<q<1$,

$$\lim_{\lambda\to\infty}[\lambda]_q=\frac{1}{1-q},$$

the above theorem does not give a rate of convergence for $P_\lambda\left(f;\cdot\right)-f\left(\cdot\right)$ in L_p−norm. However, if we choose q_λ such that $0<q_\lambda<1$ and $q_\lambda\to 1$ as $\lambda\to\infty$, we have $[\lambda]_{q_\lambda}\to 0$ as $\lambda\to\infty$. Thus we express Theorem 3.1 as follows.

Theorem 3.2. *Let $q_\lambda\in(0,1)$ such that $q_\lambda\to 1$ as $\lambda\to\infty$. If $f\in L_p\left(\mathbb{R}\right)$ for some $1\le p<\infty$, then we have*

$$\|P_\lambda\left(f;q_\lambda,\cdot\right)-f\left(\cdot\right)\|_p\le\omega_p\left(f;[\lambda]_{q_\lambda}\right)\left(1+\frac{1}{q_\lambda}\right)$$

and

$$\|W_\lambda\left(f;q_\lambda,\cdot\right)-f\left(\cdot\right)\|_p\le\omega_p\left(f;\sqrt{[\lambda]_{q_\lambda}}\right)\left(1+\sqrt{q_\lambda^{-1/2}\left(1-q_\lambda^{1/2}\right)}\right).$$

This theorem tells us that depending on the selection of q_λ, the rate of convergence of $P_\lambda\left(f;\cdot\right)$ to $f\left(\cdot\right)$ in the L_p-norm is $[\lambda]_{q_\lambda}$ that is at least so faster than λ which is the rate of convergence for the classical Picard singular integrals. Similar situation arises when approximating by $W_\lambda\left(f;\cdot\right)$ to $f\left(\cdot\right)$.

3.1.3   Convergence in Weighted Space

Now we recall the following Korovkin-type theorem in weighted L_p space given in [67].

Let ω be a positive continuous function on real axis $\mathbb{R} = (-\infty, \infty)$, satisfying the condition

$$\int_{\mathbb{R}} t^{2p} \omega(t) dt < \infty. \tag{3.6}$$

We denote by $L_{p,\omega}(\mathbb{R})$ the linear space of p-absolutely integrable functions on $\mathbb{R}$ with respect to the weight function ω, i.e., for $1 \leq p < \infty$,

$$L_{p,\omega}(\mathbb{R}) = \left\{ f : \mathbb{R} \to \mathbb{R}; \|f\|_{p,\omega} := \left\| f\omega^{\frac{1}{p}} \right\|_p = \left(\int_{\mathbb{R}} |f(t)|^p \omega(t) dt \right)^{\frac{1}{p}} < \infty \right\}.$$

Theorem 3.3. *Let $(L_n)_{n \in N}$ be a uniformly bounded sequence of linear positive operators from $L_{p,\omega}(\mathbb{R})$ to $L_{p,\omega}(\mathbb{R})$, satisfying the conditions*

$$\lim_{n \to \infty} \left\| L_n(t^i; x) - x^i \right\|_{p,\omega} = 0, \quad i = 0, 1, 2.$$

Then for every $f \in L_{p,\omega}(\mathbb{R})$,

$$\lim_{n \to \infty} \|L_n f - f\|_{p,\omega} = 0.$$

By choosing $\omega(x) = \left(\frac{1}{1+x^{6m}} \right)^p$, $p \geq 1$, and working on $L_{p,\omega}(\mathbb{R})$ space that we denote it by $L_{p,m}(\mathbb{R})$, we shall obtain direct approximation result by using Theorem 3.3. Note that this selection of $\omega(x)$ satisfies the condition (3.6). Also note that for $1 \leq p < \infty$,

$$L_{p,m}(\mathbb{R}) = \left\{ f : f : \mathbb{R} \to \mathbb{R}; \left(1 + x^{6m} \right)^{-1} f(x) \in L_p(\mathbb{R}) \right\},$$

where m is a positive integer.

Lemma 3.3. *If $f \in L_{p,m}(\mathbb{R})$ for some $1 \leq p < \infty$ and positive integer m, then*

$$\|P_\lambda(f; \cdot)\|_{p,m} \leq 2^{6m-1} \left(1 + \frac{[\lambda]_q^{6m} [6m]_q!}{q^{3m(6m+1)}} \right) \|f\|_{p,m}$$

and

$$\|W_\lambda(f; \cdot)\|_{p,m} \leq 2^{6m-1} \left(1 + [\lambda]_q^{3m} q^{-\frac{9m^2}{2}} \left(q^{1/2}; q \right)_{3m} \right) \|f\|_{p,m}$$

for $0 < q < 1$.

Proof. Using $\left(1+(x+t)^{6m}\right) \le 2^{6m-1}\left(1+x^{6m}\right)\left(1+t^{6m}\right)$ for all positive integer m and $x, t \in \mathbb{R}$ and (3.1)–(3.3), the proof is obvious.   ∎

Theorem 3.4. *Let $q_\lambda \in (0, 1)$ such that $q_\lambda \to 1$ as $\lambda \to \infty$. Then for every $f \in L_{p,m}(\mathbb{R})$,*

$$\lim_{\lambda \to 0} \|P_\lambda(f; q_\lambda, \cdot) - f\|_{p,m} = 0.$$

Proof. Using Theorem 3.3, it is sufficient to verify that the conditions

$$\lim_{\lambda \to 0} \|P_\lambda(t^i; q_\lambda, \cdot) - x^i\|_{p,m} = 0, \quad i = 0, 1, 2. \tag{3.7}$$

are satisfied. Since $P_\lambda(1; q_\lambda, \cdot) = 1$ and $P_\lambda(t; q_\lambda, \cdot) = x$, the conditions of (3.7) are fulfilled for $i = 0$ and $i = 1$.

Direct calculation shows that

$$P_\lambda(t^2; q_\lambda, \cdot) = x^2 + \frac{[2]_{q_\lambda} [\lambda]_{q_\lambda}^2}{q_\lambda^3}$$

and then we obtain

$$\|P_\lambda(t^2; q_\lambda, \cdot) - x^2\|_{p,m} = \frac{[2]_{q_\lambda} [\lambda]_{q_\lambda}^2}{q_\lambda^3} \|1\|_{p,m}.$$

This means that the condition in (3.7) for $i = 2$ also holds and by Theorem 3.3 the proof is completed.   ∎

Theorem 3.5. *Let $q_\lambda \in (0, 1)$ such that $q_\lambda \to 1$ as $\lambda \to \infty$. For every $f \in L_{p,m}(\mathbb{R})$,*

$$\lim_{\lambda \to 0} \|W_\lambda(f; q_\lambda, \cdot) - f\|_{p,m} = 0.$$

For $f \in L_{p,m}(\mathbb{R})$ with some positive integer m, we define the weighted modulus of continuity $\omega_{p,m}(f; \delta)$ as

$$\omega_{p,m}(f; \delta) = \sup_{|h| \le \delta} \left(\int_{-\infty}^{\infty} \left| \frac{f(x+h) - f(x)}{(1+h^{6m})(1+x^{6m})} \right|^p dx \right)^{1/p}$$

$$= \sup_{|h| \le \delta} \left\| \frac{f(\cdot+h) - f(\cdot)}{(1+h^{6m})} \right\|_{p,m}.$$

Now, we show that this modulus of continuity satisfies some classical properties of L_p–modulus. For $f \in L_{p,m}$, it is guaranteed that $\omega_{p,m}(f; \delta)$ is bounded as δ tends to ∞ and also, $\omega_{p,m}(f; \delta) \le 2^{6m} \|f\|_{p,m}$ for any integer m.

3.1.4   Approximation Error

The next Lemmas 3.4 and 3.5 will allow us to obtain the approximation error of generalized operators by means of the weighted modulus of continuity $\omega_{p,m}(f;\delta)$ and weighted norm $\|\cdot\|_{p,m}$.

Lemma 3.4. *Given $f \in L_{p,m}(\mathbb{R})$ and $C > 0$,*

$$\omega_{p,m}(f;C\delta) \leq 2^{6m-1}(1+C)^{6m+1}\left(1+\delta^{6m}\right)\omega_{p,m}(f;\delta) \qquad (3.8)$$

for $\delta > 0$.

Proof. For positive integer n, we can write

$$\omega_{p,m}(f;n\delta) = \sup_{|h|\leq\delta}\left\|\frac{f(\cdot+nh)-f(\cdot)}{\left(1+(nh)^{6m}\right)}\right\|_{p,m}$$

$$= \sup_{|h|\leq\delta}\left\|\sum_{k=1}^{n}\frac{f(\cdot+kh)-f(\cdot+(k-1)h)}{\left(1+(nh)^{6m}\right)}\right\|_{p,m}$$

$$\leq 2^{6m-1}\omega_{p,m}(f;\delta)\sum_{k=1}^{n}\left(1+((k-1)\delta)^{6m}\right)$$

$$\leq 2^{6m-1}n\left(1+((n-1)\delta)^{6m}\right)\omega_{p,m}(f;\delta)$$

$$\leq 2^{6m-1}n^{6m+1}\left(1+\delta^{6m}\right)\omega_{p,m}(f;\delta).$$

Using this estimation

$$\omega_{p,m}(f;C\delta) \leq 2^{6m-1}(1+\lfloor|C|\rfloor)^{6m+1}\left(1+\delta^{6m}\right)\omega_{p,m}(f;\delta)$$

$$\leq 2^{6m-1}(1+C)^{6m+1}\left(1+\delta^{6m}\right)\omega_{p,m}(f;\delta),$$

where $\lfloor|C|\rfloor$ is the greatest integer less than C.   ∎

Lemma 3.5. *If $f \in L_{p,m}(\mathbb{R})$, then $\lim_{\delta\to 0}\omega_{p,m}(f;\delta) = 0$.*

Proof. For a positive real number a, let $\chi_1^a(t)$ be a characteristic function of the interval $[a,\infty)$, $\chi_2^a(t) = 1 - \chi_1^a(t)$ and $\chi^a(t) = \chi_1^{-a}(t) \cap \chi_2^a(t)$. Since $f \in L_{p,m}$, for each $\varepsilon > 0$ there exists $a \in \mathbb{R}$ large enough such that

$$\left(\int_{-\infty}^{-a} \left| \frac{f(x)}{1+x^{6m}} \right|^p dx \right)^{\frac{1}{p}} + \left(\int_{a}^{\infty} \left| \frac{f(x)}{1+x^{6m}} \right|^p dx \right)^{\frac{1}{p}} < \frac{\varepsilon}{4}.$$

That is,

$$\left\| f \chi_2^{-a} \right\|_{p,m} + \left\| f \chi_1^{a} \right\|_{p,m} < \frac{\varepsilon}{4}.$$

Similarly, for $\delta > 0$

$$\left\| f \chi_2^{-(a+\delta)} \right\|_{p,m} + \left\| f \chi_1^{a+\delta} \right\|_{p,m} < \frac{\varepsilon}{2^{6m+1}(1+\delta^{6m})}$$

can be written. Hence for $|h| \leq \delta$

$$\left\| f(\cdot + h) \chi_2^{-(a+\delta)}(\cdot) \right\|_{p,m} + \left\| f(\cdot + h) \chi_1^{a+\delta}(\cdot) \right\|_{p,m} < \frac{\varepsilon}{4}.$$

Thus, we have

$$\omega_{p,m}(f;\delta) \leq \sup_{|h| \leq \delta} \left\| \frac{(f(\cdot + h) - f(\cdot)) \chi^{a+\delta}(\cdot)}{(1+h^{6m})} \right\|_{p,m} + \frac{\varepsilon}{2} \tag{3.9}$$

for $\delta > 0$. By the well-known Weierstrass theorem, there exist sequences $\varphi_n(x) \in C^\infty$ (the space of function having continuous derivatives of any order in the interval $[-a-2\delta, a+2\delta]$) such that

$$\lim_{n \to \infty} \left\| (f(\cdot) - \varphi_n(\cdot)) \chi^{a+2\delta}(\cdot) \right\|_{p,m} = 0.$$

That is, given $\varepsilon > 0$, there exists $n_0 \in \mathbb{N}$ such that

$$\left\| (f(\cdot) - \varphi_n(\cdot)) \chi^{a+2\delta}(\cdot) \right\|_{p,m} < \frac{\varepsilon}{2^{6m+5}} \tag{3.10}$$

whenever $n \geq n_0$ and $\delta > 0$. Thus we have

$$\left\| (f(\cdot + h) - \varphi_n(\cdot + h)) \chi^{a+\delta}(\cdot) \right\|_{p,m} \leq 2^{6m-1} \left\| (f(\cdot) - \varphi_n(\cdot)) \chi^{a+2\delta}(\cdot) \right\|_{p,m}$$

$$\leq \frac{\varepsilon}{6} \tag{3.11}$$

for $n \geq n_0$.

Applying the Minkowski inequality yields

$$\left\| \frac{(f(\cdot + h) - f(\cdot)) \chi^{a+\delta}(\cdot)}{(1 + h^{6m})} \right\|_{p,m} \leq \left\| (f(\cdot + h) - \varphi_n(\cdot + h)) \chi^{a+\delta}(\cdot) \right\|_{p,m}$$

$$+ \left\| (\varphi_n(\cdot + h) - \varphi_n(\cdot)) \chi^{a+\delta}(\cdot) \right\|_{p,m}$$

$$+ \left\| (\varphi_n(\cdot) - f(\cdot)) \chi^{a+\delta}(\cdot) \right\|_{p,m}.$$

From (3.10) and (3.11) it follows that

$$\sup_{|h| \leq \delta} \left\| \frac{(f(\cdot + h) - f(\cdot)) \chi^{a+\delta}(\cdot)}{(1 + h^{6m})} \right\|_{p,m} \leq \frac{\varepsilon}{3} + \sup_{|h| \leq \delta} \left\| (\varphi_n(\cdot + h) - \varphi_n(\cdot)) \chi^{a+\delta}(\cdot) \right\|_{p,m},$$

$$(3.12)$$

for $\delta > 0$. By the properties of $\varphi_n(x)$, for $|h| \leq \delta$ and $n \geq n_0$, we can write

$$|\varphi_n(x + h) - \varphi_n(x)| \leq \frac{\varepsilon}{6 \left\| \chi^{a+\delta} \right\|_{p,m}},$$

where $x \in [-a - 2\delta, a + 2\delta]$. Thus, we obtain

$$\sup_{|h| \leq \delta} \left\| (\varphi_n(\cdot + h) - \varphi_n(\cdot)) \chi^{a+\delta}(\cdot) \right\|_{p,m} < \frac{\varepsilon}{6}. \qquad (3.13)$$

By (3.12) and (3.13) we get

$$\sup_{|h| \leq \delta} \left\| \frac{(f(\cdot + h) - f(\cdot)) \chi^{a+\delta}(\cdot)}{(1 + h^{6m})} \right\|_{p,m} < \frac{\varepsilon}{2}, \qquad (3.14)$$

for $\delta > 0$. From (3.9) and (3.14), we get

$$\omega_{p,m}(f; \delta) < \varepsilon$$

which shows that $\lim_{\delta \to 0} \omega_{p,m}(f; \delta) = 0$. ∎

Theorem 3.6. *Let* $q_\lambda \in (0, 1)$ *such that* $q_\lambda \to 1$ *as* $\lambda \to 0$. *For every* $f \in L_{p,m}(\mathbb{R})$,

$$\left\| P_\lambda(f; q_\lambda, \cdot) - f(\cdot) \right\|_{p,m} \leq A \omega_{p,m}\left(f; [\lambda]_{q_\lambda} \right)$$

and

$$\left\| W_\lambda(f; q_\lambda, \cdot) - f(\cdot) \right\|_{p,m} \leq B \omega_{p,m}\left(f; \sqrt{[\lambda]_{q_\lambda}} \right)$$

where

$$A = 2^{12m-1}\left(1 + \frac{(6m)!\,[\lambda]_{q_\lambda}^{6m}}{q_\lambda^{3m(6m+1)}} + \frac{(6m+1)!}{q_\lambda^{(3m+1)(6m+1)}} + \frac{(12m+1)!\,[\lambda]_{q_\lambda}^{6m}}{q_\lambda^{(12m+1)(6m+1)}}\right)\left(1 + [\lambda]_{q_\lambda}^{6m}\right) \quad (3.15)$$

and

$$B = 2^{12m-1}\left(1 + [\lambda]_{q_\lambda}^{6m}\right)\left(1 + [\lambda]_{q_\lambda}^{3m} q_\lambda^{-\frac{9m^2}{2}}\left(q_\lambda^{1/2};q_\lambda\right)_{3m} + \sqrt{q_\lambda^{-\frac{(6m+1)^2}{2}}\left(q_\lambda^{1/2};q_\lambda\right)_{6m+1}}\right.$$

$$\left. + [\lambda]_{q_\lambda}^{6m}\sqrt{q_\lambda^{-\frac{(12m+1)^2}{2}}\left(q_\lambda^{1/2};q_\lambda\right)_{12m+1}}\right).$$

Proof. Part (a) of Lemma 3.1 implies that

$$P_\lambda\left(f;q_\lambda,x\right) - f\left(x\right) = \frac{(1-q_\lambda)}{2\,[\lambda]_{q_\lambda}\ln q_\lambda^{-1}}\int_{-\infty}^{\infty}\frac{(f(x+t)-f(x))}{E_{q_\lambda}\left(\frac{(1-q_\lambda)|t|}{[\lambda]_{q_\lambda}}\right)}dt.$$

Then we have

$$\|P_\lambda\left(f;q_\lambda,\cdot\right) - f\left(\cdot\right)\|_{p,\,m} \leq \frac{(1-q_\lambda)}{2[\lambda]_{q_\lambda}\ln q_\lambda^{-1}}\left(\int_{-\infty}^{\infty}\left|\int_{-\infty}^{\infty}\frac{(f(x+t)-f(x))}{E_q\left(\frac{(1-q_\lambda)|t|}{[\lambda]_{q_\lambda}}\right)(1+x^{6m})}dt\right|^p dx\right)^{1/p}$$

$$\leq \frac{(1-q_\lambda)}{2[\lambda]_{q_\lambda}\ln q_\lambda^{-1}}\int_{-\infty}^{\infty}\left(\int_{-\infty}^{\infty}\left|\frac{f(x+t)-f(x)}{(1+x^{6m})}\right|dx\right)^{1/p}\frac{dt}{E_q\left(\frac{(1-q_\lambda)|t|}{[\lambda]_{q_\lambda}}\right)}$$

$$\leq \frac{(1-q_\lambda)}{[\lambda]_{q_\lambda}\ln q_\lambda^{-1}}\int_{0}^{\infty}\omega_{p,m}\left(f;t\right)\frac{(1+t^{6m})}{E_q\left(\frac{(1-q_\lambda)t}{[\lambda]_{q_\lambda}}\right)}dt.$$

By using (3.8) and taking $C = \frac{t}{[\lambda]_{q_\lambda}}$, we have

$$\|P_\lambda\left(f;q_\lambda,\cdot\right) - f\left(\cdot\right)\|_{p,\,m} \leq \frac{2^{6m-1}(1-q_\lambda)}{[\lambda]_{q_\lambda}\ln q_\lambda^{-1}}\left(1 + [\lambda]_{q_\lambda}^{6m}\right)\omega_{p,m}\left(f;[\lambda]_{q_\lambda}\right)$$

$$\int_{0}^{\infty}\frac{\left(1+\frac{t}{[\lambda]_{q_\lambda}}\right)^{6m+1}(1+t^{6m})}{E_q\left(\frac{(1-q_\lambda)t}{[\lambda]_{q_\lambda}}\right)}dt$$

$$\leq \frac{2^{12m-1}(1-q_\lambda)}{[\lambda]_{q_\lambda} \ln q_\lambda^{-1}} \left(1+[\lambda]_{q_\lambda}^{6m}\right) \omega_{p,m}\left(f;[\lambda]_{q_\lambda}\right)$$

$$\int_0^\infty \frac{1+t^{6m}+\frac{t^{6m+1}}{[\lambda]_{q_\lambda}^{6m+1}}+\frac{t^{12m+1}}{[\lambda]_{q_\lambda}^{6m+1}}}{E_q\left(\frac{(1-q_\lambda)t}{[\lambda]_{q_\lambda}}\right)} dt.$$

From (3.1) and (3.2) it follows that

$$\|P_\lambda(f;q_\lambda,\cdot)-f(\cdot)\|_{p,\,m} \leq A\omega_{p,m}\left(f;[\lambda]_{q_\lambda}\right),$$

where A is defined as in (3.15).

For $W_\lambda(f;\cdot)$, the proof is similar. ∎

3.1.5   Global Smoothness Preservation Property

Further information on GSPP for different linear positive operators and also singular integral operators can be found in [18].

Theorem 3.7. *Let $q_\lambda \in (0,1)$ such that $q_\lambda \to 1$ as $\lambda \to \infty$. For every $f \in L_{p,m}(\mathbb{R})$ and $\delta > 0$,*

$$\omega_{p,m}(P_\lambda(f);\delta) \leq C\omega_{p,m}(f;\delta)$$

and

$$\omega_{p,m}(W_\lambda(f);\delta) \leq D\omega_{p,m}(f;\delta)$$

where

$$C = \left(1 + \frac{(6m)!\,[\lambda]_{q_\lambda}^{6m}}{q_\lambda^{3m(6m+1)}}\right) \quad and \quad D = q_\lambda^{-\frac{9m^2}{2}} \left(q_\lambda^{1/2}; q_\lambda\right)_{3m} [\lambda]_{q_\lambda}^{3m}. \tag{3.16}$$

Proof. Part (a) of Lemma 3.1 implies that

$$P_\lambda(f;q_\lambda,x+h) - P_\lambda(f;q_\lambda,x) = \frac{(1-q_\lambda)}{2\,[\lambda]_{q_\lambda} \ln q_\lambda^{-1}} \int_{-\infty}^\infty \frac{(f(x+t+h)-f(x+t))}{E_{q_\lambda}\left(\frac{(1-q_\lambda)|t|}{[\lambda]_{q_\lambda}}\right)} dt.$$

By this equality, we get

$$\left(\int_{-\infty}^{\infty}\left|\frac{P_\lambda\left(f;q_\lambda,x+h\right)-P_\lambda\left(f;q_\lambda,x\right)}{\left(1+x^{6m}\right)\left(1+h^{6m}\right)}\right|^p dx\right)^{1/p}$$

$$\leq\frac{(1-q_\lambda)}{2\left[\lambda\right]_{q_\lambda}\ln q_\lambda^{-1}}\left(\int_{-\infty}^{\infty}\left|\int_{-\infty}^{\infty}\frac{\left(f\left(x+t+h\right)-f\left(x+t\right)\right)}{E_q\left(\frac{(1-q_\lambda)|t|}{\left[\lambda\right]_{q_\lambda}}\right)\left(1+x^{6m}\right)\left(1+h^{6m}\right)}dt\right|^p dx\right)^{1/p}$$

$$\leq\frac{(1-q_\lambda)}{2\left[\lambda\right]_{q_\lambda}\ln q_\lambda^{-1}}\int_{-\infty}^{\infty}\left(\int_{-\infty}^{\infty}\left|\frac{\left(f\left(x+t+h\right)-f\left(x+t\right)\right)}{\left(1+x^{6m}\right)\left(1+h^{6m}\right)}\right|^p dx\right)^{1/p}\frac{1}{E_{q_\lambda}\left(\frac{(1-q_\lambda)|t|}{\left[\lambda\right]_{q_\lambda}}\right)}dt.$$

Using the inequality for $x, t \in \mathbb{R}$

$$1+x^{6m}\leq 2^{6m-1}\left(1+(x-t)^{6m}\right)\left(1+t^{6m}\right),$$

we have

$$\left(\int_{-\infty}^{\infty}\left|\frac{P_\lambda\left(fq_\lambda,x+h\right)-P_\lambda\left(f;q_\lambda,x\right)}{\left(1+x^{6m}\right)\left(1+h^{6m}\right)}\right|^p dx\right)^{1/p}$$

$$\leq\frac{2^{6m-1}(1-q_\lambda)}{\left[\lambda\right]_{q_\lambda}\ln q_\lambda^{-1}}\omega_{p,m}(f;h)\int_0^{\infty}\frac{\left(1+t^{6m}\right)}{E_q\left(\frac{(1-q_\lambda)|t|}{\left[\lambda\right]_{q_\lambda}}\right)}dt.$$

Besides, from (3.1) and (3.2) it follows that

$$\omega_{p,m}\left(P_\lambda\left(f\right);h\right)\leq C\omega_{p,m}\left(f;h\right),$$

where C is defined as in (3.16).

For $W_\lambda\left(f;\cdot\right)$, the proof is similar. ∎

3.2 Generalized Picard Operators

In this section, we continue the study of the generalized Picard operator $\mathcal{P}_{\lambda,\beta}$ [16] depending on nonisotropic β-distance, in the direction of weighted approximation process. For this purpose, we first define weighted n-dimensional L_p space by involving weight depending on nonisotropic distance. Then we introduce a new weighted β-Lebesgue point depending on nonisotropic distance and study pointwise approximation of $\mathcal{P}_{\lambda,\beta}$ to the unit operator at these points. Also, we compare the order of convergence at the weighted β-Lebesgue point with the order of convergence of the operators to the unit operator. Finally, we show that this type of convergence also occurs with respect to nonisotropic weighted norm.

3.2.1   Introduction

In some recent papers, various results for the q-modification of approximation operators have been increasingly studied. For the brief knowledge, it may be useful to refer to the work of Anastassiou and Aral [16] and references therein. As it is known, one of the central research directions of approximation theory is singular integral operators. Among others, we are interested in Picard singular operator in multivariate setting defined as

$$P_\lambda \left(f\right)\left(\mathbf{x}\right) = \frac{1}{\prod_{i=1}^{n}\left(2\lambda_i\right)} \int_{-\infty}^{\infty} \cdots \int_{-\infty}^{\infty} \frac{f\left(x_1+t_1,\ldots,x_n+t_n\right)}{\prod_{i=1}^{n} e^{\frac{|t_i|}{\lambda_i}}} dt_1 \ldots dt_n \qquad (3.17)$$

for $\lambda = (\lambda_1,\ldots,\lambda_n) > \mathbf{0}$, which means that each component $\lambda_i\,(i = 1,\ldots,n)$ is positive and $\mathbf{x} = (x_1,\ldots,x_n) \in \mathbb{R}^n$. For a general framework related to this operator, [15, 18, 19, 72, 73] may be referred. In [22, 23], multivariate Picard and Gauss–Weierstrass operators with kernels including nonisotropic distance were introduced, and pointwise convergence results were given. Yet, q-generalization of Picard and Gauss–Weierstrass singular integral operators has been stated, and some approximation properties in weighted space have been discussed; also complex variants of them have been studied [24, 26, 28]. Recently, another interesting improvement related to the multivariate q-Picard singular operators depending on nonisotropic norm, $\mathcal{P}_{\lambda,\beta}$, has been subsequently stated in [16]. Here, the authors have investigated pointwise convergence of the family of $\mathcal{P}_{\lambda,\beta}\left(f\right)$ to f at the so called β-Lebesgue points depending on nonisotropic β-distance. Moreover, they have introduced a suitable modulus of continuity, depending on nonisotropic distance with supremum norm to measure the rate of convergence. Also they have proved the global smoothness preservation property of these operators.

In this work, for a weight depending on nonisotropic distance, we give analogue definitions of n-dimensional nonisotropic weighted-L_p space and nonisotropic weighted β-Lebesgue point at which we obtain a pointwise convergence result for the family of $\mathcal{P}_{\lambda,\beta}\left(f\right)$ to f for f belonging to this weighted space. We also give the measure of the rate of this pointwise convergence. Convergence in the norm of this space is also discussed.

Suppose that $\mathbb{R}^n$ is the n-dimensional Euclidean space of vectors $\mathbf{x} = (x_1,\ldots,x_n)$ with real components, and let $n \in \mathbb{N}$ and $\beta = (\beta_1, \beta_2, \cdots, \beta_n) \in \mathbb{R}^n$ with each component as positive, i.e., $\beta_i > 0$ $(i = 1,\ldots,n)$. Using standard notation, we denote $|\beta| = \beta_1 + \beta_2 + \cdots + \beta_n$. Recall that the nonisotropic β-distance between $\mathbf{x}$ and $\mathbf{0}$ is defined as

$$||\mathbf{x}||_\beta = \left(|x_1|^{\frac{1}{\beta_1}} + \cdots + |x_n|^{\frac{1}{\beta_n}} \right)^{\frac{|\beta|}{n}}, \quad \mathbf{x} \in \mathbb{R}^n.$$

Note that, for $t > 0$, $||\mathbf{x}||_\beta$ is homogeneous, namely,

$$\left(\left|t^{\beta_1} x_1\right|^{\frac{1}{\beta_1}} + \cdots + \left|t^{\beta_n} x_n\right|^{\frac{1}{\beta_n}} \right)^{\frac{|\beta|}{n}} = t^{\frac{|\beta|}{n}} ||\mathbf{x}||_\beta$$

and has the following properties:

1. $||\mathbf{x}||_\beta = 0 \Leftrightarrow \mathbf{x} = \mathbf{0}$,
2. $||t^\beta \mathbf{x}||_\beta = t^{\frac{|\beta|}{n}} ||\mathbf{x}||_\beta$,
3. $||\mathbf{x} + \mathbf{y}||_\beta \leq M_\beta \left(||\mathbf{x}||_\beta + ||\mathbf{y}||_\beta \right)$,

where $\beta_{\min} = \min\{\beta_1, \beta_2, \ldots \beta_n\}$ and $M_\beta = 2^{\left(1 + \frac{1}{\beta_{\min}}\right)\frac{|\beta|}{n}}$ (see [101]).

We should note here that when $\beta_i = \frac{1}{2}$ $(i = 1, 2, \ldots, n)$, the nonisotropic β-distance $||\mathbf{x}||_\beta$ becomes the ordinary Euclidean distance $|\mathbf{x}|$ and also that $||.||_\beta$ does not satisfy the triangle inequality.

Now, we reproduce here the following result and subsequent definition from [16] (see Lemma 1 and Definition 1 of [16]):

Lemma 3.6. *For all $\lambda > 0$, $n \in \mathbb{N}$ and $\beta_i \in (0, \infty)$ $(i = 1, 2, \ldots, n)$ with $|\beta| = \beta_1 + \beta_2 + \cdots + \beta_n$, we have*

$$\frac{c(n, \beta, q)}{[\lambda]_q^{|\beta|}} \int_{\mathbb{R}^n} \mathcal{P}_\lambda(\beta, \mathbf{t}) \, d\mathbf{t} = 1,$$

where

$$\mathcal{P}_\lambda(\beta, \mathbf{t}) = 1/E_q \left(\frac{(1-q)||\mathbf{t}||_\beta}{[\lambda]_q^{\frac{|\beta|}{n}}} \right), \tag{3.18}$$

and

$$c(n, \beta, q)^{-1} = \frac{n}{2|\beta|} \omega_{\beta,n-1} \Gamma_q(n) \frac{\ln q^{-1}}{(1-q)q^{\frac{n(n-1)}{2}}}, \tag{3.19}$$

with $\Gamma_q(n)$ is given as in (3.1) and $\omega_{\beta,n-1}$ will be given by (3.23) below.

Definition 3.2. Let $f : \mathbb{R}^n \to \mathbb{R}$ be a function. For $0 < q < 1, \lambda > 0$, $n \in \mathbb{N}$ and $\beta_i \in (0, \infty)$ $(i = 1, 2, \ldots, n)$ with $|\beta| = \beta_1 + \beta_2 + \cdots + \beta_n$. The generalized q-Picard singular integral depending on nonisotropic β-distance, attached to f, is defined as

$$\mathcal{P}_{\lambda,\beta}(f;q,\mathbf{x}) \equiv \mathcal{P}_{\lambda,\beta}(f;\mathbf{x})$$

$$:= \frac{c(n, \beta, q)}{[\lambda]_q^{|\beta|}} \int_{\mathbb{R}^n} f(\mathbf{x}+\mathbf{t}) \mathcal{P}_\lambda(\beta,\mathbf{t}) dt, \qquad (3.20)$$

where $\mathcal{P}_\lambda(\beta,\mathbf{t})$ and $c(n, \beta, q)$ are defined as in (3.18) and (3.19), respectively.

The case $\beta_i = \frac{1}{2}$ $(i = 1, 2, \ldots, n)$ clearly gives the operators $\mathcal{P}_{\lambda,\frac{1}{2}}(f;q,\mathbf{x})$ introduced in [24]. Letting $q \to 1$ implies that $\mathcal{P}_{\lambda,\frac{1}{2}}(f;1,\mathbf{x})$ will be the classical multivariate Picard singular integrals (3.17).

Now, we present the following definition.

Definition 3.3. Let p, $1 \le p < \infty$, be fixed. By $L_{p,\beta}(\mathbb{R}^n)$ we denote the weighted space with nonisotropic distance of real-valued functions f defined on $\mathbb{R}^n$ for which $\frac{f(\mathbf{x})}{1+\|\mathbf{x}\|_\beta}$ is p-absolutely Lebesgue integrable on $\mathbb{R}^n$ such that the norm

$$\|f\|_{p,\beta} = \left(\int_{\mathbb{R}^n} \left| \frac{f(\mathbf{x})}{1+\|\mathbf{x}\|_\beta} \right|^p dx \right)^{\frac{1}{p}}$$

is finite.

For the case $p = \infty$, we also have

$$\|f\|_{\infty,\beta} = \sup \left\{ \left| \frac{f(\mathbf{x})}{1+\|\mathbf{x}\|_\beta} \right| : \mathbf{x} = (x_1, \ldots, x_n) \in \mathbb{R}^n \right\}.$$

Lemma 3.7. *Let $\lambda > 0$, $n \in \mathbb{N}$, and $\beta_i \in (0, \infty)$ $(i = 1, 2, \ldots n)$ with $|\beta| = \beta_1 + \beta_2 + \cdots + \beta_n$. $\mathcal{P}_{\lambda,\beta}(f)$ is a linear positive operator from the space $L_{p,\beta}(\mathbb{R}^n)$ to $L_{p,\beta}(\mathbb{R}^n)$. That is,*

$$\left\| \mathcal{P}_{\lambda,\beta}(f) \right\|_{p,\beta} \le K(n, \beta, q) \|f\|_{p,\beta},$$

where

$$K(n, \beta, q) = \max\{1, M_\beta\} \left(1 + \frac{n}{2|\beta|} c(n, \beta, q) [\lambda]_q^{\frac{|\beta|}{n}} \omega_{\beta,n-1} \Gamma_q(n+1) \frac{\ln q^{-1}}{1-q} q^{-\frac{n(n+1)}{2}} \right),$$

in which M_β is the number appeared in the property 3 of nonisotropic distance.

Proof. Using the generalized Minkowski inequality we have

$$
\left\| \mathcal{P}_{\lambda,\beta}\left(f;\mathbf{x}\right)\right\|_{p,\beta} = \left(\int\limits_{\mathbb{R}^n} \left| \frac{\mathcal{P}_{\lambda,\beta}\left(f;\mathbf{x}\right)}{1+\|\mathbf{x}\|_\beta}\right|^p dx \right)^{\frac{1}{p}}
$$

$$
= \frac{c(n,\beta,q)}{[\lambda]_q^{|\beta|}} \left(\int\limits_{\mathbb{R}^n} \left| \frac{1}{1+\|\mathbf{x}\|_\beta} \int\limits_{\mathbb{R}^n} f(\mathbf{x}+\mathbf{t})\mathcal{P}_\lambda\left(\beta,\mathbf{t}\right)d\mathbf{t}\right|^p dx \right)^{\frac{1}{p}}
$$

$$
\leq \frac{c(n,\beta,q)}{[\lambda]_q^{|\beta|}} \int\limits_{\mathbb{R}^n} \left(\int\limits_{\mathbb{R}^n} \left| \frac{f(\mathbf{x}+\mathbf{t})}{1+\|\mathbf{x}\|_\beta} dx \right|^p \right)^{\frac{1}{p}} \mathcal{P}_\lambda\left(\beta,\mathbf{t}\right)d\mathbf{t}. \qquad (3.21)
$$

From the property 3 of $\|.\|_\beta$, we have

$$
\frac{1+\|\mathbf{x}+\mathbf{t}\|_\beta}{1+\|\mathbf{x}\|_\beta} \leq \frac{1+M_\beta\left(\|\mathbf{x}\|_\beta+\|\mathbf{t}\|_\beta\right)}{1+\|\mathbf{x}\|_\beta}
$$

$$
\leq \max\{1,M_\beta\} \frac{\left(1+\|\mathbf{x}\|_\beta\right)\left(1+\|\mathbf{t}\|_\beta\right)}{1+\|\mathbf{x}\|_\beta}. \qquad (3.22)
$$

Taking into account (3.22) and Lemma 3.6, the inequality (3.21) reduces to

$$
\left\| \mathcal{P}_{\lambda,\beta}\left(f;\mathbf{x}\right)\right\|_{p,\beta}
$$

$$
\leq \max\{1,M_\beta\} \frac{c(n,\beta,q)}{[\lambda]_q^{|\beta|}} \int\limits_{\mathbb{R}^n} \left(\int\limits_{\mathbb{R}^n} \left| \frac{f(\mathbf{x}+\mathbf{t})}{1+\|\mathbf{x}+\mathbf{t}\|_\beta} dx \right|^p \right)^{\frac{1}{p}} \mathcal{P}_\lambda\left(\beta,\mathbf{t}\right)\left(1+\|\mathbf{t}\|_\beta\right)d\mathbf{t}
$$

$$
\leq \|f\|_{p,\beta} \max\{1,M_\beta\} \left(1+ \frac{c(n,\beta,q)}{[\lambda]_q^{|\beta|}} \int\limits_{\mathbb{R}^n} \mathcal{P}_\lambda\left(\beta,\mathbf{t}\right)\|\mathbf{t}\|_\beta d\mathbf{t} \right).
$$

By substitution $\mathbf{t} = [\lambda]_q^\beta \mathbf{x}$, it follows that

$$
\left\| \mathcal{P}_{\lambda,\beta}\left(f;\mathbf{x}\right)\right\|_{p,\beta} \leq \|f\|_{p,\beta} \max\{1,M_\beta\} \left(1+c(n,\beta,q)[\lambda]_q^{\frac{|\beta|}{n}} \int\limits_{\mathbb{R}^n} \frac{\|\mathbf{x}\|_\beta}{E_q\left((1-q)\|\mathbf{x}\|_\beta\right)}dx \right).
$$

Now, we shall use generalized β-spherical coordinates [101] by taking the following transformation into account:

$$x_1 = (u\cos\theta_1)^{2\beta_1}$$
$$x_2 = (u\sin\theta_1\cos\theta_2)^{2\beta_2}$$
$$\vdots$$
$$x_{n-1} = (u\sin\theta_1\sin\theta_2\cdots\sin\theta_{n-2}\cos\theta_{n-1})^{2\beta_{n-1}}$$
$$x_n = (u\sin\theta_1\sin\theta_2\cdots\sin\theta_{n-1})^{2\beta_n},$$

where $0 \le \theta_1, \theta_2, \cdots, \theta_{n-2} \le \pi, 0 \le \theta_{n-1} \le 2\pi, u \ge 0$. The Jacobian of this transformation is denoted by $J_\beta(u, \theta_1, \ldots, \theta_{n-1})$ and obtained as

$$J_\beta(u, \theta_1, \ldots, \theta_{n-1}) = u^{2|\beta|-1}\Omega_\beta(\theta),$$

where $\Omega_\beta(\theta) = 2^n \beta_1 \ldots \beta_n \prod_{j=1}^{n-1}(\cos\theta_j)^{2\beta_j-1}(\sin\theta_j)^{\sum_{k=j}^{j+1} 2\beta_k - 1}$. Clearly the integral

$$\omega_{\beta,n-1} = \int_{S^{n-1}} \Omega_\beta(\theta)\,d\theta \tag{3.23}$$

is finite. Here S^{n-1} is the unit sphere in $\mathbb{R}^n$. Thus we have

$$\left\|\mathcal{P}_{\lambda,\beta}(f;\mathbf{x})\right\|_{p,\beta}$$

$$\le \|f\|_{p,\beta} \max\left\{1, M_\beta\right\}\left(1 + c(n,\beta,q)[\lambda]_q^{\frac{|\beta|}{n}} \int_0^\infty \int_{S^{n-1}} \frac{u^{2|\beta|+\frac{2|\beta|}{n}-1}\Omega_\beta(\theta)\,d\theta\,du}{E_q\left((1-q)u^{\frac{2|\beta|}{n}}\right)}\right)$$

$$\le \|f\|_{p,\beta} \max\left\{1, M_\beta\right\}\left(1 + \frac{n}{2|\beta|}c(n,\beta,q)[\lambda]_q^{\frac{|\beta|}{n}} \omega_{\beta,n-1} \int_0^\infty \frac{u^n\,du}{E_q((1-q)u)}\right)$$

$$\le \|f\|_{p,\beta} \max\left\{1, M_\beta\right\}\left(1 + \frac{n}{2|\beta|}c(n,\beta,q)[\lambda]_q^{\frac{|\beta|}{n}} \omega_{\beta,n-1}\Gamma_q(n+1)\frac{\ln q^{-1}}{1-q}q^{-\frac{n(n+1)}{2}}\right).$$

The lemma is proved. ∎

3.2.2 Pointwise Convergence

This section provides a result related to pointwise convergence. Below, we first give the definition of the points at which pointwise convergence will be observed.

Definition 3.4. Let $f \in L_{p,\beta}(\mathbb{R}^n)$, $1 \leq p < \infty$, and $\beta_i \in (0,\infty)$ $(i = 1,\ldots,n)$ with $|\beta| = \beta_1 + \beta_2 + \ldots + \beta_n$. We say that $\mathbf{x} \in \mathbb{R}^n$ is weighted β-Lebesgue point of f provided

$$\lim_{h \to 0} \left\{ \frac{1}{h^{2|\beta|}} \int_{\|\mathbf{t}\|_\beta^{\frac{n}{2|\beta|}} < h} \left| \frac{f(\mathbf{x}+\mathbf{t}) - f(\mathbf{x})}{1 + \|\mathbf{t}\|_\beta} \right|^p d\mathbf{t} \right\}^{\frac{1}{p}} = 0.$$

Next, we give a pointwise approximation of the generalized Picard operators $P_{\lambda,\beta}(f)$ to the function $f \in L_{p,\beta}$, at any weighted β-Lebesgue point of f.

Theorem 3.8. Let $f \in L_{p,\beta}(\mathbb{R}^n)$, $1 \leq p < \infty$, with $\beta_i \in (0,\infty)$ $(i = 1,\ldots,n)$, and $|\beta| = \beta_1 + \beta_2 + \ldots + \beta_n$. Then we have

$$\lim_{\lambda \to 0} P_{\lambda,\beta}(f;q,\mathbf{x}) = f(\mathbf{x})$$

at any weighted β-Lebesgue point $\mathbf{x}$ of f.

Proof. Let $\mathbf{x}$ be a weighted β-Lebesgue point of f. Then for any $\varepsilon > 0$, there exists an $\eta > 0$ such that $h < \eta$ implies that

$$\left\{ \frac{1}{h^{2|\beta|}} \int_{\|\mathbf{t}\|_\beta^{\frac{n}{2|\beta|}} < h} \left| \frac{f(\mathbf{x}+\mathbf{t}) - f(\mathbf{x})}{1 + \|\mathbf{t}\|_\beta} \right|^p d\mathbf{t} \right\}^{\frac{1}{p}} < \varepsilon,$$

which clearly means that

$$\int_{\|\mathbf{t}\|_\beta^{\frac{n}{2|\beta|}} < h} \left| \frac{f(\mathbf{x}+\mathbf{t}) - f(\mathbf{x})}{1 + \|\mathbf{t}\|_\beta} \right|^p d\mathbf{t} < \varepsilon^p h^{2|\beta|}.$$

Transforming the last integral into the generalized β-spherical coordinates, for $h < \eta$, we get

$$\int_{\|\mathbf{t}\|_\beta^{\frac{n}{2|\beta|}} < h} \left| \frac{f(\mathbf{x}+\mathbf{t}) - f(\mathbf{x})}{1 + \|\mathbf{t}\|_\beta} \right|^p d\mathbf{t} = \int_0^h \int_{S^{n-1}} \left| \frac{f\left(\mathbf{x}+(u\theta)^\beta\right) - f(\mathbf{x})}{1 + u^{\frac{2|\beta|}{n}}} \right|^p \Omega_\beta(\theta) u^{2|\beta|-1} d\theta du$$

$$= \int_0^h u^{2|\beta|-1} g(u) du < \varepsilon^p h^{2|\beta|},$$

where

$$g(u) = \int_{S^{n-1}} \left| \frac{f\left(\mathbf{x}+(u\theta)^{\beta}\right) - f(\mathbf{x})}{1+u^{\frac{2|\beta|}{n}}} \right|^{p} \Omega_{\beta}(\theta)\, d\theta. \qquad (3.24)$$

Meanwhile, from Lemma 3.6, for all $\eta > 0$ we have

$$\left| P_{\lambda,\beta}(f;q,\mathbf{x}) - f(\mathbf{x}) \right| \leq \frac{c(n,\beta,q)}{[\lambda]_q^{|\beta|}} \int_{\|\mathbf{t}\|_{\beta}^{\frac{2|\beta|}{n}} < \eta} |f(\mathbf{x}+\mathbf{t}) - f(\mathbf{x})| P_{\lambda}(\beta,\mathbf{t})\, d\mathbf{t}$$

$$+ \frac{c(n,\beta,q)}{[\lambda]_q^{|\beta|}} \int_{\|\mathbf{t}\|_{\beta}^{\frac{2|\beta|}{n}} \geq \eta} |f(\mathbf{x}+\mathbf{t}) - f(\mathbf{x})| P_{\lambda}(\beta,\mathbf{t})\, d\mathbf{t}$$

$$= \frac{c(n,\beta,q)}{[\lambda]_q^{|\beta|}} \int_{\|\mathbf{t}\|_{\beta}^{\frac{2|\beta|}{n}} < \eta} \left| \frac{f(\mathbf{x}+\mathbf{t}) - f(\mathbf{x})}{1+\|\mathbf{t}\|_{\beta}} \right| P_{\lambda}(\beta,\mathbf{t}) \left(1+\|\mathbf{t}\|_{\beta}\right) d\mathbf{t}$$

$$+ \frac{c(n,\beta,q)}{[\lambda]_q^{|\beta|}} \int_{\|\mathbf{t}\|_{\beta}^{\frac{2|\beta|}{n}} \geq \eta} |f(\mathbf{x}+\mathbf{t}) - f(\mathbf{x})| P_{\lambda}(\beta,\mathbf{t})\, d\mathbf{t}.$$

Applying Hölder's inequality to the first integral, then we have

$$\left| P_{\lambda,\beta}(f;q,\mathbf{x}) - f(\mathbf{x}) \right| \leq \left\{ \frac{c(n,\beta,q)}{[\lambda]_q^{|\beta|}} \int_{\|\mathbf{t}\|_{\beta}^{\frac{n}{2|\beta|}} < \eta} \left| \frac{f(\mathbf{x}+\mathbf{t}) - f(\mathbf{x})}{1+\|\mathbf{t}\|_{\beta}} \right|^{p} P_{\lambda}(\beta,\mathbf{t})\, d\mathbf{t} \right\}^{\frac{1}{p}}$$

$$\times \left\{ \frac{c(n,\beta,q)}{[\lambda]_q^{|\beta|}} \int_{\|\mathbf{t}\|_{\beta}^{\frac{n}{2|\beta|}} < \eta} \left(1+\|\mathbf{t}\|_{\beta}\right)^{q} P_{\lambda}(\beta,\mathbf{t})\, d\mathbf{t} \right\}^{\frac{1}{q}}$$

$$+ \frac{c(n,\beta,q)}{[\lambda]_q^{|\beta|}} \int_{\|\mathbf{t}\|_{\beta}^{\frac{n}{2|\beta|}} \geq \eta} |f(\mathbf{x}+\mathbf{t}) - f(\mathbf{x})| P_{\lambda}(\beta,\mathbf{t})\, d\mathbf{t}$$

$$\leq \left(1+\eta^{\frac{2|\beta|}{n}}\right) \left\{ \frac{c(n,\beta,q)}{[\lambda]_q^{|\beta|}} \int_{\|\mathbf{t}\|_{\beta}^{\frac{n}{2|\beta|}} < \eta} \left| \frac{f(\mathbf{x}+\mathbf{t}) - f(\mathbf{x})}{1+\|\mathbf{t}\|_{\beta}} \right|^{p} P_{\lambda}(\beta,\mathbf{t})\, d\mathbf{t} \right\}^{\frac{1}{p}}$$

$$+ \frac{c(n,\beta,q)}{[\lambda]_q^{|\beta|}} \int_{\|\mathbf{t}\|_{\beta}^{\frac{n}{2|\beta|}} \geq \eta} |f(\mathbf{x}+\mathbf{t}) - f(\mathbf{x})| P_{\lambda}(\beta,\mathbf{t})\, d\mathbf{t}$$

$$= J_1(\lambda) + J_2(\lambda).$$

Now passing to the generalized β-spherical coordinates, $J_1(\lambda)$ gives rise to

$$J_1(\lambda) = \left(1+\eta^{\frac{2|\beta|}{n}}\right) \left\{ \int_0^{\eta} \int_{S^{n-1}} \left| \frac{f\left(\mathbf{x}+(u\theta)^{\beta}\right) - f(\mathbf{x})}{1+u^{\frac{2|\beta|}{n}}} \right|^{p} \Omega_{\beta}(\theta)\, u^{2|\beta|-1} P_{\lambda}^{0}(\beta,u)\, d\theta\, du \right\}^{\frac{1}{p}},$$

$$(3.25)$$

where

$$\mathcal{P}_{\lambda}^{0}(\beta,u) = \frac{c(n,\beta,q)}{[\lambda]_q^{|\beta|} E_q\left(\frac{(1-q)u^{\frac{2|\beta|}{n}}}{[\lambda]_q^{\frac{|\beta|}{n}}} \right)}. \qquad (3.26)$$

Therefore taking into account (3.24), (3.25) can be expressed as

$$J_1(\lambda) = \left(1 + \eta^{\frac{2|\beta|}{n}}\right) \left\{ \int_0^{\eta} g(u) u^{2|\beta|-1} \mathcal{P}_{\lambda}^0(\beta, u) \, du \right\}^{\frac{1}{p}}.$$

Integrating by parts two times gives that there exists a constant A_1, such that $J_1(\lambda) \leq \left(1 + \eta^{\frac{2|\beta|}{n}}\right) \varepsilon A_1$ (see [16]).

For $J_2(\lambda)$, from (3.22) we get

$$J_2(\lambda) = \frac{c(n,\beta,q)}{[\lambda]_q^{|\beta|}} \int_{\|\mathbf{t}\|_{\beta}^{\frac{n}{2|\beta|}} \geq \eta} |f(\mathbf{x}+\mathbf{t}) - f(\mathbf{x})| \mathcal{P}_{\lambda}(\beta, \mathbf{t}) \, d\mathbf{t}$$

$$\leq \frac{c(n,\beta,q)}{[\lambda]_q^{|\beta|}} \int_{\|\mathbf{t}\|_{\beta}^{\frac{n}{2|\beta|}} \geq \eta} \frac{|f(\mathbf{x}+\mathbf{t})|}{1 + \|\mathbf{x}+\mathbf{t}\|_{\beta}} \left(1 + \|\mathbf{x}+\mathbf{t}\|_{\beta}\right) \mathcal{P}_{\lambda}(\beta, \mathbf{t}) \, d\mathbf{t}$$

$$+ \frac{c(n,\beta,q)}{[\lambda]_q^{|\beta|}} |f(\mathbf{x})| \int_{\|\mathbf{t}\|_{\beta}^{\frac{n}{2|\beta|}} \geq \eta} \mathcal{P}_{\lambda}(\beta, \mathbf{t}) \, d\mathbf{t}$$

$$\leq \frac{c(n,\beta,q)}{[\lambda]_q^{|\beta|}} \max\{1, M_{\beta}\} \left(1 + \|\mathbf{x}\|_{\beta}\right) \int_{\|\mathbf{t}\|_{\beta}^{\frac{n}{2|\beta|}} \geq \eta} \frac{|f(\mathbf{x}+\mathbf{t})|}{1 + \|\mathbf{x}+\mathbf{t}\|_{\beta}} \left(1 + \|\mathbf{t}\|_{\beta}\right) \mathcal{P}_{\lambda}(\beta, \mathbf{t}) \, d\mathbf{t}$$

$$+ \frac{c(n,\beta,q)}{[\lambda]_q^{|\beta|}} |f(\mathbf{x})| \int_{\|\mathbf{t}\|_{\beta}^{\frac{n}{2|\beta|}} \geq \eta} \mathcal{P}_{\lambda}(\beta, \mathbf{t}) \, d\mathbf{t}.$$

Further applying Hölder's inequality $J_2(\lambda)$ gives that

$$J_2(\lambda) \leq \frac{c(n,\beta,q)}{[\lambda]_q^{|\beta|}} \max\{1, M_{\beta}\} \left(1 + \|\mathbf{x}\|_{\beta}\right)$$

$$\times \left\{ \int_{\mathbb{R}^n} \left| \frac{f(\mathbf{x}+\mathbf{t})}{1 + \|\mathbf{x}+\mathbf{t}\|_{\beta}} \right|^p \, d\mathbf{t} \right\}^{\frac{1}{p}} \left\{ \int_{\mathbb{R}^n} \left| \chi_{\eta} \left(1 + \|\mathbf{t}\|_{\beta}\right) \mathcal{P}_{\lambda}(\beta, \mathbf{t}) \right|^q \, d\mathbf{t} \right\}^{\frac{1}{q}}$$

$$+ \frac{c(n,\beta,q)}{[\lambda]_q^{|\beta|}} |f(\mathbf{x})| \left\| \chi_{\eta} \mathcal{P}_{\lambda}(\beta, \mathbf{t}) \right\|_1$$

$$= \frac{c(n,\beta,q)}{[\lambda]_q^{|\beta|}} \max\{1, M_{\beta}\} \left(1 + \|\mathbf{x}\|_{\beta}\right) \|f\|_{p,\beta}$$

$$\times \left\{ \int_{\mathbb{R}^n} \left| \chi_{\eta} \left(1 + \|\mathbf{t}\|_{\beta}\right) \mathcal{P}_{\lambda}(\beta, \mathbf{t}) \right|^{\frac{q}{p}} \left| \left(1 + \|\mathbf{t}\|_{\beta}\right) \mathcal{P}_{\lambda}(\beta, \mathbf{t}) \right| \, d\mathbf{t} \right\}^{\frac{1}{q}}$$

$$+ \frac{c(n,\beta,q)}{[\lambda]_q^{|\beta|}} |f(\mathbf{x})| \left\| \chi_{\eta} \mathcal{P}_{\lambda}(\beta, \mathbf{t}) \right\|_1,$$

where χ_{η} is the characteristic function of the set of $\mathbf{t}$ such that $\|\mathbf{t}\|^{\frac{n}{2|\beta|}} \geq \eta$.

Taking into account the fact that $\left(1+\|\mathbf{t}\|_\beta\right) P_\lambda\left(\beta,\mathbf{t}\right) \in L_{1,\beta}$, then the above inequality takes the following form:

$$J_2\left(\lambda\right) \leq \frac{c\left(n,\beta,q\right)}{[\lambda]_q^{|\beta|}} \max\left\{1,M_\beta\right\}\left(1+\|\mathbf{x}\|_\beta\right)\|f\|_{p,\beta}$$

$$\times\left\{\sup_{\|\mathbf{t}\|_\beta^{\frac{n}{2|\beta|}}\geq\eta} \left|\left(1+\|\mathbf{t}\|_\beta\right)\mathcal{P}_\lambda\left(\beta,\mathbf{t}\right)\right|^{\frac{q}{p}} \int_{\mathbb{R}^n}\chi_\eta\left|\left(1+\|\mathbf{t}\|_\beta\right)\mathcal{P}_\lambda\left(\beta,\mathbf{t}\right)\right|dt\right\}^{\frac{1}{q}}$$

$$+\frac{c\left(n,\beta,q\right)}{[\lambda]_q^{|\beta|}}\left|f\left(\mathbf{x}\right)\right|\left\|\chi_\eta\mathcal{P}_\lambda\left(\beta,\mathbf{t}\right)\right\|_1$$

$$=\max\left\{1,M_\beta\right\}\left(1+\|\mathbf{x}\|_\beta\right)\|f\|_{p,\beta}\left\{\frac{c\left(n,\beta,q\right)}{[\lambda]_q^{|\beta|}}\sup_{\|\mathbf{t}\|_\beta^{\frac{n}{2|\beta|}}\geq\eta}\left|\left(1+\|\mathbf{t}\|_\beta\right)\mathcal{P}_\lambda\left(\beta,\mathbf{t}\right)\right|\right\}^{\frac{1}{p}}$$

$$\times\left\{\frac{c\left(n,\beta,q\right)}{[\lambda]_q^{|\beta|}}\left\|\chi_\eta\left(1+\|\mathbf{t}\|_\beta\right)\mathcal{P}_\lambda\left(\beta,\mathbf{t}\right)\right\|_1\right\}^{\frac{1}{q}}+\frac{c\left(n,\beta,q\right)}{[\lambda]_q^{|\beta|}}\left|f\left(\mathbf{x}\right)\right|\left\|\chi_\eta\mathcal{P}_\lambda\left(\beta,\mathbf{t}\right)\right\|_1.$$

$$(3.27)$$

The first factor including supremum norm on the right-hand side of (3.27) tends to zero as $\lambda\to 0$; Indeed,

$$\frac{c\left(n,\beta,q\right)}{[\lambda]_q^{|\beta|}}\sup_{\|\mathbf{t}\|_\beta^{\frac{n}{2|\beta|}}\geq\eta}\left|\left(1+\|\mathbf{t}\|_\beta\right)\mathcal{P}_\lambda\left(\beta,\mathbf{t}\right)\right|$$

$$=\frac{c\left(n,\beta,q\right)}{[\lambda]_q^{|\beta|}}\sup_{\|\mathbf{t}\|_\beta^{\frac{n}{2|\beta|}}\geq\eta}\frac{\left(1+\|\mathbf{t}\|_\beta\right)}{E_q\left(\frac{(1-q)\|\mathbf{t}\|_\beta}{[\lambda]_q^{\frac{|\beta|}{n}}}\right)}$$

$$\leq\frac{c\left(n,\beta,q\right)}{[\lambda]_q^{|\beta|}}\frac{1}{1-q}\sup_{\|\mathbf{t}\|_\beta^{\frac{n}{2|\beta|}}\geq\eta}\frac{[\lambda]_q^{\frac{|\beta|}{n}(n+1)}\left(1-q+(1-q)\|\mathbf{t}\|_\beta\right)}{\prod_{k=0}^n\left([\lambda]_q^{\frac{|\beta|}{n}}+(1-q)q^k\|\mathbf{t}\|_\beta\right)}$$

$$\leq c\left(n,\beta,q\right)[\lambda]_q^{\frac{|\beta|}{n}}\left\{\frac{1+\frac{1}{1-q}\left([\lambda]_q^{\frac{|\beta|}{n}}+(1-q)\eta^{\frac{2|\beta|}{n}}\right)}{\prod_{k=0}^n\left([\lambda]_q^{\frac{|\beta|}{n}}+(1-q)q^k\eta^{\frac{2|\beta|}{n}}\right)}\right\},\qquad(3.28)$$

which clearly tends to zero as $\lambda\to 0$.

For the second factor in (3.27) we have

$$\frac{c(n,\beta,q)}{[\lambda]_q^{|\beta|}} \left\| \chi_\eta \left(1+\|\mathbf{t}\|_\beta\right) \mathcal{P}_\lambda(\beta,\mathbf{t}) \right\|_1 \le \frac{c(n,\beta,q)}{[\lambda]_q^{|\beta|}} \int_{\|\mathbf{t}\|_\beta^{\frac{n}{2|\beta|}} \ge \delta} \left(1+\|\mathbf{t}\|_\beta\right) \mathcal{P}_\lambda(\beta,\mathbf{t})\, d\mathbf{t}$$

$$= c(n,\beta,q) \int_{\|\mathbf{t}\|_\beta^{\frac{n}{2|\beta|}} \ge \frac{\delta}{\sqrt{[\lambda]_q}}} \frac{1}{E_q\left((1-q)\|\mathbf{t}\|_\beta\right)}\, d\mathbf{t}$$

$$+ c(n,\beta,q)[\lambda]_q^{\frac{|\beta|}{n}} \int_{\|\mathbf{t}\|_\beta^{\frac{n}{2|\beta|}} \ge \frac{\delta}{\sqrt{[\lambda]_q}}} \frac{\|\mathbf{t}\|_\beta}{E_q\left((1-q)\|\mathbf{t}\|_\beta\right)}\, d\mathbf{t}.$$

$$(3.29)$$

Since the function $\dfrac{1}{E_q\left((1-q)\|\mathbf{t}\|_\beta\right)}$ is integrable on $[0,\infty)$, the first and last terms of (3.29) tend to zero as $\lambda \to 0$. Finally, from the final proof of Theorem 2 in [16], the last term of (3.27) also approaches zero as $\lambda \to 0$.
Hence we obtain the assertion of the theorem. ∎

3.2.3 Order of Pointwise Convergence

Now, we shall discuss the order of pointwise convergence that we have already presented above. For this purpose we give the following generalization of the concept of weighted β-Lebesgue point.

Definition 3.5. Let $f \in L_{p,\beta}(\mathbb{R}^n)$, $1 \le p < \infty$, and $\beta_i, \gamma \in (0,\infty)$ $(i=1,\dots,n)$ with $|\beta| = \beta_1 + \beta_2 + \dots + \beta_n$. We say that $\mathbf{x} \in \mathbb{R}^n$ is weighted γ, β-Lebesgue point of f provided

$$\lim_{h \to 0} \left\{ \frac{1}{h^{2|\beta|+\gamma}} \int_{\|\mathbf{t}\|_\beta^{\frac{n}{2|\beta|}} < h} \left| \frac{f(\mathbf{x}+\mathbf{t}) - f(\mathbf{x})}{1+\|\mathbf{t}\|_\beta} \right|^p d\mathbf{t} \right\}^{\frac{1}{p}} = 0.$$

Theorem 3.9. Let $f \in L_{p,\beta}(\mathbb{R}^n)$, $1 \le p < \infty$, with $\beta_i \in (0,\infty)$ $(i=0,1,\dots,n)$, and $|\beta| = \beta_1 + \beta_2 + \dots + \beta_n$. Then we have

$$\left| \mathcal{P}_{\lambda,\beta}(f;q,\mathbf{x}) - f(\mathbf{x}) \right| = \mathcal{O}\left([\lambda]_q^{\frac{\gamma}{2p}} \right) (\lambda \to 0)$$

at every weighted γ, β-Lebesgue point $\mathbf{x}$ of f for $\gamma < \min\left\{ \frac{2|\beta|}{n}, p \right\}$.

Proof. Let $\mathbf{x}$ be a weighted γ, β-Lebesgue point of f; then for any $\varepsilon > 0$, there exists an $\eta > 0$ such that $h < \eta$ implies that

$$\int_{\|t\|_\beta^{\frac{n}{2|\beta|}} < h} \left| \frac{f(x+t) - f(x)}{1 + \|t\|_\beta} \right|^p dt < \varepsilon^p h^{2|\beta|+\gamma}. \tag{3.30}$$

Transforming into the generalized β-spherical coordinates, if $h < \eta$, we get

$$\int_{\|t\|_\beta^{\frac{n}{2|\beta|}} < h} \left| \frac{f(x+t) - f(x)}{1 + \|t\|_\beta} \right|^p dt$$

$$= \int_0^h \int_{S^{n-1}} \left| \frac{f\left(x + (u\theta)^\beta\right) - f(x)}{1 + u^{\frac{2|\beta|}{n}}} \right|^p \Omega_\beta(\theta) u^{2|\beta|-1} d\theta \, du < \varepsilon^p h^{2|\beta|+\gamma}.$$

For our future correspondence we can simplify the above expression by setting

$$G_\beta(\eta) := \int_0^\eta u^{2|\beta|-1} g(u) \, du < \varepsilon^p \eta^{2|\beta|+\gamma}, \tag{3.31}$$

where $g(u)$ is given by (3.24).

Now, as in Theorem 3.8, we need to estimate $J_1(\lambda)$ and $J_2(\lambda)$ similarly. For this aim, integrating by parts twice and taking account of (3.31), then $J_1(\lambda)$ can be estimated as

$$J_1(\lambda) \le \left(1 + \eta^{\frac{2|\beta|}{n}}\right) \left\{ \int_0^\eta g(u) u^{2|\beta|-1} P_\lambda^0(\beta, u) \, du \right\}^{\frac{1}{p}}$$

$$= \left(1 + \eta^{\frac{2|\beta|}{n}}\right) \left\{ \int_0^\eta P_\lambda^0(\beta, u) \, dG_\beta(u) \right\}^{\frac{1}{p}}$$

$$= \left(1 + \eta^{\frac{2|\beta|}{n}}\right) \left\{ P_\lambda^0(\beta, u) G_\beta(u) \Big|_0^\eta + \int_0^\eta G_\beta(u) d\left(-P_\lambda^0(\beta, u)\right) \right\}^{\frac{1}{p}}$$

$$\le \left(1 + \eta^{\frac{2|\beta|}{n}}\right) \left\{ \varepsilon^p \int_0^\eta u^{2|\beta|+\gamma-1} P_\lambda^0(\beta, u) \, du \right\}^{\frac{1}{p}}$$

$$\le \varepsilon \left(1 + \eta^{\frac{2|\beta|}{n}}\right) (2|\beta| + \gamma)^{\frac{1}{p}} \{\Delta_\lambda(\beta, \gamma)\}^{\frac{1}{p}}, \tag{3.32}$$

where $\Delta_\lambda(\beta, \gamma) := \int_0^\infty u^{2|\beta|+\gamma-1} P_\lambda^0(\beta, u) \, du$, with $P_\lambda^0(\beta, u)$ given by (3.26). Note that $\Delta_\lambda(\beta, \gamma) \to 0$ as $\lambda \to 0$. Indeed,

$$\Delta_\lambda(\beta, \gamma) = \int_0^\infty u^{2|\beta|+\gamma-1} P_\lambda^0(\beta, u) \, du$$

$$= [\lambda]_q^{\frac{\gamma}{2}} c(n, \beta, q) \frac{n}{2|\beta|} \int_0^\infty \frac{u^{n\left(1 + \frac{\gamma}{2|\beta|}\right)-1} \, du}{E_q((1-q)u)}$$

$$= [\lambda]_q^{\frac{\gamma}{2}} c(n,\beta,q) \frac{n}{2|\beta|} \ln q^{-1} \frac{c_q\left(n\left(1+\frac{\gamma}{2|\beta|}\right)\right) \Gamma_q\left(n\left(1+\frac{\gamma}{2|\beta|}\right)\right)}{(1-q)q^{\frac{n\left(1+\frac{\gamma}{2|\beta|}\right)\left\{n\left(1+\frac{\gamma}{2|\beta|}\right)-1\right\}}{2}}}$$

$$= \mathcal{O}\left([\lambda]_q^{\frac{\gamma}{2}}\right) \quad (\lambda \to 0).$$

For $J_2(\lambda)$, from (3.27) we have

$$J_2(\lambda) = \max\{1, M_\beta\}\left(1+\|\mathbf{x}\|_\beta\right)\|f\|_{p,\beta}\left\{\frac{c(n,\beta,q)}{[\lambda]_q^{|\beta|}} \sup_{\|\mathbf{t}\|_\beta^{\frac{2|\beta|}{\beta}} \geq \eta}\left|\left(1+\|\mathbf{t}\|_\beta\right)\mathcal{P}_\lambda(\beta,\mathbf{t})\right|\right\}^{\frac{1}{p}}$$

$$\times \left\{\frac{c(n,\beta,q)}{[\lambda]_q^{|\beta|}}\left\|\chi_\eta\left(1+\|\mathbf{t}\|_\beta\right)\mathcal{P}_\lambda(\beta,\mathbf{t})\right\|_1\right\}^{\frac{1}{q}} + \frac{c(n,\beta,q)}{[\lambda]_q^{|\beta|}}|f(\mathbf{x})|\left\|\chi_\eta\mathcal{P}_\lambda(\beta,\mathbf{t})\right\|_1,$$

where

$$\frac{c(n,\beta,q)}{[\lambda]_q^{|\beta|}} \sup_{\|\mathbf{t}\|_\beta^{\frac{n}{2|\beta|}} \geq \eta}\left|\left(1+\|\mathbf{t}\|_\beta\right)\mathcal{P}_\lambda(\beta,\mathbf{t})\right| = o\left([\lambda]_q^{\frac{\gamma}{q}}\right) \quad (\lambda \to 0). \qquad (3.33)$$

Indeed, from (3.28) we can reach to the following Inequality:

$$\frac{c(n,\beta,q)}{[\lambda]_q^{\frac{\gamma}{q}}[\lambda]_q^{|\beta|}} \sup_{\|\mathbf{t}\|_\beta^{\frac{n}{2|\beta|}} \geq \eta}\left|\left(1+\|\mathbf{t}\|_\beta\right)\mathcal{P}_\lambda(\beta,\mathbf{t})\right|$$

$$\leq c(n,\beta,q)[\lambda]_q^{\frac{|\beta|}{n}-\frac{\gamma}{2}}\left\{\frac{1+\frac{1}{1-q}\left([\lambda]_q^{\frac{|\beta|}{n}}+(1-q)\eta^{\frac{2|\beta|}{n}}\right)}{\prod_{k=0}^n\left([\lambda]_q^{\frac{|\beta|}{n}}+(1-q)q^k\eta^{\frac{2|\beta|}{n}}\right)}\right\},$$

which gives that (3.33) holds for $\gamma < \frac{2|\beta|}{n}$. For the last term of $J_2(\lambda)$ we get that

$$\frac{c(n,\beta,q)}{[\lambda]_q^{|\beta|}}|f(\mathbf{x})|\left\|\chi_\eta\mathcal{P}_\lambda(\beta,\mathbf{t})\right\|_1 = o\left([\lambda]_q^{\frac{\gamma}{2p}}\right) \quad (\lambda \to 0). \qquad (3.34)$$

Actually, by taking into account the fact that $\|\mathbf{t}\|_\beta^{\frac{n}{2|\beta|}} \geq \eta$ and making the substitution $\mathbf{t} = [\lambda]_q^{\frac{\beta}{n}}\mathbf{x}$, it follows that

$$\frac{c(n,\beta,q)}{[\lambda]_q^{\frac{\gamma}{q}}[\lambda]_q^{|\beta|}}\,|f(\mathbf{x})|\,\big\|\chi_\eta P_\lambda\,(\beta,\mathbf{t})\big\|_1 = \frac{c(n,\beta,q)}{[\lambda]_q^{\frac{\gamma}{2P}}[\lambda]_q^{|\beta|}}\,|f(\mathbf{x})|\int_{\|\mathbf{t}\|_\beta^{\frac{n}{2|\beta|}}\geq\eta}\frac{dt}{E_q\left(\frac{(1-q)\|\mathbf{t}\|_\beta}{[\lambda]_q^{\frac{|\beta|}{n}}}\right)}$$

$$\leq \frac{c(n,\beta,q)}{\eta[\lambda]_q^{\frac{\gamma}{2P}}[\lambda]_q^{|\beta|}}\,|f(\mathbf{x})|\int_{\|\mathbf{t}\|_\beta^{\frac{n}{2|\beta|}}\geq\eta}\frac{\|\mathbf{t}\|_\beta^{\frac{n}{2|\beta|}}\,dt}{E_q\left(\frac{(1-q)\|\mathbf{t}\|_\beta}{[\lambda]_q^{\frac{|\beta|}{n}}}\right)}$$

$$= \frac{c(n,\beta,q)}{\eta}\,|f(\mathbf{x})|\int_{\|\mathbf{t}\|_\beta^{\frac{n}{2|\beta|}}\geq\frac{\delta}{\sqrt{[\lambda]_q}}}\frac{[\lambda]_q^{\frac{P-\gamma}{2P}}\|\mathbf{t}\|_\beta^{\frac{n}{2|\beta|}}\,dt}{E_q\left((1-q)\,\|\mathbf{t}\|_\beta\right)}$$

tends to zero as $\lambda \to 0$, which indicates that (3.34) is satisfied. Finally for the other factor of $J_2(\lambda)$, we get

$$\lim_{\lambda\to 0}\frac{c(n,\beta,q)}{[\lambda]_q^{|\beta|}}\left\|\chi_\eta\left(1+\|\mathbf{t}\|_\beta\right)P_\lambda\,(\beta,\mathbf{t})\right\|_1 = 0$$

as in (3.29) and this completes the proof. ∎

3.2.4   Norm Convergence

In this section, we give a convergence result in the norm of $L_{p,\beta}(\mathbb{R}^n)$.

Theorem 3.10. *Let* $f \in L_{p,\beta}(\mathbb{R}^n)$, $1 \leq p < \infty$, *with* $\beta_i \in (0,\infty)\,(i=0,1,\cdots,n)$ *and* $|\beta| = \beta_1 + \beta_2 + \cdots + \beta_n$. *If the following condition*

$$\lim_{h\to 0}\frac{1}{h^{2|\beta|}}\int_{\|\mathbf{t}\|_\beta^{\frac{n}{2|\beta|}}<h}\|f(\mathbf{x}+\mathbf{t})-f(\mathbf{x})\|_{p,\beta}\,d\mathbf{t} = 0, \tag{3.35}$$

is satisfied, then we have

$$\lim_{\lambda\to 0}\big\|\mathcal{P}_{\lambda,\beta}\,(f;\mathbf{x})-f(\mathbf{x})\big\|_{p,\beta} = 0.$$

Proof. From (3.35), we have for any $\varepsilon > 0$, there exists a $\delta > 0$ such that $h < \delta$ implies that

$$\int_{\|\mathbf{t}\|_\beta^{\frac{n}{2|\beta|}}<h}\|f(\mathbf{x}+\mathbf{t})-f(\mathbf{x})\|_{p,\beta}\,d\mathbf{t} < \varepsilon h^{2|\beta|}.$$

Transforming the integral in the above inequality into generalized β-spherical coordinates, then we get for $h < \delta$

$$\int\limits_{\|\mathbf{t}\|_\beta^{\frac{n}{2|\beta|}} < h} \|f(\mathbf{x}+\mathbf{t}) - f(\mathbf{x})\|_{p,\beta} \, d\mathbf{t}$$

$$= \int_0^h \int_{S^{n-1}} \left\| f\left(\mathbf{x}+(u\theta)^\beta\right) - f(\mathbf{x}) \right\|_{p,\beta} \Omega_\beta(\theta) u^{2|\beta|-1} d\theta \, du < \varepsilon h^{2|\beta|};$$

hence, for simplicity, using the similar setting as in the proof of Theorem 2 of [16], we denote

$$k(u) = \int_{S^{n-1}} \left\| f\left(\mathbf{x}+(u\theta)^\beta\right) - f(\mathbf{x}) \right\|_{p,\beta} \Omega_\beta(\theta) \, d\theta \qquad (3.36)$$

and considering $k(u)$, we get that

$$\int\limits_{\|\mathbf{t}\|_\beta^{\frac{n}{2|\beta|}} < h} \|f(\mathbf{x}+\mathbf{t}) - f(\mathbf{x})\|_{p,\beta} \, d\mathbf{t} = \int_0^h u^{2|\beta|-1} k(u) \, du < \varepsilon h^{2|\beta|}.$$

Using generalized Minkowski inequality we conclude that

$$\left\| \mathcal{P}_{\lambda,\beta}(f;\mathbf{x}) - f(\mathbf{x}) \right\|_{p,\beta} = \left(\int_{\mathbb{R}^n} \left| \frac{\mathcal{P}_{\lambda,\beta}(f;\mathbf{x}) - f(\mathbf{x})}{1 + \|\mathbf{x}\|_\beta} \right|^p dx \right)^{\frac{1}{p}}$$

$$= \frac{c(n,\beta,q)}{[\lambda]_q^{|\beta|}} \left(\int_{\mathbb{R}^n} \left| \frac{1}{1 + \|\mathbf{x}\|_\beta} \int_{\mathbb{R}^n} [f(\mathbf{x}+\mathbf{t}) - f(\mathbf{x})] \mathcal{P}_\lambda(\beta,\mathbf{t}) dt \right|^p dx \right)^{\frac{1}{p}}$$

$$\leq \frac{c(n,\beta,q)}{[\lambda]_q^{|\beta|}} \int_{\mathbb{R}^n} \left(\int_{\mathbb{R}^n} \left| \frac{[f(\mathbf{x}+\mathbf{t}) - f(\mathbf{x})]}{1 + \|\mathbf{x}\|_\beta} \right|^p dx \right)^{\frac{1}{p}} \mathcal{P}_\lambda(\beta,\mathbf{t}) dt$$

$$= \frac{c(n,\beta,q)}{[\lambda]_q^{|\beta|}} \int_{\mathbb{R}^n} \|f(\mathbf{x}+\mathbf{t}) - f(\mathbf{x})\|_{p,\beta} \mathcal{P}_\lambda(\beta,\mathbf{t}) dt$$

$$= \frac{c(n,\beta,q)}{[\lambda]_q^{|\beta|}} \int\limits_{\|\mathbf{t}\|_\beta^{\frac{n}{2|\beta|}} < \delta} \|f(\mathbf{x}+\mathbf{t}) - f(\mathbf{x})\|_{p,\beta} \mathcal{P}_\lambda(\beta,\mathbf{t}) dt$$

$$+ \frac{c(n,\beta,q)}{[\lambda]_q^{|\beta|}} \int\limits_{\|\mathbf{t}\|_\beta^{\frac{n}{2|\beta|}} \geq \delta} \|f(\mathbf{x}+\mathbf{t}) - f(\mathbf{x})\|_{p,\beta} \mathcal{P}_\lambda(\beta,\mathbf{t}) dt$$

$$= L_1(\lambda) + L_2(\lambda).$$

Therefore using generalized β-spherical coordinates, $L_1(\lambda)$ can be estimated as

$$L_1(\lambda) = \frac{c(n,\beta,q)}{[\lambda]_q^{|\beta|}} \int\limits_{\|\mathbf{t}\|_\beta^{\frac{n}{2|\beta|}} < \delta} \|f(\mathbf{x}+\mathbf{t}) - f(\mathbf{x})\|_{p,\beta}\, \mathcal{P}_\lambda(\beta,\mathbf{t})\, d\mathbf{t}$$

$$\leq \int_0^h k(u)\, u^{2|\beta|-1} \mathcal{P}_\lambda^0(\beta,u)\, du,$$

where $k(u)$ is given by (3.36). Using two times integration by parts, we easily obtained that there exists a constant C_1, such that $L_1(\lambda) \leq \varepsilon C_1$.

Since

$$\|f(\mathbf{x}+\mathbf{t}) - f(\mathbf{x})\|_{p,\beta} \leq \left(\int_{\mathbb{R}^n} \left| \frac{[f(\mathbf{x}+\mathbf{t}) - f(\mathbf{x})]}{1 + \|\mathbf{x}\|_\beta} \right|^p d\mathbf{x} \right)^{\frac{1}{p}}$$

$$\leq \left(\int_{\mathbb{R}^n} \left| \frac{f(\mathbf{x}+\mathbf{t})}{1 + \|\mathbf{x}\|_\beta} \right|^p d\mathbf{x} \right)^{\frac{1}{p}}$$

$$+ \left(\int_{\mathbb{R}^n} \left| \frac{f(\mathbf{x})}{1 + \|\mathbf{x}\|_\beta} \right|^p d\mathbf{x} \right)^{\frac{1}{p}}$$

$$\leq \max\{1, M_\beta\} \left(1 + \|\mathbf{t}\|_\beta \right) \|f\|_{p,\beta} + \|f\|_{p,\beta},$$

then, from (3.29) $L_2(\lambda)$ tends to zero as $\lambda \to 0$. The theorem is proved. ∎

3.3 q-Meyer–König–Zeller–Durrmeyer Operators

Trif [150] studied some approximation properties of the operators $\hat{M}_{n,q} f(x)$. Very recently Heping [97] established some approximation properties based on q-hypergeometric series of these operators. Also Dogru and Gupta [55] proposed some other bivariate q-Meyer–König and Zeller operators having different test function and established some approximation properties. This section is based on [84].

3.3.1   Introduction

Trif [150] introduced the q-Meyer–König and Zeller operators as

$$\hat{M}_{n,q}f(x) = \sum_{k=0}^{\infty} m_{n,k,q}(x)f\left(\frac{[k]_q}{[n+k]_q}\right)$$

$$\hat{M}_{n,q}f(1) = f(1),$$

where

$$m_{n,k,q}(x) = \prod_{j=0}^{n}(1-q^j x)\begin{bmatrix} n+k \\ k \end{bmatrix}_q x^k.$$

Govil and Gupta [84] introduced a new sequence of the Durrmeyer type integrating q-Meyer–König–Zeller operators as follows:

Definition 3.6. For $f \in C[0,1]$, we define the q-Meyer–König–Zeller–Durrmeyer operators (q-MKZD operators) as

$$M_{n,q}(f) \equiv M_{n,q}(f;x) = [n]_q \sum_{k=1}^{\infty} m_{n,k,q}(x)q^{1-k}\int_0^1 \frac{m_{n,k-1,q}(qt)}{(1-q^n t)(1-q^{n+1}t)}f(t)d_q t + m_{n,0,q}(x)f(0)$$

$$= \sum_{k=0}^{\infty} W_{n,k}(f)m_{n,k,q}(x). \tag{3.37}$$

Alternately we can rewrite the operators (3.37) as

$$M_{n,q}(f;x) = \sum_{k=1}^{\infty} m_{n,k,q}(x)\int_0^1 \frac{1}{B_q(n,k)}t^{k-1}(qt;q)_{n-1}f(t)d_q t + m_{n,0,q}(x)f(0), 0 \leq x < 1,$$

where $m_{n,k,q}(x) = \frac{(q;q)_{n+k}}{(q;q)_k(q;q)_n}(x;q)_{n+1}x^k$.

We may note here that the Meyer–König–Zeller basis function $m_{n,k,q}(x)$ considered in (3.37) and $\hat{M}_{n,q}f(x)$ are the two alternative forms having the same value.

3.3.2   Estimation of Moments

Theorem 3.11. For all $x \in [0,1]$, we have the following identities:

$$M_{n,q}(1;x) = 1, M_{n,q}(t;x) = x$$

and for all integers $n \geq 3$, we have

$$M_{n,q}(t^2;x) = x^2 + \frac{[2]_q x(1-x)(1-q^n x)}{[n-1]_q} - E_{n,q}(x),$$

where

$$0 \leq E_{n,q}(x) \leq \frac{x[2]_q[3]_q q^{n-1}}{[n-1]_q[n-2]_q}(1-x)(1-qx)(1-q^n x).$$

Proof. Clearly by definition, we have $M_{n,q}(1;x) = 1$. Also by easy computation, we have

$$\int_0^1 \frac{1}{B_q(n,k)} t^{k+s-1}(qt;q)_{n-1} d_q t = \frac{[n+k-1]_q!\,[k+s-1]_q!}{[k-1]_q!\,[n+k+s-1]_q!}.$$

Thus

$$M_{n,q}(t;x) = (x;q)_{n+1} \sum_{k=1}^{\infty} \begin{bmatrix} n+k \\ k \end{bmatrix}_q x^k \frac{[n+k-1]_q!\,[k]_q!}{[k-1]_q!\,[n+k]_q!}$$

$$= (x;q)_{n+1} \sum_{k=1}^{\infty} \begin{bmatrix} n+k \\ k \end{bmatrix}_q x^k \frac{[k]_q}{[n+k]_q}$$

$$= (x;q)_{n+1} \sum_{k=1}^{\infty} \begin{bmatrix} n+k-1 \\ k-1 \end{bmatrix}_q x^k = x.$$

Next for the estimation of $M_{n,q}(t^2;x)$, we fix an integer $n \geq 3$ as well as $x \in [0,1)$ (the result is trivial for $x = 1$). We proceed as follows:

$$M_{n,q}(t^2;x) = (x;q)_{n+1} \sum_{k=1}^{\infty} \begin{bmatrix} n+k \\ k \end{bmatrix}_q x^k \frac{[n+k-1]_q!\,[k+1]_q!}{[k-1]_q!\,[n+k+1]_q!}$$

$$= (x;q)_{n+1} \sum_{k=1}^{\infty} \begin{bmatrix} n+k-1 \\ k-1 \end{bmatrix}_q x^k \frac{[k+1]_q}{[n+k+1]_q}$$

$$= x(x;q)_{n+1} \sum_{k=0}^{\infty} \begin{bmatrix} n+k \\ k \end{bmatrix}_q x^k \frac{[k+2]_q}{[n+k+2]_q}.$$

On the other hand, as above we have

$$x^2 = x(x;q)_{n+1} \sum_{k=1}^{\infty} \begin{bmatrix} n+k-1 \\ k-1 \end{bmatrix}_q x^k,$$

and on using this, we get

$$M_{n,q}(t^2;x) - x^2$$

$$= x(x;q)_{n+1} \sum_{k=1}^{\infty} \frac{1}{[n]_q[n+k+2]_q} \begin{bmatrix} n+k-1 \\ k \end{bmatrix}_q x^k ([n+k]_q[k+2]_q - [k]_q[n+k+2]_q)$$

$$+ (x;q)_{n+1} \frac{[2]_q x}{[n+2]_q}$$

$$= x(x;q)_{n+1} \sum_{k=1}^{\infty} \frac{1}{[n]_q[n+k+2]_q} \begin{bmatrix} n+k-1 \\ k \end{bmatrix}_q x^k$$

$$([n+k]_q(1+q+q^2[k]_q) - [k]_q(1+q+q^2[n+k]_q)) + (x;q)_{n+1} \frac{[2]_q x}{[n+2]_q}$$

$$= x(x;q)_{n+1} \sum_{k=1}^{\infty} \frac{1}{[n]_q[n+k+2]_q} \begin{bmatrix} n+k-1 \\ k \end{bmatrix}_q x^k ((1+q)\{[n+k]_q - [k]_q\})$$

$$+ (x;q)_{n+1} \frac{(1+q)x}{[n+2]_q}$$

$$= x(x;q)_{n+1} \sum_{k=1}^{\infty} \frac{1}{[n]_q[n+k+2]_q} \begin{bmatrix} n+k-1 \\ k \end{bmatrix}_q x^k \left((1+q)q^k[n]_q \right)$$

$$+ (x;q)_{n+1} \frac{(1+q)x}{[n+2]_q}$$

$$= x(x;q)_{n+1} \sum_{k=0}^{\infty} \frac{1}{[n+k+2]_q} \begin{bmatrix} n+k-1 \\ k \end{bmatrix}_q x^k \left((1+q)q^k \right).$$

Thus

$$M_{n,q}(t^2;x) = x^2 + \frac{x(1+q)(x;q)_{n+1}}{[n-1]_q} \sum_{k=0}^{\infty} q^k \frac{[n+k-1]_q}{[n+k+2]_q} \begin{bmatrix} n+k-2 \\ k \end{bmatrix}_q x^k.$$

Since

$$\frac{[n+k-1]_q}{[n+k+2]_q} = 1 - \frac{[3]_q q^{n+k-1}}{[n+k+2]_q},$$

we get

$$M_{n,q}(t^2;x) = x^2 + \frac{x(1+q)(x;q)_{n+1}}{[n-1]_q} \sum_{k=0}^{\infty} \begin{bmatrix} n+k-2 \\ k \end{bmatrix}_q (qx)^k$$

$$- \frac{x(1+q)[3]_q q^{n-1}(x;q)_{n+1}}{[n-1]_q} \sum_{k=0}^{\infty} \frac{1}{[n+k+2]_q} \begin{bmatrix} n+k-2 \\ k \end{bmatrix}_q (q^2 x)^k.$$

By the Corollary 1.1(c), we have

$$\sum_{k=0}^{\infty} \begin{bmatrix} n+k-2 \\ k \end{bmatrix}_q (qx)^k = \frac{1}{(qx;q)_{n-1}} = \frac{(1-x)(1-q^n x)}{(x;q)_{n+1}}.$$

Hence

$$M_{n,q}(t^2;x) = x^2 + \frac{[2]_q x(1-x)(1-q^n x)}{[n-1]_q} - E_{n,q}(x),$$

where

$$0 \le E_{n,q}(x) = \frac{x(1+q)[3]_q q^{n-1}(x;q)_{n+1}}{[n-1]_q} \sum_{k=0}^{\infty} \frac{1}{[n+k+2]_q} \begin{bmatrix} n+k-2 \\ k \end{bmatrix}_q (q^2 x)^k$$

$$\le \frac{x(1+q)[3]_q q^{n-1}(x;q)_{n+1}}{[n-1]_q} \sum_{k=0}^{\infty} \frac{1}{[n+k-2]_q} \begin{bmatrix} n+k-2 \\ k \end{bmatrix}_q (q^2 x)^k$$

$$= \frac{x(1+q)[3]_q q^{n-1}(x;q)_{n+1}}{[n-1]_q[n-2]_q} \sum_{k=0}^{\infty} \begin{bmatrix} n+k-3 \\ k \end{bmatrix}_q (q^2 x)^k$$

$$= \frac{x(1+q)[3]_q q^{n-1}(x;q)_{n+1}}{[n-1]_q[n-2]_q} \frac{1}{(q^2 x;q)_{n-2}}$$

$$= \frac{x[2]_q[3]_q q^{n-1}}{[n-1]_q[n-2]_q}(1-x)(1-qx)(1-q^n x).$$

This completes the proof of Theorem 3.11.   ∎

Definition 3.7. For $q \in (0,1)$, we define $M_{\infty,q}(f,1) = f(1)$ and for $x \in [0,1)$ and $m_{\infty,k,q}(x) = \frac{x^k}{(q;q)_k}(x;q)_{\infty}$,

$$M_{\infty,q}(f) \equiv M_{\infty,q}(f;x) = \frac{1}{1-q} \sum_{k=1}^{\infty} m_{\infty,k,q}(x) q^{1-k} \int_0^1 m_{\infty,k-1,q}(qt) f(t) d_q t + m_{\infty,0,q}(x) f(0)$$

$$= \sum_{k=0}^{\infty} W_{\infty,k}(f) m_{n,k,q}(x). \tag{3.38}$$

Lemma 3.8. For $q \in (0,1)$, we have

$$M_{\infty,q}(1;x) = 1, \qquad M_{\infty,q}(t;x) = x.$$

For $x \in (0,1)$ and $s \ge 2$, we have the following recurrence relation:

$$M_{\infty,q}(t^s;x) = M_{\infty,q}(t^{s-1};x) - q^{s-1}(1-x)M_{\infty,q}(t^{s-1};qx).$$

Proof. By simple computation, we have

$$\sum_{k=0}^{\infty} m_{\infty,k,q}(x) = 1, \quad \sum_{k=0}^{\infty} (1 - q^k) m_{\infty,k,q}(x) = x.$$

Thus by Definition 3.7, we have

$$M_{\infty,q}(1;x) = 1, \qquad M_{\infty,q}(t;x) = x.$$

Also it can be easily observed that

$$\int_0^1 t^s m_{\infty,k,q}(qt) d_q t = \frac{q^k}{(q;q)_k} \int_0^1 t^{k+s}(qt;q)_\infty d_q t$$

$$= \frac{q^k}{(q;q)_k} [k+s]_q! (1-q)^{k+s+1} = (1-q)^{s+1} \frac{q^k [k+s]_q!}{[k]_q!}.$$

Using the formula (see [100, pp. 76–79]) $\int_0^1 t^k (qt;q)_\infty d_q t = [k]_q!(1-q)^{k+1}$ and $q^k m_{\infty,k,q}(x) = (1-x) m_{\infty,k,q}(qx)$, we get

$$M_{\infty,q}(t^s;x) = \frac{1}{1-q} \sum_{k=1}^{\infty} m_{\infty,k,q}(x) q^{1-k} \frac{(1-q)^{k+s} q^{k-1} [k-1+s]_q!}{(q;q)_{k-1}}$$

$$= \sum_{k=1}^{\infty} (1-q^k)(1-q^{k+1}) \cdots (1-q^{k+s-1}) m_{\infty k,q}(x).$$

$$= \sum_{k=1}^{\infty} (1-q^k) \cdots (1-q^{k+s-2}) m_{\infty k,q}(x) - q^{s-1} \sum_{k=1}^{\infty} (1-q^k) \cdots (1-q^{k+s-2}) q^k m_{\infty,k,q}(x).$$

$$= M_{\infty,q}(t^{s-1};x) - q^{s-1}(1-x) \sum_{k=1}^{\infty} (1-q^k) \cdots (1-q^{k+s-2}) m_{\infty,k,q}(qx).$$

$$= M_{\infty,q}(t^{s-1};x) - q^{s-1}(1-x) M_{\infty,q}(t^{s-1};qx).$$

This completes the proof of recurrence relation.   ■

Remark 3.1. Using the recurrence relation of the above lemma, we obtain

$$M_{\infty,q}(t^2;x) = x - q^2 x(1-x),$$

$$M_{\infty,q}(t^3;x) = x - q^2 x(1-x) - q^3 x(1-x) + q^5 x(1-x)(1-qx),$$

and

$$M_{\infty,q}(t^4,x) = x - q^2 [3]_q x(1-x) + q^5 [3]_q x(1-x)(1-qx) - q^9 x(1-x)(1-qx)(1-q^2 x).$$

Remark 3.2. For the limiting operators, if $T_{\infty,q,m}(x) = M_{\infty,q}((t-x)^m;x)$, then by Lemma 3.8, it can be easily seen that

$$T_{\infty,q,0}(x) = 1, T_{\infty,q,1}(x) = 0, T_{\infty,q,2}(x) = x(1-x)(1-q^2),$$

$$T_{\infty,q,3}(x) = x(1-x)[1-q^2-q^3+q^5(1-qx)-2x+3q^2x]$$

and

$$T_{\infty,q,4}(x) = x(1-x)[q^5[3]_q(1-qx)-[3]_qq^2-q^9(1-qx)(1-q^2x)$$
$$+4q^2x(1-x)+4q^3x-4q^5x(1-qx)-2q^2x^2]+x-4x^2+6x^3-3x^4.$$

3.3.3   Convergence

Theorem 3.12. *Let* $q_n \in (0,1)$. *Then the sequence* $\{M_{n,q_n}(f)\}$ *converges to* f *uniformly on* $[0,1]$ *for each* $f \in C[0,1]$ *if and only if* $\lim_{n\to\infty} q_n = 1$.

Proof. Since the operators M_{n,q_n} are linear positive operators defined in $C[0,1]$, the well-known theorem due to Korovkin (see [113, pp. 8–9]) implies that $M_{n,q_n}(f;x)$ converges to $f(x)$ uniformly, $[x \in [0,1); n \to \infty]$ for any $f \in C[0,1]$ if and only if

$$M_{n,q_n}(t^i;x) \to x^i \ \ i = 1,2 \ \ [x \in [0,1); n \to \infty]. \tag{3.39}$$

If $q_n \to 1$, then $[n]_{q_n} \to \infty$; hence, (3.39) follows from Theorem 3.11. On the other hand, if we assume that for any $f \in C[0,1]$, $M_{n,q_n}(f,x)$ converges to $f(x)$ uniformly $[x \in [0,1); n \to \infty]$, then $q_n \to 1$. In fact, if the sequence (q_n) does not tend to 1, then by Theorem 3.1, $E_{n,q}(x) \to 0$ as $n \to \infty$ for all $x \in [0,1)$. Also for $0 < q < 1$, we have $[n-1]_q \to \frac{1}{1-q}$ as $n \to \infty$; thus,

$$M_{n,q}(t^2;x) \to x^2 + [2]_q x(1-x)(1-q) \nrightarrow x^2 \ \ (n \to \infty),$$

which leads to a contradiction. Hence, $q_n \to 1$, and the proof of Theorem 3.12 is thus complete. ∎

For $f \in C[0,1]$, $t > 0$, we define the modulus of continuity $\omega(f,t)$ as follows:

$$\omega(f,t) := \sup_{\substack{|x-y|\leq t \\ x,y\in[0,1]}} |f(x)-f(y)|.$$

Theorem 3.13. *For any* $f \in C[0,1]$, *we have*

$$|M_{\infty,q}(f)-f(x)| \leq 2\omega(f,2^{-1}\sqrt{1-q^2}),$$

where $\omega(f,\delta)$ *denotes the modulus of continuity of the function* f *on the segment* $[0,1]$.

The proof of Theorem 3.13 follows along the lines of [151, p. 13]; we omit the details.

Lemma 3.9. *Let* $f \in C[0,1]$ *and* $f(1) = 0$. *Then we have*

$$|W_{n,k}(f)| \leq W_{n,k}(|f|) \leq \omega(f,q^n)(1+q^{k-n}),$$

and for any n,k, *we have*

$$|W_{\infty,k}(f)| \leq W_{\infty,k}(|f|) \leq \omega(f,q^n)(1+q^{k-n}).$$

Proof. By the well-known property of modulus of continuity (see [113, p. 20])

$$\omega(f,\lambda t) \leq (1+\lambda)\omega(f,t), \quad \lambda > 0,$$

we get

$$|f(t)| = |f(t) - f(1)| \leq \omega(f, 1-t) \leq \omega(f,q^n)(1+(1-t)/q^n).$$

Thus,

$$|W_{n,k}(f)| \leq W_{n,k}(|f|) := [n]_q \int_0^1 q^{1-k}|f(t)| \frac{m_{n,k-1,q}(qt)}{(1-q^n t)(1-q^{n+1} t)} d_q t$$

$$\leq [n]_q \int_0^1 q^{1-k}\omega(f,q^n)(1+(1-t)/q^n) \frac{m_{n,k-1,q}(qt)}{(1-q^n t)(1-q^{n+1} t)} d_q t$$

$$= \omega(f,q^n)\left[1+q^{-n}\left(1-\frac{[k]_q}{[n+k]_q}\right)\right]$$

$$= \omega(f,q^n)\left(1+\frac{q^k(1-q^n)}{q^n(1-q^{n+k})}\right) \leq \omega(f,q^n)(1+q^{k-n}).$$

Similarly,

$$|W_{\infty,k}(f)| \leq W_{\infty,k}(|f|) := \frac{q^{1-k}}{1-q} \int_0^1 |f(t)|m_{\infty,k-1,q}(qt)d_q t$$

$$\leq \omega(f,q^n)\frac{q^{1-k}}{1-q} \int_0^1 (1+(1-t)/q^n)m_{\infty,k-1,q}(qt)d_q t$$

$$= \omega(f,q^n)(1+(1-(1-q^k))/q^n) = \omega(f,q^n)(1+q^{k-n}).$$

Lemma 3.9 is proved.   ∎

Theorem 3.14. *Let* $0 < q < 1$. *Then for each* $f \in C[0,1]$ *the sequence* $\{M_{n,q}(f;x)\}$ *converges to* $M_{\infty,q}(f;x)$ *uniformly on* $[0,1]$. *Furthermore,*

$$\|M_{n,q}(f) - M_{\infty,q}(f)\| \le C_q\, \omega(f,q^n), \tag{3.40}$$

where $C_q = \frac{11-2q}{1-q}$.

Proof. The operators $M_{n,q}$ and $M_{\infty,q}$ reproduce constant functions, that is,

$$M_{n,q}(1;x) = M_{\infty,q}(1;x) = 1.$$

Note that, without loss of generality, we may assume that $f(1) = 0$. If $x = 1$, then by Lemma 3.9, we have

$$|M_{n,q}(f;1) - M_{\infty,q}(f;1)| = |W_{n,n}(f) - f(1)| = |W_{n,n}(f)| \le 2\omega(f,q^n).$$

For $x \in [0,1)$, by the definitions of $M_{n,q}(f;x)$ and $M_{\infty,q}(f;x)$, we know that

$$|M_{n,q}(f;x) - M_{\infty,q}(f;x)| = \left| \sum_{k=0}^{n} W_{n,k}(f)m_{n,k,q}(x) - \sum_{k=0}^{\infty} A_{\infty k}(f)m_{\infty,k,q}(x) \right|$$

$$\le \sum_{k=0}^{n} |W_{n,k}(f) - W_{\infty,k}(f)|m_{n,k,q}(x) + \sum_{k=0}^{n} |W_{\infty,k}(f)||m_{n,k,q}(x) - m_{\infty,k,q}(x)|$$

$$+ \sum_{k=n+1}^{\infty} |W_{\infty,k}(f))|m_{\infty,k,q}(x) =: I_1 + I_2 + I_3.$$

First we have

$$|m_{n,k,q}(x) - m_{\infty,k,q}(x)| := \left| \begin{bmatrix} n+k \\ k \end{bmatrix}_q x^k(x;q)_{n+1} - \frac{x^k}{(q;q)_k}(x;q)_\infty \right|$$

$$= \left| \begin{bmatrix} n+k \\ k \end{bmatrix}_q x^k((x;q)_{n+1} - (x;q)_\infty) + x^k(x;q)_\infty \left(\begin{bmatrix} n+k \\ k \end{bmatrix}_q - \frac{1}{(q;q)_k} \right) \right|$$

$$\le m_{n,k,q}(x)\left|1 - \prod_{j=n+1}^{\infty}(1-q^j x)\right| + m_{\infty,k,q}(x)\left|\prod_{j=n}^{n+k}(1-q^j) - 1\right|$$

$$\le \frac{q^{n-k}}{1-q}(m_{n,k,q}(x) + m_{\infty,k,q}(x)),$$

where in the last formula, we use the following inequality, which can be easily proved by the induction on n (see [100]):

$$1 - \prod_{s=1}^{n}(1-a_s) \le \sum_{s=1}^{n} a_s, \quad (a_1,\ldots,a_n \in (0,1),\ n = 1,2,\ldots,\infty).$$

Using the above inequality we get

$$|W_{n,k}(f) - W_{\infty,k}(f)| \leq \int_0^1 q^{1-k}|f(t)| \left|[n]_q \frac{m_{n,k-1,q}(qt)}{(1-q^n t)(1-q^{n+1}t)} - \frac{1}{1-q}P_{\infty,k-1}(q;qt)\right| d_q t$$

$$\leq \int_0^1 q^{1-k}|f(t)| \left|[n]_q - \frac{1}{1-q}\right| m_{\infty,k-1,q}(qt) d_q t$$

$$+ \int_0^1 q^{1-k}|f(t)|[n]_q \left|\frac{m_{n,k-1,q}(qt)}{(1-q^n t)(1-q^{n+1}t)} - m_{\infty,k-1,q}(qt)\right| d_q t$$

$$\leq \frac{q^{n+1}}{1-q} \int_0^1 q^{1-k}|f(t)| m_{\infty,k-1,q}(qt) d_q t$$

$$+ \frac{q^{n-k}}{1-q} \int_0^1 q^{1-k}|f(t)|[n]_q (m_{n,k-1,q}(qt) + m_{\infty,k-1,q}(qt)) d_q t$$

$$= q^{n+1} W_{\infty,k}(|f|) + \frac{q^{n-k}}{1-q} W_{n,k}(|f|) + q^{n-k}[n]_q W_{\infty,k}(|f|)$$

$$\leq q^{n+1}\omega(f,q^n)(1+q^{k-n}) + 2\frac{q^{n-k}}{1-q}\omega(f,q^n)(1+q^{k-n}) \leq \frac{5\omega(f,q^n)}{1-q}.$$

Now we estimate I_1 and I_3. We have

$$I_1 \leq \frac{5\omega(f,q^n)}{1-q} \sum_{k=0}^n m_{n,k,q}(x) = \frac{5\omega(f,q^n)}{1-q},$$

and

$$I_3 \leq \omega(f,q^n) \sum_{k=n+1}^\infty (1+q^{k-n})m_{\infty,k,q}(x) \leq 2\omega(f,q^n) \sum_{k=n+1}^\infty P_{\infty,k}(q;x) \leq 2\omega(f,q^n).$$

Finally we estimate I_2, as follows:

$$I_2 \leq \sum_{k=0}^n \omega(f,q^n)(1+q^{k-n})\frac{q^{n-k}}{1-q}(m_{n,k,q}(x) + m_{\infty,k,q}(x))$$

$$\leq \frac{2\omega(f,q^n)}{1-q} \sum_{k=0}^n (m_{n,k,q}(x) + m_{\infty,k,q}(x)) \leq \frac{4\omega(f,q^n)}{1-q}.$$

Combining the estimates of I_1, I_2, I_3 for each $x \in [0,1)$, we conclude that

$$|M_{n,q}(f;x) - M_{\infty,q}(f;x)| \leq \frac{11-2q}{1-q}\omega(f,q^n).$$

This completes the proof of Theorem 3.14.   ∎

Remark 3.3. As a special case when $f(x) = x^2, 0 < q < 1$, we have

$$\|M_{n,q}(f) - M_{\infty,q}(f)\| \geq c_1 q^n \geq c_2\, \omega(f, q^n),$$

where c_1 and c_2 are positive constants independent of n. Hence, the estimate (3.40) is sharp as the sequence q^n in (3.40) cannot be replaced by any other sequence, which decreases to zero more rapidly as $n \to \infty$.

Lemma 3.10. *Let L be a positive linear operator on $C[0,1]$ which reproduces linear functions. If $L(t^2, x) > x^2$ for all $x \in (0,1)$, then $L(f) = f$ if and only if f is linear.*

As an application of Lemma 3.10, we have the following theorem for our operator:

Theorem 3.15. *Let $0 < q < 1$ be fixed and let $f \in C[0,1]$. Then $M_{\infty,q}(f; x) = f(x)$ for all $x \in [0,1]$ if and only if f is linear.*

Theorem 3.16. *For any $f \in C[0,1]$, the sequence $\{M_{\infty,q}(f)\}$ converges to f uniformly on $[0,1]$ as $q \to 1-$.*

Proof. The proof is standard. We know that the operators $M_{\infty,q}$ are positive linear operators on $C[0,1]$ and reproduce linear functions. Also, by Lemma 3.8, we have $M_{\infty,q}(t; x) = x$

$$M_{\infty,q}(t^2; x) = (1 - q^2)x + q^2 x^2 \to x^2$$

uniformly on $[0,1]$ as $q \to 1-$. Thus $M_{\infty,q}(t^i; x) \to x^i, i = 0, 1, 2$, and by the well-known theorem due to Korovkin, we obtain the desired result. ∎

Theorem 3.17. *If $f \in C^{(2)}[0,1]$, then*

$$\left| M_{\infty,q}(f; x) - f(x) - \frac{1-q}{2} f^{(2)}(x) x(1-x) \right| \leq \frac{1}{2} \omega(f^{(2)}, \sqrt{1-q}) \left[T_{\infty,q,2}(x) + \frac{T_{\infty,q,4}(x)}{1-q} \right],$$

where $T_{\infty,q,2}(x)$ and $T_{\infty,q,4}(x)$ are given in Remark 3.2.

Proof. Let $x \in [0,1]$ and $f \in C^{(2)}[0,1]$ and then by Taylor's expansion, we have

$$f(t) = f(x) + \frac{f^{(1)}(x)}{1!}(t-x) + \frac{f^{(2)}(x)}{2!}(t-x)^2 + \frac{f^{(2)}(\xi) - f^{(2)}(x)}{2}(t-x)^2,$$

where ξ lies between x and t. By Definition 3.7, we have

$$\left| M_{\infty,q}(f; x) - f(x) - \frac{1-q}{2} f^{(2)}(x) x(1-x) \right| \leq M_{\infty,q}\left(\frac{f^{(2)}(\xi) - f^{(2)}(x)}{2}(t-x)^2; x \right).$$

Also we have

$$|f^{(2)}(\xi) - f^{(2)}(x)| \leq \omega(f^{(2)}, |\xi - x|) \leq \omega(f^{(2)}, \sqrt{1-q})\left(1 + \frac{(t-x)^2}{\sqrt{1-q}}\right).$$

Hence,

$$M_{\infty,q}\left(\frac{f^{(2)}(\xi) - f^{(2)}(x)}{2}(t-x)^2; x\right) \leq \frac{1}{2}\omega\left(f^{(2)}, \sqrt{1-q}\right)\left[T_{\infty,q,2}(x) + \frac{T_{\infty,q,4}(x)}{1-q}\right],$$

and this completes the proof of Theorem 3.17.   ∎

Our next direct theorem is in terms of the second-order modulus of continuity, for this firstly we introduce some basic definitions:
For $\delta > 0$ and $W^2 = \{g \in C[0,1] : g', g'' \in C[0,1]\}$, the $K-$functional are defined as

$$K_2(f,\delta) = \inf\{\|f - g\| + \eta\|g''\| : g \in W^2\},$$

where norm-$\|.\|$ is the uniform norm on $C[0,1]$. Following [50, p. 177], there exists a positive constant $C > 0$ such that

$$K_2(f,\delta) \leq C\omega_2(f, \sqrt{\delta}), \tag{3.41}$$

where the second-order modulus of smoothness for $f \in C[0,1]$ is defined as

$$\omega_2(f, \sqrt{\delta}) = \sup_{0 < h \leq \sqrt{\delta}} \sup_{x, x+h \in [0,1]} |f(x + 2h) - 2f(x+h) + f(x)|.$$

Theorem 3.18. *Let $n > 1$ be a natural number and let $q_0 = q_0(n) \in (0,1)$. Then for $f \in C[0,1], q \in (q_0, 1)$, there exists an absolute constant $C > 0$ such that*

$$|M_{n,q}(f;x) - f(x)| \leq C\omega_2\left(f, \sqrt{\delta}\right),$$

where $\delta = \frac{[2]_q x(1-x)(1-q^n x)}{[n-1]_q}$.

Proof. Let $x \in [0,1]$ and $g \in W^2$. Applying Taylor's formula

$$g(t) = g(x) + (t-x)\, g'(x) + \int_x^t (t-u)\, g''(u)\, du,$$

and using Theorem 3.11, we obtain

$$M_{n,q}(g;x) - g(x) = M_{n,q}\left(\int_x^t (t-u)\, g''(u)\, du; x\right).$$

On the other hand

$$\left| \int_x^t (t-u)\, g''(u)\, du \right| \le (t-x)^2 \|g''\|.$$

Thus

$$|M_{n,q}(g;x) - g(x)| \le M_{n,q}(t-x)^2;x)\|g''\|.$$

Also by Theorem 3.11, we have

$$|M_{n,q}(f;x)| \le [n]_q \sum_{k=1}^{\infty} m_{n,k,q}(x) q^{1-k} \int_0^1 \frac{m_{n,k-1,q}(qt)}{(1-q^n t)(1-q^{n+1} t)} |f(t)| d_q t + m_{n,0,q}(x)|f(0)| \le \|f\|.$$

Therefore,

$$|M_{n,q}(f;x) - f(x)| \le |M_{n,q}(f-g;x) - (f-g)(x)| + |M_{n,q}(g;x) - g(x)|$$

$$\le 2\|f-g\| + \frac{[2]_q x(1-x)(1-q^n x)}{[n-1]_q} \|g''\|.$$

Taking the infimum on the right side over all $g \in W^2$ and using (3.41), we get

$$|M_{n,q}(f;x) - f(x)| \le 2K_2\left(f, \frac{[2]_q x(1-x)(1-q^n x)}{[n-1]_q} \right)$$

$$\le C\omega_2(f;\sqrt{\delta}),$$

and the proof of Theorem 3.18 is thus complete.   ∎

Remark 3.4. It may be observed that for $q = q_0(n) \to 1$ as $n \to \infty$, the sequence $\{M_{n,q}(f)\}$ converges to f uniformly on $[0,1]$ for each $f \in C[0,1]$, because

$$\lim_{n\to\infty} [n+1]_q = \lim_{n\to\infty} \frac{1 - (q_0(n))^{n+1}}{1 - q_0(n)} = \infty,$$

if $\lim_{n\to\infty} q_0(n) = 1$.

Chapter 4
q-Bernstein-Type Integral Operators

4.1 Introduction

In order to approximate integrable functions on the interval $[0,1]$, Kantorovich gave modified Bernstein polynomials. Later in the year 1967 Durrmeyer [58] considered a more general integral modification of the classical Bernstein polynomials, which were studied first by Derriennic [47]. Also some other generalizations of the Bernstein polynomials are available in the literature. The other most popular generalization as considered by Goodman and Sharma [82], namely, genuine Bernstein–Durrmeyer operators. In this chapter we discuss the q analogues of various integral modifications of Bernstein polynomials. The results were discussed in recent papers [45, 62, 86, 89, 92, 94, 121], etc.

4.2 q-Bernstein–Kantorovich Operators

Recently, Dalmanoglu [45] proposed the q-Kantorovich–Bernstein operators as

$$K_{n,q}(f,x) = [n+1]_q \sum_{k=0}^{n} p_{n,k}(q;x) \int_{[k]_q/[n+1]_q}^{[k+1]_q/[n+1]_q} f(t)d_q t, \ x \in [0,1] \qquad (4.1)$$

where

$$p_{n,k}(q;x) := \begin{bmatrix} n \\ k \end{bmatrix}_q x^k \prod_{s=0}^{n-k-1} (1 - q^s x).$$

In case $q = 1$, the operators (4.1) reduce to well-known Bernstein–Kantorovich operators

$$K_n(f,x) = (n+1) \sum_{k=0}^{n} p_{n,k}(x) \int_{k/(n+1)}^{(k+1)/(n+1)} f(t)dt, \ x \in [0,1]$$

A. Aral et al., *Applications of q-Calculus in Operator Theory*, 113
DOI 10.1007/978-1-4614-6946-9_4, © Springer Science+Business Media New York 2013

where $p_{n,k}(x)$ is the Bernstein basis function given by

$$p_{n,k}(x) := \binom{n}{k} x^k (1-x)^{n-k}.$$

4.2.1   Direct Results

For the operators (4.1), Dalmanoglu [45] obtained the following theorems:

Theorem 4.1. *If the sequence (q_n) satisfies the conditions $\lim_{n\to\infty} q_n = 1$ and $\lim_{n\to\infty} \frac{1}{[n]_{q_n}} = 0$ and $0 < q_n < 1$, then*

$$\|K_{n,q}(f,x) - f\| \to 0, n \to \infty,$$

for every $f \in C[0,a]$, $0 < a < 1$.

Proof. First, we have

$$K_{n,q}(1,x) = [n+1]_q \sum_{k=0}^{n} q^{-k} \begin{bmatrix} n \\ k \end{bmatrix}_q x^k \prod_{s=0}^{n-k-1} (1-q^s x) \int_{[k]_q/[n+1]_q}^{[k+1]_q/[n+1]_q} d_q t.$$

Also by definition of q-integral

$$\int_{[k]_q/[n+1]_q}^{[k+1]_q/[n+1]_q} d_q t = \int_{0}^{[k+1]_q/[n+1]_q} d_q t - \int_{0}^{[k]_q/[n+1]_q} d_q t$$

$$= (1-q) \frac{[k+1]_q}{[n+1]_q} \sum_{j=0}^{\infty} q^j - (1-q) \frac{[k]_q}{[n+1]_q} \sum_{j=0}^{\infty} q^j$$

$$= \frac{1-q}{[n+1]_q} ([k+1]_q - [k]_q) \sum_{j=0}^{\infty} q^j = \frac{q^k}{[n+1]_q}.$$

Thus $K_{n,q}(1,x) = 1$. Next

$$K_{n,q}(t,x) = [n+1]_q \sum_{k=0}^{n} q^{-k} \begin{bmatrix} n \\ k \end{bmatrix}_q x^k \prod_{s=0}^{n-k-1} (1-q^s x) \int_{[k]_q/[n+1]_q}^{[k+1]_q/[n+1]_q} t\, d_q t.$$

Again by definition of q-integral

$$\int_{[k]_q/[n+1]_q}^{[k+1]_q/[n+1]_q} t d_q t = \int_0^{[k+1]_q/[n+1]_q} t d_q t - \int_0^{[k]_q/[n+1]_q} t d_q t$$

$$= (1-q) \frac{[k+1]_q}{[n+1]_q} \sum_{j=0}^{\infty} q^{2j} \frac{[k+1]_q}{[n+1]_q} - (1-q) \frac{[k]_q}{[n+1]_q} \sum_{j=0}^{\infty} q^{2j} \frac{[k]_q}{[n+1]_q}$$

$$= \frac{1-q}{[n+1]_q^2} ([k+1]_q^2 - [k]_q^2) \sum_{j=0}^{\infty} q^{2j} = \frac{q^k}{[n+1]_q^2} \frac{1}{1+q} ([k]_q(1+q) + 1).$$

Therefore

$$K_{n,q}(t,x) = [n+1]_q \sum_{k=0}^{n} \begin{bmatrix} n \\ k \end{bmatrix}_q x^k \prod_{s=0}^{n-k-1} (1-q^s x) \frac{1}{[n+1]_q^2} \frac{1}{1+q} ([k]_q(1+q) + 1)$$

$$\frac{[n]_q}{[n+1]_q} x + \frac{1}{1+q} \frac{1}{[n+1]_q}.$$

To estimate $K_{n,q}(t^2, x)$, we have

$$\int_{[k]_q/[n+1]_q}^{[k+1]_q/[n+1]_q} t^2 d_q t = \int_0^{[k+1]_q/[n+1]_q} t^2 d_q t - \int_0^{[k]_q/[n+1]_q} t^2 d_q t$$

$$= \frac{1}{[n+1]_q^3} \frac{1}{1+q+q^2} (q^k [k+1]_q^2 + [k]_q [k+1]_q + [k]_q^2).$$

Therefore using $[k+1]_q = q[k]_q + 1$ and using the similar methods as above, we have

$$K_{n,q}(t^2,x) = \frac{[n]_q [n-1]_q}{[n+1]_q^2} \frac{q^3+q^2+q}{1+q+q^2} x^2 + \frac{[n]_q}{[n+1]_q^2} \frac{q^2+3q+2}{1+q+q^2} x + \frac{1}{[n+1]_q^2} \frac{1}{1+q+q^2}.$$

Replacing q by a sequence $\{q_n\}$ such that $\lim_{n\to\infty} q_n = 1$, it is easily seen that $K_{n,q}(t^i,x), i=0,1,2$ converges uniformly to t^i. Thus the result follows by Korovkin's theorem. ∎

Theorem 4.2. *If the sequence (q_n) satisfies the conditions $\lim_{n\to\infty} q_n = 1$ and $\lim_{n\to\infty} \frac{1}{[n]q_n} = 0$ and $0 < q_n < 1$, then*

$$|K_{n,q}(f,x) - f(x)| \le 2\omega(f,\sqrt{\delta_n}),$$

for all $f \in C[0,a]$ and $\delta_n = K_{n,q}((t-x)^2,x)$.

Proof. Let $f \in C[0,a]$. From the linearity and monotonicity of $K_{n,q}(f,x)$, we can write

$$|K_{n,q}(f,x) - f(x)| \le K_{n,q}(|f(t) - f(x)|,x)$$

$$= [n+1]_q \sum_{k=0}^{n} q^{-k} \begin{bmatrix} n \\ k \end{bmatrix}_q x^k \prod_{s=0}^{n-k-1} (1 - q^s x) \int_{[k]_q/[n+1]_q}^{[k+1]_q/[n+1]_q} |f(t) - f(x)| d_q t.$$

On the other hand

$$|f(t) - f(x)| \le \omega(f, |t - x|).$$

If $|t - x| < \delta$, it is obvious that

$$|f(t) - f(x)| \le \left(1 + \frac{(t-x)^2}{\delta^2}\right) \omega(f, \delta) \tag{4.2}$$

If $|t - x| > \delta$, we use the property of modulus of continuity

$$\omega(f, \lambda \delta) \le (1 + \lambda) \omega(f, \delta) \le (1 + \lambda^2) \omega(f, \delta), \lambda \in R^+$$

as $\lambda = \frac{|t-x|}{\delta}$. Therefore, we have

$$|f(t) - f(x)| \le \left(1 + \frac{(t-x)^2}{\delta^2}\right) \omega(f, \delta) \tag{4.3}$$

for $|t - x| > \delta$. Consequently by (4.2) and (4.3), we get

$$|K_{n,q}(f,x) - f(x)| \le [n+1]_q \sum_{k=0}^{n} q^{-k} \begin{bmatrix} n \\ k \end{bmatrix}_q x^k$$

$$\prod_{s=0}^{n-k-1} (1 - q^s x) \int_{[k]_q/[n+1]_q}^{[k+1]_q/[n+1]_q} \left(1 + \frac{(t-x)^2}{\delta^2}\right) \omega(f, \delta) d_q t$$

$$= \left(K_{n,q}(1,x) + \frac{1}{\delta^2} K_{n,q}((t-x)^2,x)\right) \omega(f, \delta).$$

Taking $q = (q_n)$ satisfies the conditions $\lim_{n \to \infty} q_n = 1$, $\lim_{n \to \infty} \frac{1}{[n]_{q_n}} = 0$, and $0 < q_n < 1$, using the methods of Theorem 4.1, that

$$\lim_{n \to \infty} K_{n,q_n}((t-x)^2,x) = 0,$$

letting $\delta_n = K_{n,q_n}((t-x)^2,x)$ and taking $\delta = \sqrt{\delta_n}$, we finally get the desired result. This completes the proof of theorem. ∎

4.3 *q*-Bernstein–Durrmeyer Operators

For $f \in C[0,1], x \in [0,1], n = 1,2,,,,;0 < q < 1$, very recently Gupta [86] defined the *q*-Durrmeyer-type operators as

$$D_{n,q}(f,x) \equiv (D_{n,q}f)(x) = [n+1]_q \sum_{k=0}^{n} q^{-k} p_{n,k}(q;x) \int_0^1 f(t) p_{n,k}(q;qt) d_q t \quad (4.4)$$

where

$$p_{n,k}(q;x) := \begin{bmatrix} n \\ k \end{bmatrix}_q x^k \prod_{s=0}^{n-k-1} (1 - q^s x).$$

It can be easily verified that in the case $q = 1$, the operators defined by (4.4) reduce to the well-known Bernstein–Durrmeyer operators

$$D_n(f,x) = (n+1) \sum_{k=0}^{n} p_{n,k}(x) \int_0^1 f(t) p_{n,k}(t) dt,$$

where

$$p_{n,k}(x) := \binom{n}{k} x^k (1-x)^{n-k}.$$

4.3.1 *Auxiliary Results*

In the sequel, we shall need the following auxiliary results:

Lemma 4.1. *For* $n,k \geq 0$*, we have*

$$D_q(1-x)_q^{n-k} = -[n-k]_q(1-qx)_q^{n+k-1}, \quad (4.5)$$

Proof. Using the *q*-derivative operator, we can write

$$D_q(1-x)_q^{n-k} = \frac{1}{(q-1)x} \left(\prod_{j=0}^{n-k-1} (1-q^{j+1}x) - \prod_{j=0}^{n-k-1} (1-q^j x) \right)$$

$$= -\frac{(q^{n-k}-1)}{(q-1)} \prod_{j=0}^{n-k-2} (1+q^{j+1}x)$$

$$= -[n-k]_q(1-qx)_q^{n-k-1}. \qquad \blacksquare$$

Remark 4.1. By using (4.5) and $D_q x^k = [k]_q x^{k-1}$, we get

$$D_q(x^k(1-x)_q^{n-k}) = [k]_q x^{k-1}(1-x)_q^{n-k} - q^k x^k [n-k]_q (1-qx)_q^{n-k-1}$$
$$= x^{k-1}(1-qx)_q^{n-k-1}((1-x)[k]_q - q^k x[n-k]_q)$$
$$= x^{k-1}(1-qx)_q^{n-k-1}([k]_q - [n]_q x).$$

Hence, we obtain

$$x(1-x)D_q\left(x^k(1-x)_q^{n-k}\right) = x^k(1-x)_q^{n-k}[n]_q\left(\frac{[k]_q}{[n]_q} - x\right). \tag{4.6}$$

Lemma 4.2. *We have the following equalities:*

$$x(1-x)D_q(p_{n,k}(q;x)) = [n]_q p_{n,k}(q;x)\left(\frac{[k]_q}{[n]_q} - x\right), \tag{4.7}$$

$$t(1-qt)D_q(p_{n,k}(q;qt)) = [n]_q p_{n,k}(q;qt)\left(\frac{[k]_q}{[n]_q} - qt\right). \tag{4.8}$$

Proof. Above equalities can be obtained by direct computations using definition of operator and (4.6). ∎

Theorem 4.3 ([92]). *If m-th $(m > 0, m \in \mathbb{N})$ order moments of operator (4.4) is defined as*

$$D_{n,m}^q(x) := D_{n,q}(t^m, x) = [n+1]_q \sum_{k=0}^{n} q^{-k} p_{n,k}(q;x) \int_0^1 p_{n,k}(q;qt)t^m d_q t, x \in [0,1],$$

then $D_{n,0}^q(x) = 1$ and for $n > m+2$, we have the following recurrence relation:
$$[n+m+2]D_{n,m+1}^q(x)$$

$$= ([m+1]_q + q^{m+1}x[n]_q)D_{n,m}^q(x) + x(1-x)q^{m+1}D_q(D_{n,m}^q(x)). \tag{4.9}$$

Proof. By (4.7), we have
$$x(1-x)D_q(D_{n,m}^q(x))$$

$$= [n+1]_q \sum_{k=0}^{n} q^{-k}x(1-x)D_q(p_{n,k}(q;x)) \int_0^1 p_{n,k}(q;qt)t^m d_q t$$

$$= [n+1]_q[n]_q \sum_{k=0}^{n} q^{-k}p_{n,k}(q;x) \int_0^1 \left(\frac{[k]_q}{[n]_q} - qt\right) p_{n,k}(q;qt)t^m d_q t$$

$$+ q[n+1]_q[n]_q \sum_{k=0}^{n} q^{-k}x(1-x)D_q(p_{n,k}(q;x)) \int_0^1 p_{n,k}(q;qt)t^{m+1} d_q t$$

$$- x[n+1]_q[n]_q \sum_{k=0}^{n} q^{-k} x(1-x) D_q(p_{n,k}(q;x)) \int_0^1 p_{n,k}(q;qt) t^m d_q t$$

$$= I + [n]_q D_{n,m+1}^q(x) - x[n]_q D_{n,m}^q(x),$$

Set

$$u(t) = \frac{t^{m+1}}{q^{m+1}} - \frac{t^{m+2}}{q^{m+1}},$$

by *q*-integral by parts, we get
$\int_0^1 u(qt) D_q(p_{n,k}(q;qt)) d_q t$

$$= [u(t) p_{n,k}(q;qt)]_0^1 - \frac{1}{q^{m+1}} \int_0^1 p_{n,k}(q;qt)([m+1]_q t^m - [m+2]_q t^{m+1}) d_q t$$

$$= -\frac{1}{q^{m+1}} \int_0^1 p_{n,k}(q;qt)([m+1]_q t^m - [m+2]_q t^{m+1}) d_q t,$$

therefore

$$I = -\frac{1}{q^{m+1}} \left([m+1]_q D_{n,m}^q(x) - [m+2]_q D_{n,m+1}^q(x)\right)$$

by combining the above two equations, we can write

$$q^{m+1} x(1-x) D_q(D_{n,m}^q(x)) = -\left([m+1]_q D_{n,m}^q(x) - [m+2]_q D_{n,m+1}^q(x)\right)$$
$$+ q^{m+1}\left([n]_q D_{n,m+1}^q(x) - x[n]_q D_{n,m}^q(x)\right).$$

Hence we get the result. ∎

Corollary 4.1. *We have*

$$D_{n,1}^q(x) = \frac{(1+qx[n]_q)}{[n+2]_q}, \tag{4.10}$$

$$D_{n,2}^q(x) = \frac{q^3 x^2 [n]_q([n]_q - 1) + (1+q)^2 qx[n]_q + 1 + q}{[n+2]_q[n+3]_q}. \tag{4.11}$$

The corollary follows from (4.9).

Lemma 4.3. *For $f \in C[0,1]$, we have $\|D_{n,q}f\| \le \|f\|$.*

Proof. By definition (4.4) and using Theorem 4.3, we have

$$|D_{n,q}(f;x)| \le [n+1]_q \sum_{k=0}^{n} q^{-k} p_{n,k}(q;x) \int_0^1 |f(t)| p_{n,k}(q;qt) d_q t$$

$$\le \|f\| D_{n,q}(1;x) = \|f\|. \qquad \blacksquare$$

Lemma 4.4. *Let $n > 3$ be a given natural number and let $q_0 = q_0(n) \in (0,1)$ be the least number such that $q^{n+2} - q^{n+1} - 2q^n - 2q^{n-1} - \cdots - 2q^3 - q^2 + q + 2 < 0$ for every $q \in (q_0, 1)$. Then*

$$D_{n,q}((t-x)^2, x) \leq \frac{2}{[n+2]_q} \left(\varphi^2(x) + \frac{1}{[n+3]_q} \right),$$

where $\varphi^2(x) = x(1-x)$, $x \in [0,1]$.

Proof. In view of Theorem 4.3, we obtain

$$D_{n,q}((t-x)^2, x) = x^2 \cdot \frac{q^3 [n]_q ([n]_q - 1) - 2q[n]_q [n+3]_q + [n+2]_q [n+3]_q}{[n+2]_q [n+3]_q}$$

$$+ x \cdot \frac{q(1+q)^2 [n]_q - 2[n+3]_q}{[n+2]_q [n+3]_q} + \frac{1+q}{[n+2]_q [n+3]_q}$$

By direct computations, using the definition of the q-integers, we get

$$q(1+q)^2 [n]_q - 2[n+3]_q = q(1+q)^2 (1+q+\cdots+q^{n-1}) - 2(1+q+\cdots+q^{n+2})$$

$$= -q^{n+2} + q^{n+1} + 2q^n + 2q^{n-1} + \cdots + 2q^3 + q^2 - q - 2 > 0,$$

for every $q \in (q_0, 1)$. Furthermore

$$q(1+q)^2 [n]_q - 2[n+3]_q \leq 4[n] - q - 2[n+3]_q$$

$$= 4([n+3]_q - q^n - q^{n+1} - q^{n+2}) - 2[n+3]_q$$

$$\leq 4[n+3]_q - 2[n+3]_q = 2[n+3]_q$$

and

$$q(1+q)^2 [n]_q - 2[n+3]_q + q^3 [n]_q ([n]_q - 1) - 2q[n]_q [n+3]_q + [n+2]_q [n+3]_q$$

$$= q(1+q)^2 [n]_q - 2(1+q+q^2+q^3 [n]_q) + q^3 [n]_q^2 - q^3 [n]_q$$

$$- 2q[n]_q (1+q+q^2+q^3 [n]_q) + (1+q+q^2 [n]_q)(1+q+q^2+q^3 [n]_q)$$

$$= q^3 (1-q)^2 [n]_q^2 - (q-q^2+2q^3-2q^4)[n]_q - (1-q^3)$$

$$= q^3 (1-q)^2 \cdot \left(\frac{1-q^n}{1-q} \right)^2 - q(1-q)(1+2q^2) \cdot \frac{1-q^n}{1-q} - (1-q^3)$$

$$= q^{2n+3} + q^{n+1} - q - 1 \leq 0.$$

In conclusion, for $x \in [0,1]$, we have
$$D_{n,q}((t-x)^2, x)$$

$$
= \frac{q(1+q)^2[n]_q - 2[n+3]_q}{[n+2]_q[n+3]_q} \cdot x(1-x) + \left(\frac{q(1+q)^2[n]_q - 2[n+3]_q}{[n+2]_q[n+3]_q} \right.
$$

$$
\left. + \frac{q^3[n]_q([n]_q - 1) - 2q[n]_q[n+3]_q + [n+2]_q[n+3]_q}{[n+2]_q[n+3]_q} \right) \cdot x^2 + \frac{1+q}{[n+2]_q[n+3]_q}
$$

$$
\leq \frac{2[n+3]_q}{[n+2]_q[n+3]_q} \cdot \varphi^2(x) + \frac{2}{[n+2]_q[n+3]_q} \leq \frac{2}{[n+2]_q} \cdot \left(\varphi^2(x) + \frac{1}{[n+3]_q} \right),
$$

which was to be proved. ∎

For $\delta > 0$ and $W^2 = \{ g \in C[0,1] : g', g'' \in C[0,1] \}$, the K-functional are defined as

$$K_2(f, \delta) = \inf\{ \|f - g\| + \eta \|g''\| : g \in W^2 \},$$

where norm-$\|.\|$ is the uniform norm on $C[0,1]$. Following [50], there exists a positive constant $C > 0$ such that

$$K_2(f, \delta) \leq C \omega_2(f, \sqrt{\delta}), \tag{4.12}$$

where the second-order modulus of smoothness for $f \in C[0,1]$ is defined as

$$\omega_2(f, \sqrt{\delta}) = \sup_{0 < h \leq \sqrt{\delta}} \sup_{x, x+h \in [0,1]} |f(x+h) - f(x)|.$$

We define the usual modulus of continuity for $f \in C[0,1]$ as

$$\omega(f, \delta) = \sup_{0 < h \leq \delta} \sup_{x, x+h \in [0,1]} |f(x+h) - f(x)|.$$

4.3.2 Direct Results

Our first main result is the following local theorem:

Theorem 4.4. *Let $n > 3$ be a natural number and let $q_0 = q_0(n) \in (0,1)$ be defined as in Lemma 4.4. Then there exists an absolute constant $C > 0$ such that*

$$|D_{n,q}(f,x) - f(x)| \leq C \omega_2 \left(f, [n+2]_q^{-1/2} \delta_n(x) \right) + \omega \left(f, \frac{1-x}{[n+2]_q} \right),$$

where $f \in C[0,1]$, $\delta_n^2(x) = \varphi^2(x) + \frac{1}{[n+3]_q}$, $x \in [0,1]$, and $q \in (q_0, 1)$.

Proof. For $f \in C[0,1]$ we define

$$\tilde{D}_{n,q}(f,x) = D_{n,q}(f,x) + f(x) - f\left(\frac{1+q[n]_q x}{[n+2]_q}\right).$$

Then, by Corollary 4.1, we find

$$\tilde{D}_{n,q}(1,x) = D_{n,q}(1,x) = 1 \tag{4.13}$$

and

$$\tilde{D}_{n,q}(t,x) = D_{n,q}(t,x) + x - \frac{1+q[n]_q x}{[n+2]_q} = x. \tag{4.14}$$

Using Taylor's formula

$$g(t) = g(x) + (t-x)\, g'(x) + \int_x^t (t-u)\, g''^2,$$

we obtain

$$\tilde{D}_{n,q}(g,x) = g(x) + \tilde{D}_{n,q}\left(\int_x^t (t-u)\, g''(u)\, du, x\right)$$

$$= g(x) + D_{n,q}\left(\int_x^t (t-u)\, g''(u)\, du, x\right)$$

$$- \int_x^{\frac{1+q[n]_q x}{[n+2]_q}} \left(\frac{1+q[n]_q x}{[n+2]_q} - u\right) g''(u)\, du$$

Hence $|\tilde{D}_{n,q}(g,x) - g(x)| \le$

$$\le D_{n,q}\left(\left|\int_x^t |t-u| \cdot |g''(u)|\, du\right|, x\right) + \left|\int_x^{\frac{1+q[n]_q x}{[n+2]_q}} \left|\frac{1+q[n]_q x}{[n+2]_q} - u\right| \cdot |g''(u)|\, du\right|$$

$$\le D_{n,q}((t-x)^2, x) \cdot \|g''\| + \left(\frac{1+q[n]_q x}{[n+2]_q} - x\right)^2 \cdot \|g''\| \tag{4.15}$$

On the other hand

$$D_{n,q}((t-x)^2, x) + \left(\frac{1+q[n]_q x}{[n+2]_q} - x\right)^2 \le$$

$$\le \frac{2}{[n+2]_q}\left(\varphi^2(x) + \frac{1}{[n+3]_q}\right) + \left(\frac{1 - ([n+2]_q - q[n]_q)x}{[n+2]_q}\right)^2, \tag{4.16}$$

by Lemma 4.4. Because $[n+2]_q - q[n]_q = (1+q+\ldots+q^{n+1}) - q(1+q+\ldots+q^{n-1}) = 1+q^{n+1}$, we have

$$1 \leq [n+2]_q - q[n]_q \leq 2 \tag{4.17}$$

Then using (4.17), we have

$$\left(\frac{1-([n+2]_q-q[n]_q)x}{[n+2]_q}\right)^2 \cdot \delta_n^{-2}(x) \leq$$

$$= \frac{1-2([n+2]_q-q[n]_q)x+([n+2]_q-q[n]_q)^2x^2}{[n+2]_q^2} \cdot \frac{[n]_q}{[n]_q x(1-x)+1}$$

$$\leq \frac{1-2x+4x^2}{[n+2]_q} \cdot \frac{[n]_q}{[n+2]_q} \cdot \frac{1}{[n]_q x(1-x)+1} \leq \frac{3}{[n+2]_q}, \tag{4.18}$$

for $n = 1,2,\ldots$ and $0 < q < 1$. In conclusion, by (4.16) and (4.18), we get

$$D_{n,q}((t-x)^2,x)+\left(\frac{1+q[n]_q x}{[n+2]_q}-x\right)^2 \leq \frac{5}{[n+2]_q}\cdot\delta_n^2(x), \tag{4.19}$$

where $x \in [0,1]$. Hence, by (4.15),

$$|\tilde{D}_{n,q}(g,x)-g(x)| \leq \frac{5}{[n+2]_q}\cdot\delta_n^2(x)\cdot\|g''\|, \tag{4.20}$$

where $n > 3$ and $x \in [0,1]$. Furthermore, by Theorem 4.3, we have

$$|\tilde{D}_{n,q}(f,x)| \leq |D_{n,q}(f,x)|+|f(x)|+\left|f\left(\frac{1+q[n]_q x}{[n+2]_q}\right)\right| \leq 3\|f\|.$$

Thus

$$\|\tilde{D}_{n,q}(f,x)\| \leq 3\,\|f\|, \tag{4.21}$$

for all $f \in C[0,1]$.

Now, for $f \in C[0,1]$ and $g \in W^2$, we obtain

$$|D_{n,q}(f,x)-f(x)| \leq$$

$$= \left|\tilde{D}_{n,q}(f,x)-f(x)+f\left(\frac{1+q[n]_q x}{[n+2]_q}\right)-f(x)\right|$$

$$\leq |\tilde{D}_{n,q}(f-g,x)|+|\tilde{D}_{n,q}(g,x)-g(x)|+|g(x)-f(x)|+\left|f\left(\frac{1+q[n]_q x}{[n+2]_q}\right)-f(x)\right|$$

$$\leq 4 \|f - g\| + \frac{5}{[n+2]} \cdot \delta_n^2(x) \cdot \|g''\| + \omega \left(f, \left| \frac{1 - ([n+2]_q - q[n]_q)x}{[n+2]_q} \right| \right)$$

$$\leq 5 \left(\|f - g\| + \frac{1}{[n+2]_q} \cdot \delta_n^2(x) \cdot \|g''\| \right) + \omega \left(f, \frac{1-x}{[n+2]_q} \right),$$

where we used (4.20) and (4.21). Taking the infimum on the right hand side over all $g \in W^2$, we obtain

$$|D_{n,q}(f,x) - f(x)| \leq 5 K_2 \left(f, \frac{1}{[n+2]_q} \delta_n^2(x) \right) + \omega \left(f, \frac{1-x}{[n+2]_q} \right).$$

In view of (4.12), we find

$$|D_{n,q}(f,x) - f(x)| \leq C \omega_2 \left(f, [n+2]_q^{-1/2} \delta_n(x) \right) + \omega \left(f, \frac{1-x}{[n+2]_q} \right),$$

this completes the proof of the theorem.   ∎

For the next theorem we shall use some notations: for $f \in C[0,1]$ and $\varphi(x) = \sqrt{x(1-x)}$, $x \in [0,1]$, let

$$\omega_2^{\varphi}(f, \sqrt{\delta}) = \sup_{0 < h \leq \sqrt{\delta}} \sup_{x \pm h\varphi \in [0,1]} |f(x + h\varphi(x)) - 2f(x) + f(x - h\varphi(x))|$$

be the second-order Ditzian–Totik modulus of smoothness, and let

$$\overline{K}_{2,\varphi}(f, \delta) = \inf\{\|f - g\| + \delta \|\varphi^2 g''\|^2 \|g''^2(\varphi)\}$$

be the corresponding K-functional, where

$$W^2(\varphi) = \{g \in C[0,1] : g' \in AC_{loc}[0,1], \varphi^2 g'' \in C[0,1]\}$$

and $g' \in AC_{loc}[0,1]$ means that g is differentiable and g' is absolutely continuous on every closed interval $[a,b] \subset [0,1]$. It is well known (see [51, p. 24, Theorem 1.3.1]) that

$$\overline{K}_{2,\varphi}(f, \delta) \leq C \omega_2^{\varphi}(f, \sqrt{\delta}) \tag{4.22}$$

for some absolute constant $C > 0$. Moreover, the Ditzian–Totik moduli of first order is given by

$$\omega_{\psi}(f, \delta) = \sup_{0 < h \leq \delta} \sup_{x, x \pm h\psi(x) \in [0,1]} |f(x + h\psi(x)) - f(x)|,$$

where ψ is an admissible step-weight function on $[0,1]$.
Now we state our next main result.

Theorem 4.5. *Let $n > 3$ be a natural number and let $q_0 = q_0(n) \in (0,1)$ be defined as in Lemma 4.3. Then there exists an absolute constant $C > 0$ such that*

$$\|D_{n,q}f - f\| \leq C\,\omega_2^{\varphi}(f, [n+2]_q^{-1/2}) + \omega_{\psi}(f, [n+2]_q^{-1}),$$

where $f \in C[0,1]$, $q \in (q_0,1)$, and $\psi(x) = 1 - x$, $x \in [0,1]$.

Proof. Again, let

$$\tilde{D}_{n,q}(f,x) = D_{n,q}(f,x) + f(x) - f\left(\frac{1+q[n]_q x}{[n+2]_q}\right),$$

where $f \in C[0,1]$. Using Taylor's formula:

$$g(t) = g(x) + (t-x)\,g'(x) + \int_x^t (t-u)\,g''^2(\varphi),$$

the formulas (4.13) and (4.14), we obtain

$$\tilde{D}_{n,q}(g,x) = g(x) + D_{n,q}\left(\int_x^t (t-u)\,g''(u)\,du, x\right) - \int_x^{\frac{1+q[n]_q x}{[n+2]-q}}\left(\frac{1+q[n]_q x}{[n+2]_q} - u\right)g''(u)\,du$$

Hence

$$|\tilde{D}_{n,q}(g,x) - g(x)|$$

$$\leq D_{n,q}\left(\left|\int_x^t |t-u| \cdot |g''(u)|\,du\right|, x\right) + \left|\int_x^{\frac{1+q[n]_q x}{[n+2]}}\left|\frac{1+q[n]_q x}{[n+2]_q} - u\right| \cdot |g''(u)|\,du\right|$$

$$\tag{4.23}$$

Because the function δ_n^2 is concave on $[0,1]$, we have for $u = t + \tau(x-t)$, $\tau \in [0,1]$, the estimate

$$\frac{|t-u|}{\delta_n^2(u)} = \frac{\tau|x-t|}{\delta_n^2(t+\tau(x-t))} \leq \frac{\tau|x-t|}{\delta_n^2(t) + \tau(\delta_n^2(x) - \delta_n^2(t))} \leq \frac{|t-x|}{\delta_n^2(x)}.$$

Hence, by (4.23), we find

$$|\tilde{D}_{n,q}(g,x) - g(x)| \leq$$

$$\leq D_{n,q}\left(\left|\int_x^t \frac{|t-u|}{\delta_n^2(u)}\,du\right|, x\right) \cdot \|\delta_n^2 g''\| + \left|\int_x^{\frac{1+q[n]_q x}{[n+2]_q}}\frac{\left|\frac{1+q[n]_q x}{[n+2]_q} - u\right|}{\delta_n^2(u)}\,du\right| \cdot \|\delta_n^2 g''\|$$

$$\leq \frac{1}{\delta_n^2(x)} \cdot D_{n,q}((t-x)^2, x) \cdot \|\delta_n^2 g''\| + \frac{1}{\delta_n^2(x)} \cdot \left(\frac{1+q[n]_q x}{[n+2]_q} - x\right)^2 \cdot \|\delta_n^2 g''\|$$

In view of (4.19) and

$$\delta_n^2(x) \cdot |g''^2(x)g''(x)| + \frac{1}{[n+3]_q} \cdot |g''^2 g''\| + \frac{1}{[n+3]_q} \cdot \|g''\|,$$

where $x \in [0,1]$, we get

$$|\tilde{D}_{n,q}(g,x) - g(x)| \le \frac{5}{[n+2]_q} \cdot \left(\|\varphi^2 g''\| + \frac{1}{[n+3]_q} \cdot \|g''\| \right) \qquad (4.24)$$

Using $[n]_q \le [n+2]_q$, (4.21), and (4.24), we find for $f \in C[0,1]$,

$|D_{n,q}(f,x) - f(x)| \le$

$$\le |\tilde{D}_{n,q}(f-g,x)| + |\tilde{D}_{n,q}(g,x) - g(x)| + |g(x) - f(x)| + \left| f\left(\frac{1+q[n]_q x}{[n+2]_q} \right) - f(x) \right|$$

$$\le 4 \|f-g\| + \frac{5}{[n+2]_q} \cdot \|\varphi^2 g''\| + \frac{5}{[n+2]_q} \cdot \|g''\| + \left| f\left(\frac{1+q[n]_q x}{[n+2]_q} \right) - f(x) \right|$$

Taking the infimum on the right hand side over all $g \in W^2(\varphi)$, we obtain

$$|D_{n,q}(f,x) - f(x)| \le 5 \overline{K}_{2,\varphi}\left(f, \frac{1}{[n+2]_q} \right) + \left| f\left(\frac{1+q[n]_q x}{[n+2]_q} \right) - f(x) \right| \qquad (4.25)$$

On the other hand

$$\left| f\left(\frac{1+q[n]x}{[n+2]} \right) - f(x) \right| =$$

$$= \left| f\left(x + \psi(x) \cdot \frac{1-([n+2]_q - q[n]_q)x}{[n+2]_q \psi(x)} \right) - f(x) \right|$$

$$\le \sup_{t, t+\psi(t) \cdot (1-([n+2]_q - q[n]_q)x)/[n+2]_q \in [0,1]} \left| f\left(t + \psi(t) \cdot \frac{1-([n+2]_q - q[n]_q)x}{[n+2]_q \psi(x)} \right) - f(t) \right|$$

$$\le \omega_\psi\left(f, \frac{|1-([n+2]_q - q[n]_q)x|}{[n+2]_q \psi(x)} \right) \le \omega_\psi\left(f, \frac{1-x}{[n+2]_q \psi(x)} \right) = \omega_\psi\left(f, \frac{1}{[n+2]_q} \right).$$

Hence, by (4.25) and (4.22), we get

$$\|D_{n,q} f - f\| \le C \, \omega_2^\varphi(f, [n+2]_q^{-1/2}) + \omega_\psi(f, [n+2]_q^{-1}),$$

$x \in [0,1]$, which completes the proof of the theorem.   ∎

Remark 4.2. In [86] it is proved for $q = q(n) \to 1$ as $n \to \infty$ that the sequence $\{D_{n,q}f\}$ converges to f uniformly on $[0,1]$ for each $f \in C[0,1]$. The same result follows from Theorem 4.5, because

$$\lim_{n\to\infty}[n+2]_{qn} = \lim_{n\to\infty}\frac{1-(q(n))^{n+2}}{1-q(n)} = \infty,$$

if $\lim_{n\to\infty}q(n) = 1$.

4.3.3   Applications to Random and Fuzzy Approximation

Let $(X, ||.||)$ be a normed space over K, where $K = R$ or $K = C$. Similar to the case of real-valued functions can be introduced the following concepts.

Definition 4.1 (Gal [74]).

(i) For $f : [0,1] \to X$, the first-order Ditzian–Totik modulus of continuity $\omega_\psi(f,\delta)$ and the second-order Ditzian–Totik modulus of smoothness $\omega_2^\varphi(f,\delta)$ are respectively defined as

$$\omega_\psi(f,\delta) = \sup_{0<h\leq\delta} \sup_{x,x\pm h\psi(x)\in[0,1]} ||f(x+h\psi(x)) - f(x)||,$$

and

$$\omega_2^\varphi(f,\delta) =$$
$$\sup\{\sup\{||f(x+h\varphi(x)) - 2f(x) + f(x-h\varphi(x))||, x \in I_{2,h}\}, h \in [0,\delta]\}$$

where $I_{2,h} = \left[-\frac{1-h^2}{1+h^2}, \frac{1-h^2}{1+h^2}\right]$, $\varphi(x) = \sqrt{x(1-x)}$, $\psi(x) = 1-x, 0 < \delta \leq 1$.

(ii) $f : [0,a] \to X$ is called q-integrable $(0 < q < 1)$ on $[0,a]$ if there exists $I \in X$ denoted by $I := \int_0^a f(u)d_q u$ with the property

$$\lim_{n\to\infty}\left\| I - (1-q)\sum_{k=1}^n q^k f(aq^k) \right\| = 0.$$

Remark 4.3. Let $(X, ||.||)$ be a Banach space. If $f : [0,a] \to X$ is continuous on $[0,a]$, then it is q-integrable. Indeed, denoting $S_n(f) = (1-q)\sum_{k=1}^n q^k f(aq^k)$, we get $S_{n+p}(f) - S_n(f) = (1-q)\sum_{k=n}^{n+p} q^k f(aq^k)$ and since $||f(x)||$ is bounded (by continuity) by a positive constant denoted by M, for all $n, p \in \mathbb{N}$ it follows

$$||S_{n+p}(f) - S_n(f)|| \leq M(1-q)\sum_{k=n}^{n+p}q^k \leq M(1-q)q^n\sum_{j=0}^\infty q^j = Mq^n,$$

which shows that $(S_n(f))_{n\in\mathbb{N}}$ is a Cauchy sequence. Since X is a Banach space, it follows that this sequence is convergent and therefore f is q-integrable.

Definition 4.2 (see Gupta [86] for real-valued functions). For $f : [0,1] \to X, 0 < q < 1$, q-integrable on $[0,1]$, the q-Durrmeyer operators attached to f can be defined as

$$D_{n,q}(f,x) \equiv (D_{n,q}f)(x) = [n+1]\sum_{k=0}^{n} q^{-k}p_{n,k}(q;x)\int_0^1 f(u)p_{n,k}(q;qu)d_qu \quad (4.26)$$

where

$$p_{n,k}(q;x) := \begin{bmatrix} n \\ k \end{bmatrix} x^k(x;q)_{n-k}.$$

Theorem 4.6 (see, e.g., [124], p. 183). *Let $(X, \|.\|)$ be a normed space over K, where $K = R$ or $K = C$ and denote by $X^* = \{x^* : X \to K, x^*$ is linear and continuous$\}$. Then*

$$\|x\| = \sup\{|x^*(x)| : x^* \in X^*, \|x^*\| < 1\}.$$

Gal and Gupta [77] established the following theorem:

Theorem 4.7. *Let $(X, \|\cdot\|)$ be a Banach space and suppose that $f : [0,1] \to X$ is continuous on $[0,1]$. Then under the conditions on q as given in Lemma 4.4, we have*

$$\|D_{n,q}f - f\|_u \le C\,\omega_2^\varphi(f, [n+2]^{-1/2}) + \omega_\psi(f, [n+2]^{-1}),$$

where $\|f\|_u = \sup\{\|f(x)\| : x \in [0,1]\}$.

Proof. Let $x^* \in X^*, 0 < \||x^*\|| \le 1$ and define $g : [0,1] \to \mathbb{R}, g(x) = x^*(f(x))$. Obviously g is continuous on $[0,1]$. First, we have

$$
\begin{aligned}
\omega_\psi\left(g, \frac{1}{[n+2]}\right) &= \sup_{0<h\le 1/[n+2]} \sup_{x,x\pm h\psi(x)\in[0,1]} |x^*[f(x+h\psi(x)) - f(x)]| \\
&\le \sup_{0<h\le 1/[n+2]} \sup_{x,x\pm h\psi(x)\in[0,1]} \||x^*\|| \cdot \|[f(x+h\psi(x)) - f(x)]\| \\
&\le \sup_{0<h\le 1/[n+2]} \sup_{x,x\pm h\psi(x)\in[0,1]} \|[f(x+h\psi(x)) - f(x)]\| \\
&= \omega_\psi\left(f, \frac{1}{[n+2]}\right),
\end{aligned}
$$

and

$$\omega_2^\varphi(g,[n+2]^{-1/2})$$

$$= \sup\{\sup\{|x^*[f(x+h\varphi(x))-2f(x)+f(x-h\varphi(x))]|,\ x\in I_{2,h}\},h\in[0,[n+2]^{-1/2}]\}$$

$$\le \sup\{\sup\{\||x^*\||\cdot\|f(x+h\varphi(x))-2f(x)+f(x-h\varphi(x))\|,x\in I_{2,h}\},h\in[0,[n+2]^{-1/2}]\}$$

$$\le \omega_2^\varphi(f,[n+2]^{-1/2}).$$

Now, by Theorem 4.5, for all $x\in[0,1]$ and $n\in\mathbb{N}$, we have

$$|D_{n,q}g(x)-g(x)|\le C[\ \omega_2^\varphi(g,[n+2]^{-1/2})\ +\ \omega_\psi(g,[n+2]^{-1})].$$

But by the linearity and the continuity of x^* (the continuity allows to x^* to commutes with the integral), we easily get $D_{n,q}g(x)-g(x)=x^*[D_{n,q}f(x)-f(x)]$, which combined with the above inequalities lead to

$$|x^*[D_{n,q}f(x)-f(x)]|\le C[\ \omega_2^\varphi(f,[n+2]^{-1/2})\ +\ \omega_\psi(f,[n+2]^{-1})],$$

for all $x\in[0,1]$. Passing to supremum with $\||x^*\||\le 1$ and taking into account Theorem 4.6, it follows

$$\|D_{n,q}f(x)-f(x)]\|\le C[\ \omega_2^\varphi(f,[n+2]^{-1/2})\ +\ \omega_\psi(f,[n+2]^{-1})],$$

for all $x\in[0,1]$, which proves the theorem. ∎

Some applications to the approximation of random functions by q-Durrmeyer random polynomials and of fuzzy-number-valued functions by q-Durrmeyer fuzzy polynomials were discussed in [77] as

If (S,B,P) is a probability space (P is the probability), then the set of almost sure (a.s.) finite real random variables is denoted by $L(S,B,P)$ and it is a Banach space with respect to the norm $\|g\|=\int_S|g(t)|dP(t)$. Here, for $g_1,g_2\in L(S,B,P)$, we consider $g_1=g_2$ if $g_1(t)=g_2(t)$, a.s. $t\in S$.

A random function defined on $[0,1]$ is a mapping $f:[0,1]\to L(S,B,P)$ and we denote $f(x)(t)\in\mathbb{R}$ by $f(x,t)$. For this kind of f, the q-Durrmeyer random polynomials are defined by

$$(D_{n,q}f)(x,t)=[n+1]\sum_{k=0}^{n}q^{-k}p_{n,k}(q;x)\int_0^1 f(u,t)p_{n,k}(q;qu)d_qu.$$

Corollary 4.2. *If* $f:[0,1]\to L(S,B,P)$ *is continuous on* $[0,1]$, *then*

$$\|D_{n,q}f-f\|_u\le C\,\omega_2^\varphi(f,[n+2]^{-1/2})\ +\ \omega_\psi(f,[n+2]^{-1}),$$

where $\|f\|_u=\sup\{\|f(x)\|;x\in[0,1]\}=\sup\{\int_S|f(x,t)|dP(t);x\in[0,1]\}$.

Given a set $X \neq \emptyset$, a fuzzy subset of X is a mapping $u : X \rightarrow [0,1]$, and obviously any classical subset A of X can be considered as a fuzzy subset of X defined by $\chi_A : X \rightarrow [0,1]$, $\chi_A(x) = 1$, if $x \in A$, $\chi_A(x) = 0$ if $x \in X \setminus A$. (see, e.g., Zadeh [154]).

Let us denote by $\mathbb{R}_{\mathcal{F}}$ the class of fuzzy subsets of real axis $\mathbb{R}$ (i.e., $u : \mathbb{R} \rightarrow [0,1]$), satisfying the following properties:

(i) $\forall u \in \mathbb{R}_{\mathcal{F}}$, u is normal, i.e., $\exists x_u \in \mathbb{R}$ with $u(x_u) = 1$.
(ii) $\forall u \in \mathbb{R}_{\mathcal{F}}$, u is convex fuzzy set (i.e., $u(tx + (1-t)y) \geq \min\{u(x), u(y)\}$, $\forall t \in [0,1]$, $x, y \in \mathbb{R}$).
(iii) $\forall u \in \mathbb{R}_{\mathcal{F}}$, u is upper semicontinuous on $\mathbb{R}$.
(iv) $\overline{\{x \in \mathbb{R} : u(x) > 0\}}$ is compact, where $\overline{A}$ denotes the closure of A.

Then $\mathbb{R}_{\mathcal{F}}$ is called the space of fuzzy real numbers (see, e.g., Dubois–Prade [56]).

Remark 4.4. Obviously $\mathbb{R} \subset \mathbb{R}_{\mathcal{F}}$, because any real number $x_0 \in \mathbb{R}$ can be described as the fuzzy number whose value is 1 for $x = x_0$ and 0 otherwise.

For $0 < r \leq 1$ and $u \in \mathbb{R}_{\mathcal{F}}$, define $[u]^r = \{x \in \mathbb{R}; u(x) \geq r\}$ and $[u]^0 = \overline{\{x \in \mathbb{R}; u(x) > 0\}}$. Then it is well known that for each $r \in [0,1]$, $[u]^r$ is a bounded closed interval. For $u, v \in \mathbb{R}_{\mathcal{F}}$ and $\lambda \in \mathbb{R}$, we have the sum $u \oplus v$ and the product $\lambda \odot u$ defined by $[u \oplus v]^r = [u]^r + [v]^r$, $[\lambda \odot u]^r = \lambda [u]^r$, $\forall r \in [0,1]$, where $[u]^r + [v]^r$ means the usual addition of two intervals (as subsets of $\mathbb{R}$) and $\lambda [u]^r$ means the usual product between a scalar and a subset of $\mathbb{R}$ (see, e.g., Dubois–Prade [56], Congxin–Zengtai [44]).

Let $D : \mathbb{R}_{\mathcal{F}} \times \mathbb{R}_{\mathcal{F}} \rightarrow \mathbb{R}_+ \cup \{0\}$ by

$$D(u,v) = \sup_{r \in [0,1]} \max \left\{ \left| u_-^r - v_-^r \right|, \left| u_+^r - v_+^r \right| \right\},$$

where $[u]^r = \left[u_-^r, u_+^r \right]$, $[v]^r = \left[v_-^r, v_+^r \right]$. The following properties are known (Dubois–Prade [56]):

$D(u \oplus w, v \oplus w) = D(u,v)$, $\forall u, v, w \in \mathbb{R}_{\mathcal{F}}$
$D(k \odot u, k \odot v) = |k| D(u,v)$, $\forall u, v \in \mathbb{R}_{\mathcal{F}}, \forall k \in \mathbb{R}$;
$D(u \oplus v, w \oplus e) \leq D(u,w) + D(v,e)$, $\forall u, v, w, e \in \mathbb{R}_{\mathcal{F}}$ and $(\mathbb{R}_{\mathcal{F}}, D)$ is a complete metric space.

Also, we need the following concept of q-integral. A function $f : [0,a] \rightarrow \mathbb{R}_{\mathcal{F}}$, $[0,a] \subset \mathbb{R}$ will be called q-integrable on $[0,a]$, if there exists $I \in \mathbb{R}_{\mathcal{F}}$, denoted by $I = \int_0^a f(u) d_q u$ with the property

$$\lim_{n \to \infty} D[I, (1-q) \odot \Sigma^{*n}_{k=1} q^k \odot f(aq^k)]\| = 0.$$

Here the sum Σ^* is considered with respect to the operation $\oplus$.

Remark 4.5. If $f : [0,a] \rightarrow \mathbb{R}_{\mathcal{F}}$ is continuous on $[0,a]$, then it is q-integrable. Indeed, denoting $S_n(f) = (1-q) \odot \Sigma^{*n}_{k=1} q^k \odot f(aq^k)$, from the above properties of the metric D, we can write

$$D[S_n(f), S_{n+p}(f)] = (1-q)D[0_{\mathbb{R}_\mathcal{F}}, \Sigma^{*n+p}_{k=n} q^k \odot f(aq^k)] \leq$$

$$(1-q)\sum_{k=n}^{n+p} q^k D[0_{\mathbb{R}_\mathcal{F}}, f(aq^k)] \leq M(1-q)\sum_{k=n}^{n+p} q^k,$$

where the continuity implies that f is bounded and that there exists $M > 0$ such that $D[0_{\mathbb{R}_\mathcal{F}}, f(x)] \leq M$ for all $x \in [0,a]$. In continuation, taking into account that $(\mathbb{R}_\mathcal{F}, D)$ is a complete metric space, the reasonings are similar to those in the Remark 4.3.

Theorem 4.8 (see [44]). $\mathbb{R}_\mathcal{F}$ *can be embedded in* $\mathbb{B} = \bar{C}[0,1] \times \bar{C}[0,1]$, *where* $\bar{C}[0,1]$ *is the class of all real-valued bounded functions* $f : [0,1] \to \mathbb{R}$ *such that* f *is left continuous for any* $x \in (0,1]$, f *has right limit for any* $x \in [0,1)$, *and* f *is right continuous at* 0. *With the norm* $\|\cdot\| = \sup_{x \in [0,1]} |f(x)|$, $\bar{C}[0,1]$ *is a Banach space. Denote* $\|\cdot\|_\mathbb{B}$ *the usual product norm, i.e.,* $\|(f,g)\|_\mathbb{B} = \max\{\|f\|, \|g\|\}$. *Let us denote the embedding by* $j : \mathbb{R}_\mathcal{F} \to \mathbb{B}$, $j(u) = (u_-, u_+)$. *Then* $j(\mathbb{R}_\mathcal{F})$ *is a closed convex cone in* $\mathbb{B}$ *and* j *satisfies the following properties:*

(i) $j(s \odot u \oplus t \odot v) = s \cdot j(u) + t \cdot j(v)$ *for all* $u, v \in \mathbb{R}_\mathcal{F}$ *and* $s, t \geq 0$ *(here "·" and "+" denote the scalar multiplication and addition in* $\mathbb{B}$)
(ii) $D(u,v) = \|j(u) - j(v)\|_\mathbb{B}$ *(i.e.,* j *embeds* $\mathbb{R}_\mathcal{F}$ *in* $\mathbb{B}$ *isometrically)*

Let $f : [0,1] \to \mathbb{R}_\mathcal{F}$ be a continuous fuzzy-number-valued function. The fuzzy q-Durrmeyer polynomials attached to f can be defined by

$$(D_{n,q}f)(x) = [n+1]\sum_{k=0}^{n} q^{-k} p_{n,k}(q;x) \odot \int_0^1 p_{n,k}(q;qu) \odot f(u)d_qu.$$

Also, let us define the following moduli of continuity and smoothness of f :

$$\omega_\psi(f,\delta) = \sup_{0<h\leq\delta, x\pm h\psi(x)\in[0,1]} \sup D[f(x+h\psi(x)), f(x)],$$

$$\omega_2^\phi(f;\delta) = \sup\{D[f(x+h\phi(x)) \oplus f(x-h\phi(x)), 2 \odot f(x)];$$

$$x, x+h\phi(x), x-h\phi(x) \in [0,1], 0 \leq h \leq \delta\}.$$

Here $\phi^2(x) = x(1-x)$, $\psi(x) = 1-x$.

Theorem 4.9. *Let* $f : [0,1] \to \mathbb{R}_\mathcal{F}$ *be continuous on* $[0,1]$. *There exist the absolute constant* C, *such that for all* $n \in \mathbb{N}$ *we have*

$$\sup\{D[(D_{n,q}f)(x), f(x)]; x \in [0,1]\} \leq C\,\omega_2^\phi(f, [n+2]^{-1/2}) + \omega_\psi(f, [n+2]^{-1}).$$

4.4   Discretely Defined *q*-Durrmeyer Operators

For $f \in C[0,1]$, Gupta and Wang [94] proposed the following *q*-Durrmeyer operators as

$$M_{n,q}(f;x) = [n+1]_q \sum_{k=1}^{n} q^{1-k} p_{n,k}(q;x) \int_0^1 f(t) p_{n,k-1}(q;qt) d_q t + f(0) p_{n,0}(q;x)$$

(4.27)

It can be easily verified that in the case $q = 1$, the operators defined by (4.27) reduce to the Durrmeyer-type operators recently introduced and studied in [3].

4.4.1   Moment Estimation

By the definition of *q*-Beta function, we have

$$\int_0^1 t^s p_{n,k}(q;qt) d_q t = \begin{bmatrix} n \\ k \end{bmatrix} q^k \int_0^1 t^{k+s}(1-qt)_q^{n-k} d_q t$$

$$= \frac{q^k [n]_q!}{[k]_q![n-k]_q!} \frac{[k+s]_q![n-k]_q!}{[k+s+n-k+1]_q!} = \frac{q^k [n]_q![k+s]_q!}{[n+s+1]_q![k]_q!}$$

(4.28)

and

$$\int_0^1 t^s p_{\infty,k}(q;qt) d_q t = \frac{q^k}{(1-q)^k [k]_q!} \int_0^1 t^{k+s}(1-qt)_q^\infty d_q t$$

$$= \frac{q^k}{(1-q)^k [k]_q!}[k+s]_q!(1-q)^{k+s+1} = (1-q)^{s+1} \frac{q^k [k+s]_q!}{[k]_q!}.$$

(4.29)

Lemma 4.5. *We have*

$$M_{n,q}(1;x) = 1, \quad M_{n,q}(t;x) = x \frac{[n]_q}{[n+2]_q}$$

and

$$M_{n,q}(t^2;x) = \frac{(1+q)x[n]_q}{[n+3]_q[n+2]_q} + x^2 \frac{q[n]_q([n]_q-1)}{[n+3]_q[n+2]_q}.$$

Proof. In order to prove the theorem we shall use the following identities:

$$\sum_{k=0}^{n} p_{n,k}(q;x) = 1, \quad \sum_{k=0}^{n} \frac{[k]_q}{[n]_q} p_{n,k}(q;x) = x,$$

$$\sum_{k=0}^{n} \left(\frac{[k]_q}{[n]_q}\right)^2 p_{nk}(q;x) = x^2 + \frac{x(1-x)}{[n]_q}.$$

By (4.28) and (4.29), it can easily be verified that $M_{n,q}(1;x) = 1$. Next, using the above, we have

$$M_{n,q}(t;x) = [n+1]_q \sum_{k=1}^{n} q^{1-k} p_{n,k}(q;x) \frac{q^{k-1}[n]_q![k]_q}{[n+2]_q!}$$

$$= \frac{1}{[n+2]_q} \sum_{k=1}^{n} [k]_q p_{n,k}(q;x) = x \frac{[n]_q}{[n+2]_q}.$$

Finally, using $[a+1]_q = 1 + q[a]_q$, we have

$$M_{n,q}(t^2;x) = \frac{1}{[n+3]_q[n+2]_q} \sum_{k=1}^{n} p_{n,k}(q;x)[k+1]_q[k]_q$$

$$= \frac{1}{[n+3]_q[n+2]_q} \left\{ \sum_{k=1}^{n} p_{n,k}(q;x)(1+q[k]_q)[k]_q \right\}$$

$$= \frac{1}{[n+3]_q[n+2]_q} \left\{ \sum_{k=1}^{n} p_{n,k}(q;x)[k]_q + q \sum_{k=1}^{n} p_{n,k}(q;x)[k]_q^2 \right\}$$

$$= \frac{1}{[n+3]_q[n+2]_q} \left\{ x[n]_q + q(x^2[n]_q^2 + x(1-x)[n]_q) \right\}$$

$$= \frac{x[n]_q(1+q)}{[n+3]_q[n+2]_q} + \frac{q^2 x^2}{[n+3]_q[n+2]_q} \left[\frac{[n]_q^2 - [n]_q}{q} \right].$$

Thus,

$$M_{n,q}(t^2;x) = \frac{x[n]_q(1+q)}{[n+3]_q[n+2]_q} + \frac{qx^2[n]_q([n]_q - 1)}{[n+3]_q[n+2]_q}.$$

This completes the proof of the lemma. ∎

Remark 4.6. By simple computation, it can easily be verified that

$$M_{n,q}(t^r;x) = \frac{[n+1]_q!}{[n+r+1]_q!} \sum_{k=1}^{n} [k]_q[k+1]_q \cdots [k+r-1]_q p_{n,k}(q;x), \quad r \geq 1.$$

Using $[k+s]_q = [s]_q + q^s[k]_q$, we get

$$[k]_q[k+1]_q \cdots [k+r-1]_q = \prod_{s=0}^{r-1}([s]_q + q^s[k]_q) = \sum_{s=1}^{r} c_s(r)[k]_q^s,$$

where $c_s(r) > 0$, $s = 1, 2, \ldots, r$ are the constants independent of k. Hence

$$M_{n,q}(t^r; x) = \frac{[n+1]_q!}{[n+r+1]_q!} \sum_{s=1}^{r} c_s(r) \sum_{k=1}^{n} [k]_q^s p_{n,k}(q;x) = \frac{[n+1]_q!}{[n+r+1]_q!} \sum_{s=1}^{r} c_s(r)[n]_q^s B_{n,q}(t^s; x).$$

Since $c_s(r) > 0$ for $s = 1, 2, \ldots, r$ and $B_{n,q}(t^s; x)$ is a polynomial of degree $\leq \min(s,n)$ (see [7]), we get $M_{n,q}(t^r; x)$ is a polynomial of degree $\leq \min(r,n)$.

4.4.2   Rate of Approximation

Theorem 4.10. *Let $q_n \in (0,1)$. Then the sequence $\{M_{n,q_n}(f)\}$ converges to f uniformly on $[0,1]$ for each $f \in C[0,1]$ if and only if $\lim_{n\to\infty} q_n = 1$.*

Proof. Since the operators M_{n,q_n} are positive linear operators on $C[0,1]$ and preserve constant functions, the well-known Korovkin theorem [113] implies that $M_{n,q_n}(f;x)$ converges to $f(x)$ uniformly on $[0,1]$ as $n \to \infty$ for any $f \in C[0,1]$ if and only if

$$M_{n,q_n}(t^i; x) \to x^i \quad (i = 1, 2), \tag{4.30}$$

uniformly on $[0,1]$ as $n \to \infty$. If $q_n \to 1$, then $[n]_{q_n} \to \infty$ (see [151]) and for $s = 1, 2, 3$, $\lim_{n\to\infty} \frac{[n+s]_{q_n}}{[n]_{q_n}} = 1$, hence (4.30) follows from Lemma 4.5.

On the other hand, if we assume that for any $f \in C[0,1]$, $M_{n,q_n}(f,x)$ converges to $f(x)$ uniformly on $[0,1]$ as $n \to \infty$, then $q_n \to 1$. In fact, if the sequence (q_n) does not tend to 1, then it must contain a subsequence (q_{n_k}) such that $q_{n_k} \in (0,1)$, $q_{n_k} \to q_0 \in [0,1)$ as $k \to \infty$. Thus, $\frac{1}{[n_k+s]_{q_{n_k}}} = \frac{1-q_{n_k}}{1-(q_{n_k})^{n_k+s}} \to (1-q_0)$ as $k \to \infty$, $s = 0, 1, 2, 3$. Taking $n = n_k$, $q = q_{n_k}$ in $M_{n,q}(t^2; x)$, by Lemma 4.5, we get

$$M_{n_k, q_{n_k}}(t^2; x) \to x(1-q_0^2) + x^2 q_0^2 \not\to x^2 \quad (k \to \infty),$$

which leads to a contradiction. Hence, $q_n \to 1$.
This completes the proof of Theorem 4.10. ∎

Let $q \in (0,1)$ be fixed. We define $M_{\infty,q}(f,1) = f(1)$ and for $x \in [0,1)$

$$M_{\infty,q}(f,x) := \frac{1}{1-q} \sum_{k=1}^{\infty} p_{\infty,k}(q;x) q^{1-k} \int_0^1 f(t) p_{\infty,k-1}(q;qt) d_q t + f(0) p_{\infty,0}(q;x)$$

$$=: \sum_{k=0}^{\infty} A_{\infty k}(f) p_{\infty,k}(q;x). \tag{4.31}$$

Using (4.29), (4.31), and the fact that (see [125])

$$\sum_{k=0}^{\infty} p_{\infty,k}(q;x) = 1, \quad \sum_{k=0}^{\infty}(1-q^k)p_{\infty,k}(q;x) = x$$

and

$$\sum_{k=0}^{\infty}(1-q^k)^2 p_{\infty,k}(q;x) = x^2 + (1-q)x(1-x),$$

it is easy to prove that

$$M_{\infty,q}(1;x) = 1, \quad M_{\infty,q}(t;x) = x,$$

and

$$M_{\infty,q}(t^2;x) = \sum_{k=0}^{\infty}(1-q^k)(1-q^{k+1})p_{\infty,k}(q;x)$$

$$= (1-q)x + q(x^2 + (1-q)x(1-x)) = (1-q^2)x + q^2 x^2.$$

For $f \in C[0,1]$, $t > 0$, we define the modulus of continuity $\omega(f,t)$ as follows:

$$\omega(f,t) := \sup_{\substack{|x-y| \le t \\ x,y \in [0,1]}} |f(x) - f(y)|.$$

Lemma 4.6. *Let $f \in C[0,1]$ and $f(1) = 0$. Then we have*

$$|A_{nk}(f)| \le A_{nk}(|f|) \le \omega(f,q^n)(1+q^{k-n})$$

and

$$|A_{\infty k}(f)| \le A_{\infty k}(|f|) \le \omega(f,q^n)(1+q^{k-n}).$$

Proof. By the well-known property of modulus of continuity (see [4], pp. 20)

$$\omega(f,\lambda t) \le (1+\lambda)\omega(f,t), \quad \lambda > 0,$$

we get

$$|f(t)| = |f(t) - f(1)| \le \omega(f,1-t) \le \omega(f,q^n)(1 + (1-t)/q^n).$$

Thus,

$$|A_{nk}(f)| \leq A_{nk}(|f|) := [n+1]_q \int_0^1 q^{1-k}|f(t)|p_{n,k-1}(q;qt)d_qt$$

$$\leq [n+1]_q \int_0^1 q^{1-k}\omega(f,q^n)(1+(1-t)/q^n)p_{n,k-1}(q;qt)d_qt$$

$$= \omega(f,q^n)(1+q^{-n}(1-\frac{[k]_q}{[n+2]_q}))$$

$$= \omega(f,q^n)\left(1+\frac{q^k(1-q^{n+2-k})}{q^n(1-q^{n+2})}\right) \leq \omega(f,q^n)(1+q^{k-n}).$$

Similarly,

$$|A_{\infty k}(f)| \leq A_{\infty k}(|f|) := \frac{q^{1-k}}{1-q}\int_0^1 |f(t)|p_{\infty,k-1}(q;qt)d_qt$$

$$\leq \omega(f,q^n)\frac{q^{1-k}}{1-q}\int_0^1 (1+(1-t)/q^n)p_{\infty,k-1}(q;qt)d_qt$$

$$= \omega(f,q^n)(1+(1-(1-q^k))/q^n) = \omega(f,q^n)(1+q^{k-n}).$$

Lemma 4.6 is proved.   ∎

Theorem 4.11. *Let $0 < q < 1$. Then for each $f \in C[0,1]$ the sequence $\{M_{n,q}(f;x)\}$ converges to $M_{\infty,q}(f;x)$ uniformly on $[0,1]$. Furthermore,*

$$\|M_{n,q}(f) - M_{\infty,q}(f)\| \leq C_q \,\omega(f,q^n). \tag{4.32}$$

Remark 4.7. When $f(x) = x^2$, we have

$$\|M_{n,q}(f) - M_{\infty,q}(f)\| \geq c_1 q^n \geq c_2 \,\omega(f,q^n),$$

where $c_1, c_2 > 0$ are the constants independent of n. Hence, the estimate (4.32) is sharp in the following sense: The sequence q^n in (4.32) cannot be replaced by any other sequence decreasing to zero more rapidly as $n \to \infty$.

Proof. The operators $M_{n,q}$ and $M_{\infty,q}$ preserve constant functions, that is,

$$M_{n,q}(1,x) = M_{\infty,q}(1,x) = 1.$$

Without loss of generality, we assume that $f(1) = 0$. If $x = 1$, then by Lemma 4.1, we have

$$|M_{n,q}(f;1) - M_{\infty,q}(f;1)| = |A_{nn}(f) - f(1)| = |A_{nn}(f)| \leq 2\omega(f,q^n).$$

For $x \in [0,1)$, by the definitions of $M_{n,q}(f;x)$ and $M_{\infty,q}(f;x)$, we know that

$$|M_{n,q}(f;x) - M_{\infty,q}(f;x)| = \left| \sum_{k=0}^{n} A_{nk}(f) p_{n,k}(q;x) - \sum_{k=0}^{\infty} A_{\infty k}(f) p_{\infty,k}(q;x) \right|$$

$$\leq \sum_{k=0}^{n} |A_{nk}(f) - A_{\infty k}(f)| p_{n,k}(q;x) + \sum_{k=0}^{n} |A_{\infty k}(f)| |p_{n,k}(q;x) - p_{\infty,k}(q;x)|$$

$$+ \sum_{k=n+1}^{\infty} |A_{\infty k}(f)| p_{\infty,k}(q;x) =: I_1 + I_2 + I_3.$$

First we have

$$|p_{n,k}(q;x) - p_{\infty,k}(q;x)| = \left| \begin{bmatrix} n \\ k \end{bmatrix}_q x^k \prod_{s=0}^{n-k-1}(1 - q^s x) - \frac{x^k}{(1-q)^k [k]_q!} \prod_{s=0}^{\infty}(1 - q^s x) \right|$$

$$= \left| \begin{bmatrix} n \\ k \end{bmatrix}_q x^k \left(\prod_{s=0}^{n-k-1}(1 - q^s x) - \prod_{s=0}^{\infty}(1 - q^s x) \right) \right.$$

$$\left. + x^k \prod_{s=0}^{\infty}(1 - q^s x) \left(\begin{bmatrix} n \\ k \end{bmatrix}_q - \frac{1}{(1-q)^k [k]_q!} \right) \right|$$

$$\leq p_{n,k}(q;x) \left| 1 - \prod_{s=n-k}^{\infty} \lim (1 - q^s x) \right|$$

$$+ p_{\infty k}(q;x) \left| \prod_{s=n-k+1}^{n}(1 - q^s) - 1 \right|$$

$$\leq \frac{q^{n-k}}{1-q}(p_{n,k}(q;x) + p_{\infty k}(q;x)),$$

where in the last formula, we use the following inequality, which can be easily proved by the induction on n (see [100]):

$$1 - \prod_{s=1}^{n}(1 - a_s) \leq \sum_{s=1}^{n} a_s, \quad (a_1, \ldots, a_n \in (0,1), \ n = 1,2,\ldots,\infty).$$

Using the above inequality we get

$$|A_{nk}(f) - A_{\infty k}(f)| \leq \int_0^1 q^{1-k} |f(t)| |[n+1]_q p_{n,k-1}(q;qt) - \frac{1}{1-q} p_{\infty,k-1}(q;qt)| d_q t$$

$$\leq \int_0^1 q^{1-k} |f(t)| \left| [n+1]_q - \frac{1}{1-q} \right| p_{\infty,k-1}(q;qt) d_q t$$

$$+ \int_0^1 q^{1-k}|f(t)|[n+1]_q \Big| p_{n,k-1}(q;qt) - p_{\infty,k-1}(q;qt) \Big| d_q t$$

$$\leq \frac{q^{n+1}}{1-q} \int_0^1 q^{1-k}|f(t)| p_{\infty,k-1}(q;qt) d_q t$$

$$+ \frac{q^{n-k}}{1-q} \int_0^1 q^{1-k}|f(t)|[n+1](p_{n,k-1}(q;qt) + p_{\infty,k-1}(q;qt)) d_q t$$

$$= q^{n+1} A_{\infty k}(|f|) + \frac{q^{n-k}}{1-q} A_{nk}(|f|) + q^{n-k}[n+1]_q A_{\infty k}(|f|)$$

$$\leq q^{n+1} \omega(f,q^n)(1+q^{k-n}) + 2\frac{q^{n-k}}{1-q}\omega(f,q^n)(1+q^{k-n}) \leq \frac{5\omega(f,q^n)}{1-q}.$$

Now we estimate I_1 and I_3. We have

$$I_1 \leq \frac{5\omega(f,q^n)}{1-q} \sum_{k=0}^n p_{n,k}(q;x) = \frac{5\omega(f,q^n)}{1-q}.$$

and

$$I_3 \leq \omega(f,q^n) \sum_{k=n+1}^\infty (1+q^{k-n}) p_{\infty,k}(q;x) \leq 2\omega(f,q^n) \sum_{k=n+1}^\infty p_{\infty,k}(q;x) \leq 2\omega(f,q^n).$$

Finally we estimate I_2 as follows:

$$I_2 \leq \sum_{k=0}^n \omega(f,q^n)(1+q^{k-n})\frac{q^{n-k}}{1-q}(p_{n,k}(q;x) + p_{\infty,k}(q;x))$$

$$\leq \frac{2\omega(f,q^n)}{1-q} \sum_{k=0}^n (p_{n,k}(q;x) + p_{\infty,k}(q;x)) \leq \frac{4\omega(f,q^n)}{1-q}.$$

We conclude that for $x \in [0,1)$,

$$|M_{n,q}(f;x) - M_{\infty,q}(f;x)| \leq C_q \omega(f,q^n),$$

where $C_q = 2 + \frac{9}{1-q}$. This completes the proof of Theorem 4.11. ∎

Since $M_{\infty,q}(t^2,x) = (1-q^2)x + q^2 x^2 > x^2$ for $0 < q < 1$, as a consequence of Lemma 3.10, we have the following:

Theorem 4.12. *Let $0 < q < 1$ be fixed and let $f \in C[0,1]$. Then $M_{\infty,q}(f;x) = f(x)$ for all $x \in [0,1]$ if and only if f is linear.*

Remark 4.8. Let $0 < q < 1$ be fixed and let $f \in C[0,1]$. Then by Theorem 4.11 and Theorem 4.12, it can easily be verified that the sequence $\{M_{n,q}(f;x)\}$ does not approximate $f(x)$ unless f is linear. This is completely in contrast to the classical Bernstein polynomials, by which $\{B_{n,1}(f;x)\}$ approximates $f(x)$ for any $f \in C[0,1]$.

At last, we discuss approximating property of the operators $M_{\infty,q}$.

Theorem 4.13. *For any $f \in C[0,1]$, $\{M_{\infty,q}(f)\}$ converges to f uniformly on $[0,1]$ as $q \to 1-$.*

Proof. The proof is standard. We know that the operators $M_{\infty,q}$ are positive linear operators on $C[0,1]$ and reproduce linear functions. Also,

$$M_{\infty,q}(t^2;x) = (1-q^2)x + q^2x^2 \to x^2$$

uniformly on $[0,1]$ as $q \to 1-$. Theorem 4.5 follows from the Korovkin theorem. ∎

4.5 Genuine q-Bernstein–Durrmeyer Operators

For $f \in C[0,1]$, Mahmudov and Sabancigil [121] defined the following genuine q-Bernstein–Durrmeyer operators as

$$U_{n,q}(f;x) = [n-1]_q \sum_{k=1}^{n-1} q^{1-k} p_{n,k}(q;x) \int_0^1 f(t) p_{n-2,k-1}(q;qt) d_q t$$

$$+ f(0) p_{n,0}(q;x) + f(1) p_{n,n}(q;x)$$

$$=: \sum_{k=0}^{n} A_{nk}(f) p_{n,k}(q;x), \quad 0 \le x \le 1. \tag{4.33}$$

It can be easily verified that in the case $q = 1$, the operators defined by (4.33) reduce to the genuine Bernstein–Durrmeyer operators [82].

4.5.1 Moments

Lemma 4.7 ([121]). *We have*

$$U_{n,q}(1;x) = 1, U_{n,q}(t;x) = x$$

$$U_{n,q}(t^2;x) = \frac{(1+q)x(1-x)}{[n+1]_q} + x^2$$

and

$$U_{n,q}((t-x)^2;x) = \frac{(1+q)x(1-x)}{[n+1]_q} \le \frac{2}{[n+1]_q} x(1-x).$$

Lemma 4.8 ([121]). $U_{n,q}(t;x)$ *is a polynomial of degree less than or equal to* $\min\{m,n\}$.

Proof. By simple computation,

$$U_{n,q}(t^m;x) = [n-1]_q \sum_{k=1}^{n-1} q^{1-k} p_{n,k}(q;x) \int_0^1 f(t) p_{n-2,k-1}(q;qt) t^m d_q t + p_{n,n}(q;x)$$

$$= [n-1]_q \sum_{k=1}^{n-1} p_{n,k}(q;x) \frac{[n-2]_q![k+m-1]_q!}{[k-1]_q![n+m-1]_q!} + p_{n,n}(q;x)$$

$$= \frac{[n-1]_q!}{[n+m-1]_q!} \sum_{k=1}^{n-1} p_{n,k}(q;x) \frac{[k+m-1]_q!}{[k-1]_q!} + p_{n,n}(q;x)$$

$$= \frac{[n-1]_q!}{[n+m-1]_q!} \sum_{k=1}^{n} p_{n,k}(q;x)[k]_q[k+1]_q \cdots [k+m-1]_q + p_{n,n}(q;x).$$

Next using

$$[k]_q[k+1]_q \cdots [k+m-1]_q = \prod_{s=0}^{m-1} (q^s[k]_q + [s]_q) = \sum_{s=1}^{m} c_c(m)[k]_q^s,$$

where $c_s(m) > 0, s = 1,2,3,\cdots,m$ are the constants independent of k, we get

$$U_{n,q}(t^m;x) = \frac{[n-1]_q!}{[n+m-1]_q!} \sum_{k=1}^{n} \sum_{s=1}^{m} c_s(m)[n]_q^s B_{n,q}(t^s;x),$$

where $B_{n,q}$ is the q Bernstein operator. Since $B_{n,q}(t^s;x)$ is a polynomial of degree less than or equal to $\min\{s,n\}$ and $c_s(m) > 0, s = 1,2,3,\ldots,m$, it follows that $U_{n,q}(t^m;x)$ is a polynomial of degree less than or equal to $\min\{m,n\}$. ∎

4.5.2 Direct Results

The following theorems were established by [121]:

Theorem 4.14. *Let* $0 < q_n < 1$. *Then the sequence* $\{U_{n,q}(f;x)\}$ *converges to* f *uniformly on* $[0,1]$ *for each* $f \in C[0,1]$, *if and only if* $\lim_{n \to \infty} q_n = 1$.

Theorem 4.15. *Let* $0 < q < 1$ *and* $n > 3$. *Then for each* $f \in C[0,1]$ *the sequence* $\{U_{n,q}(f;x)\}$ *converges to* $f(x)$ *uniformly on* $[0,1]$. *Furthermore*

$$||U_{n,q}(f;.) - U_{\infty,q}(f;.)|| \le c_q \omega(f, q^{n-2}),$$

where $c_q = \frac{10}{1-q} + 4$ *and* $||.||$ *is the uniform norm on* $[0,1]$.

Theorem 4.16. *There exists an absolute constant $C > 0$ such that*

$$|U_{n,q}(f;x) - f(x)| \le C\,\omega_2\left(f, \sqrt{\frac{x(1-x)}{[n+1]_q}}\right),$$

where $f \in C[0,1]$, $0 < q < 1$, and $x \in [0,1]$.

Proof. Using Taylor's formula

$$g(t) = g(x) + (t-x)\,g'(x) + \int_x^t (t-u)\,g''^2[0,1],$$

we obtain

$$U_{n,q}(g;x) = g(x) + U_{n,q}\left(\int_x^t (t-u)\,g''(u)\,du;x\right), g \in C^2[0,1]$$

Hence

$$|U_{n,q}(g;x) - g(x)| \le U_{n,q}\left(\left|\int_x^t |t-u|\cdot|g''(u)|\,du\right|,x\right)$$

$$\le U_{n,q}((t-x)^2;x)\cdot\|g''\| \le \|g''\|\frac{2}{[n+1]_q}x(1-x).$$

Now for $f \in C[0,1]$ and $g \in C^2[0,1]$ and with the fact $\|U_{n,q}(f;;.)\| \le \|f\|$, we obtain

$$|U_{n,q}(f;x) - g(x)| \le |U_{n,q}(f-g;x)| + |U_{n,q}(g;x) - g(x)| + \|f(x) - g(x)\|$$

$$\le 2\,\|f-g\| + \|g''\|\frac{2}{[n+1]_q}x(1-x).$$

Taking the infimum on the right hand side over all $g \in C^2[0,1]$, we obtain

$$|U_{n,q}(f;x) - f(x)| \le 2K_2\left(f, \frac{1}{[n+1]_q}x(1-x)\right). \tag{4.34}$$

The desired results follow from (4.12), (4.34). This completes the proof of the theorem. ∎

4.6 q-Bernstein Jacobi Operators

In the year 2005, Derriennic [48] introduced the generalization of modified Bernstein polynomials for q-Jacobi weights using the q-Bernstein basis functions. For $q \in (0,1)$ and $\alpha, \beta > -1$

$$L_{n,q}^{\alpha,\beta}(f;x) = \sum_{k=0}^n f_{n,k,q}^{\alpha,\beta} p_{n,k}(q;x) \tag{4.35}$$

where

$$p_{n,k}(q;x) := \begin{bmatrix} n \\ k \end{bmatrix}_q x^k \prod_{s=0}^{n-k-1}(1-q^s x)$$

and

$$f_{n,k,q}^{\alpha,\beta} = \frac{\int_0^1 t^{k+\alpha}(1-qt)_q^{n-k+\beta} f(q^{\beta+1}t)d_q t}{\int_0^1 t^{k+\alpha}(1-qt)_q^{n-k+\beta}d_q t}.$$

It is observed in [48] that for any $n \in N$, $L_{n,q}^{\alpha,\beta}(f;x)$ is linear and positive and preserves the constant functions.

It is self adjoint. It preserves the degree of polynomials of degree $\leq n$.

The polynomial $L_{n,q}^{\alpha,\beta}(f;x)$ is well defined if there exists $\gamma \geq 0$ such that $x^\gamma f(x)$ is bounded on $(0,A]$ for some $A \in 90,1]$ and $\alpha > \gamma - 1$. Indeed $x^\alpha f(x)$ is then q-integrable for the weight $w_q^{\alpha,\beta}(x) = x^\alpha(1-qx)_q^\beta$. Thus we call that f is said to satisfy the condition $C(\alpha)$. Also $< f,g >_q^{\alpha,\beta}$ is well defined if the product fg satisfies $C(\alpha)$, particularly if f^2 and g^2 do it, where

$$< f,g >_q^{\alpha,\beta} = \int_0^{q^{\beta+1}} t^\alpha(1-q^{-\beta}t)_q^\beta f(t)g(t)d_q t$$

and

$$< f,g >_q^{\alpha,\beta} = q^{(\alpha+1)(\beta+1)} \int_0^1 t^\alpha(1-qt)_q^\beta f(q^{\beta+1}t)g(q^{\beta+1}t)d_q t.$$

4.6.1   Basic Results

Proposition 4.1. *If f verifies the condition $C(\alpha)$, we have*

$$D_q L_{n,q}^{\alpha,\beta}(f;x) = \frac{[n]_q}{[n+\alpha+\beta+2]_q}q^{\alpha+\beta+2}L_{n-1,q}^{\alpha+1,\beta+1}D_q\left(f\left(\frac{\cdot}{q}\right);qx\right), x \in [0,1]$$

Proposition 4.2. *For any $m,n \in N, x \in [0,1]$ and $q \in [1/2,1]$ if*

$$T_{n,m,q}(x) = \sum_{k=0}^n p_{n,k}(q;x)\frac{\int_0^1 t^{k+\alpha}(1-qt)_q^{n-k+\beta}(x-t)^m d_q t}{\int_0^1 t^{k+\alpha}(1-qt)_q^{n-k+\beta}d_q t}.$$

Lemma 4.9. *For any $m,n \in N, x \in [0,1]$ and $q \in [1/2,1]$ if*

$$T_{n,m,q}^1(x) = \sum_{k=0}^n p_{n,k}(q;x)\frac{\int_0^1 t^{k+\alpha}(1-qt)_q^{n-k+\beta}(x-t)^m d_q t}{\int_0^1 t^{k+\alpha}(1-qt)_q^{n-k+\beta}d_q t}.$$

Then for $m \geq 2$, the following recurrence formula holds

$$[n+m+\alpha+\beta+2]_q q^{-\alpha-2m-1} T_{n,m+1,q}^1(x)$$

$$= (-x(1-x)D_q T_{n,m,q}^1(x) + T_{n,m,q}^1(x)(p_{1,m}(x) + x(1-q)[n+\alpha+\beta]_q[m+1]_q q^{1-\alpha-m})$$

$$= +T_{n,m-1,q}^1(x)p_{2,m}(x) + T_{n,m-2,q}^1(x)p_{3,m}(x)(1-q),$$

where the polynomials $p_{i,m}(x), i = 1,2,3$ are uniformly bounded with regard to n and q.

Lemma 4.10. *For any $m \in N, x \in [0,1]$ and $q \in [1/2,1]$, the expansion of $(x-t)^m$ on the Newton basis at the points $x/q^i, i = 0,1,2,\ldots.m-1$ is*

$$(x-t)^m = \sum_{k=1}^m d_{m,k}(1-q)^{m-k}(x-t)_q^k, \tag{4.36}$$

where the coefficient $d_{m,k}$ verify $|d_{m,k}| \leq d_m, k = 1,2,\ldots,m$ and d_m does not depend on x,t,q.

Remark 4.9. From Lemmas 4.9 and 4.10, we have for any m there exists a constant $K_m > 0$ independent of n and q, such that

$$\sup_{x \in [0,1]} |T_{n,m,q}(x)| \leq \begin{cases} \frac{K_m}{[n]_q^{m/2}}, & \text{if } m \text{ is even} \\ \frac{K_m}{[n]_q^{(m+1)/2}}, & \text{if } m \text{ is odd}. \end{cases}$$

Remark 4.10. The sequence (q_n) has the property S if and only if there exists $n \in N$ and $c > 0$ such that for any $n > N, 1 - q_n < c/n$.

4.6.2 Convergence

Theorem 4.17. *If f is continuous at the point $x \in (0,1)$, then*

$$\lim_{n \to \infty} L_{n,q_n}^{\alpha,\beta}(f;x) = f(x)$$

in the following cases:

1. *If f is bounded on $[0,1]$ and the sequence (q_n) is such that $\lim_{n \to \infty} q_n = 1$*
2. *If there exist real numbers $\alpha', \beta' \geq 0$ and a real $k' > 0$ such that, for any $x \in (0,1), |x^{\alpha'}(1-x)^{\beta'} f(x)| \leq k', \alpha' < \alpha+1, \beta' < \beta+1$ and the sequence (q_n) owns the property S*

Theorem 4.18. *If the function f admits a second derivative at the point $x \in [0,1]$, then as in cases 1 and 2 of Theorem 4.17, we have*

$$\lim_{n \to \infty} [n]_{q_n} [L_{n,q}^{\alpha,\beta}(f;x) - f(x)] = \frac{d}{dx} \frac{\left(x^{\alpha+1}(1-x)^{\beta+1} f'x\right)}{x^\alpha (1-x)^\beta} \qquad (4.37)$$

Proof. By Taylor's formula, we have

$$f(t) = f(x) + (t-x)f'(x) + \frac{(t-x)^2}{2!} f''(x) + (t-x)^2 \varepsilon(t-x),$$

where $\lim_{u \to 0} \varepsilon(u) = 0$. Thus

$$L_{n,q_n}^{\alpha,\beta}(f;x) - f(x) = -f'(x) T_{n,1,q_n}(x) + \frac{f''(x)}{2!} T_{n,2,q_n}(x) + R_n(x),$$

where $R_n(x) = L_{n,q_n}^{\alpha,\beta}((t-x)^2 \varepsilon(t-x);x)$. Using $\lim_{q \to 1} [a]_q = a$ for any $a \in R$. Using Lemmas 4.9 and 4.10, we have $\lim_{[n]_{q_n} \to \infty} [n]_{q_n} T_{n,1,q_n}(x) = (\alpha+\beta+2)x - \alpha - 1$ and $\lim_{[n]_{q_n} \to \infty} [n]_{q_n} T_{n,2,q_n}(x) = 2x(1-x)$. The result follows immediately if we show that $\lim_{[n]_{q_n} \to \infty} [n]_{q_n} R_n(x) = 0$. Proceeding along the same manner as in Theorem 4.17. For any $\eta > 0$ we can find a $\delta > 0$ such that for n large enough $\varepsilon(t-x) < \eta$ if $|x - q_n^{\beta+1} t| < \delta$.

We obtain the inequality $|(t-x)^2 \varepsilon(t-x)| \leq \eta(x-t)^2 + (\rho_x + |f(t)|) I_{x,\delta}(q^{-(\beta+1)}t)$ for any $t \in (0,1)$ where ρ_x is independent of t and δ. We deduce

$$[n]_{q_n} |R_n(x)| \leq \begin{cases} [n]_{q_n} \left(\eta T_{n,2,q_n}(x) + (\rho_x + k) T_{n,4,q_n}(x)/\delta^4 \right), & \text{in case 1} \\ [n]_{q_n} \left(\eta T_{n,2,q_n}(x) + \rho_x T_{n,4,q_n}(x)/\delta^4 \right) + k'nE_n(x,\delta), & \text{in case 1} \end{cases}$$

The right hand side tends to $2\eta x(1-x)$ when n (hence $[n]_{q_n}$) tends to infinity is as small as wanted.   ∎

Chapter 5
q-Summation–Integral Operators

5.1 q-Baskakov–Durrmeyer Operators

Aral and Gupta [32], proposed a q-analogue of the Baskakov operators and investigated its approximation properties. In continuation of their work they introduced Durrmeyer-type modification of q-Baskakov operators. These operators, opposed to Bernstein–Durrmeyer operators, are defined to approximate a function f on $[0, \infty)$. The Durrmeyer-type modification of the q-Bernstein operators was first introduced in [48]. Some results on the approximation of functions by the q-Bernstein–Durrmeyer operators were recently studied in [94]. In [62], some direct local and global approximation theorems were given for the q-Bernstein–Durrmeyer operators. We may also mention that some article related to Baskakov–Durrmeyer operators and different generalizations of them given in [61, 83, 153].

The main motivation of this section is to present a local approximation theorem and a rate of convergence of these new operators as well as their weighted approximation properties. The resulting approximation processes turn out to have an order of approximation at least as good as the classical Baskakov–Durrmeyer operators in certain subspace of continuous functions.

Recently, in [32], we introduced the following q-generalization of the classical Baskakov operators. For $f \in C[0, \infty)$, $q > 0$ and each positive integer n, the q-Baskakov operators are defined as

$$\mathcal{B}_{n,q}(f,x) = \sum_{k=0}^{\infty} \begin{bmatrix} n+k-1 \\ k \end{bmatrix}_q q^{\frac{k(k-1)}{2}} \frac{x^k}{(1+x)_q^{n+k}} f\left(\frac{[k]_q}{q^{k-1}[n]_q} \right)$$

$$= \sum_{k=0}^{\infty} b_{n,k}^q(x) f\left(\frac{[k]_q}{q^{k-1}[n]_q} \right). \tag{5.1}$$

A. Aral et al., *Applications of q-Calculus in Operator Theory*,
DOI 10.1007/978-1-4614-6946-9_5, © Springer Science+Business Media New York 2013

Lemma 5.1 ([32]). *For* $\mathcal{B}_{n,q}(t^m, x)$, $m = 0, 1, 2$, *one has the following:*

$$\mathcal{B}_{n,q}(1, x) = 1.$$

$$\mathcal{B}_{n,q}(t, x) = x,$$

$$\mathcal{B}_{n,q}(t^2, x) = x^2 + \frac{x}{[n]_q}\left(1 + \frac{1}{q}x\right).$$

5.1.1   Construction of Operators

For every $n \in \mathbb{N}$, $q \in (0, 1)$, the positive linear operator $\mathcal{D}_n^q$ is defined by

$$\mathcal{D}_n^q(f(t), x) := [n-1]_q \sum_{k=0}^{\infty} \mathcal{P}_{n,k}^q(x) \int_0^{\infty/A} \mathcal{P}_{n,k}^q(t) f(t) \, d_q t, \tag{5.2}$$

where

$$\mathcal{P}_{n,k}^q(x) := \begin{bmatrix} n+k-1 \\ k \end{bmatrix}_q q^{\frac{k^2}{2}} \frac{x^k}{(1+x)_q^{n+k}}$$

for $x \in [0, \infty)$ and for every real-valued continuous and bounded function f on $[0, \infty)$ (see [31]).

These operators satisfy linearity property. Also it can be observed that in case $q = 1$ the above operators reduce to the Baskakov–Durrmeyer operators discussed in [139] and [142]. Also see [144] for similar type of operators.

Lemma 5.2. *The following equalities hold:*

(i) $\mathcal{D}_n^q(1, x) = 1.$

(ii) $\mathcal{D}_n^q(t, x) = \left(1 + \frac{[2]_q}{q^2[n-2]_q}\right)x + \frac{1}{q[n-2]_q}$, *for* $n > 2$.

(iii) $\mathcal{D}_n^q(t^2, x) = \left(1 + \frac{[3]_q}{q^3[n-3]_q} + \frac{[2]_q}{q^2[n-2]_q} + \frac{q[2]_q[3]_q + [n]_q}{q^6[n-2]_q[n-3]_q}\right)x^2$.

$$+ \frac{[n]_q + q(1 + [2]_q)[n]_q}{q^5[n-2]_q[n-3]_q}x + \frac{[2]_q}{q^3[n-2]_q[n-3]_q}, \text{ for } n > 3.$$

Proof. The operators $\mathcal{D}_n^q$ are well defined on the function 1, t, t^2. Then for every $n > 3$ and $x \in [0, \infty)$, we obtain

$$\mathcal{D}_n^q(1, x) = [n-1]_q \sum_{k=1}^{\infty} \mathcal{P}_{n,k}^q(x) \int_0^{\infty/A} \mathcal{P}_{n,k}^q(t) \, d_q t$$

$$= [n-1]_q \sum_{k=0}^{\infty} \mathcal{P}_{n,k}^q(x) \begin{bmatrix} n+k-1 \\ k \end{bmatrix}_q q^{\frac{k^2}{2}} \int_0^{\infty/A} \frac{t^k}{(1+t)_q^{n+k}} \, d_q t.$$

Using (1.15) and (1.17), we can write

$$D_n^q(1,x) = [n-1]_q \sum_{k=0}^{\infty} \mathcal{P}_{n,k}^q(x) \begin{bmatrix} n+k-1 \\ k \end{bmatrix}_q q^{\frac{k^2}{2}} \frac{B_q(k+1,n-1)}{K(A,k+1)}$$

$$= [n-1]_q \sum_{k=0}^{\infty} \mathcal{P}_{n,k}^q(x) q^{\frac{k^2}{2}} \frac{[n+k-1]_q!}{[n-1]_q![k]_q!} \frac{[k]_q![n-2]_q!}{[n+k-1]_q! q^{\frac{k(k+1)}{2}}}$$

$$= [n-1]_q \sum_{k=0}^{\infty} \mathcal{P}_{n,k}^q(x) \frac{q^{-\frac{k}{2}}}{[n-1]_q}$$

$$= \sum_{k=0}^{\infty} \begin{bmatrix} n+k-1 \\ k \end{bmatrix}_q q^{\frac{k(k-1)}{2}} \frac{x^k}{(1+x)_q^{n+k}}$$

$$= \mathcal{B}_n^q(1,x) = 1,$$

where $\mathcal{B}_n^q(f,x)$ is the q-Baskakov operator defined by (5.1).

Similarly

$$D_n^q(t,x) = [n-1]_q \sum_{k=0}^{\infty} \mathcal{P}_{n,k}^q(x) \int_0^{\infty/A} \mathcal{P}_{n,k}^q(t) t\, d_q t$$

$$= [n-1]_q \sum_{k=0}^{\infty} \mathcal{P}_{n,k}^q(x) \begin{bmatrix} n+k-1 \\ k \end{bmatrix}_q q^{\frac{k^2}{2}} \int_0^{\infty/A} \frac{t^{k+1}}{(1+t)_q^{n+k}} d_q t$$

$$= [n-1]_q \sum_{k=0}^{\infty} \mathcal{P}_{n,k}^q(x) \begin{bmatrix} n+k-1 \\ k \end{bmatrix}_q q^{\frac{k^2}{2}} \frac{B_q(k+2,n-2)}{K(A,k+2)}$$

$$= \sum_{k=0}^{\infty} \begin{bmatrix} n+k-1 \\ k \end{bmatrix}_q q^{\frac{k^2-3k-2}{2}} \frac{[k+1]_q}{[n-2]_q} \frac{x^k}{(1+x)_q^{n+k}}.$$

Using the equality $[k+1]_q = [k]_q + q^k$

$$D_n^q(t,x) = \sum_{k=0}^{\infty} \begin{bmatrix} n+k-1 \\ k \end{bmatrix}_q q^{\frac{k(k-1)}{2}} q^{-k+1} q^{-2} \frac{[k]_q}{[n-2]_q} \frac{x^k}{(1+x)_q^{n+k}}$$

$$+ \sum_{k=0}^{\infty} \begin{bmatrix} n+k-1 \\ k \end{bmatrix}_q q^{\frac{k(k-1)}{2}} q^{-k+1} q^{-2} \frac{q^k}{[n-2]_q} \frac{x^k}{(1+x)_q^{n+k}}$$

$$= \frac{[n]_q}{q^2 [n-2]_q} \sum_{k=0}^{\infty} \begin{bmatrix} n+k-1 \\ k \end{bmatrix}_q q^{\frac{k(k-1)}{2}} q^{-k+1} \frac{[k]_q}{[n]_q} \frac{x^k}{(1+x)_q^{n+k}}$$

$$+ \frac{1}{q[n-2]_q} \sum_{k=0}^{\infty} \begin{bmatrix} n+k-1 \\ k \end{bmatrix}_q q^{\frac{k(k-1)}{2}} \frac{x^k}{(1+x)_q^{n+k}}$$

$$= \frac{[n]_q}{q^2 [n-2]_q} \mathcal{B}_n^q(t,x) + \frac{1}{q[n-2]_q} \mathcal{B}_n^q(1,x).$$

From Lemma 5.1, we can write

$$\mathcal{D}_n^q(t,x) = \frac{[n]_q}{q^2[n-2]_q}x + \frac{1}{q[n-2]_q}$$

$$= \left(1 + \frac{[2]_q}{q^2[n-2]_q}\right)x + \frac{1}{q[n-2]_q}.$$

Finally

$$\mathcal{D}_n^q(t^2,x) = [n-1]_q\sum_{k=0}^{\infty}\mathcal{P}_{n,k}^q(x)\int_0^{\infty/A}\mathcal{P}_{n,k}^q(t)t^2 d_qt$$

$$= [n-1]_q\sum_{k=0}^{\infty}\mathcal{P}_{n,k}^q(x)\begin{bmatrix}n+k-1\\k\end{bmatrix}_q q^{\frac{k^2}{2}}\int_0^{\infty/A}\frac{t^{k+2}}{(1+t)_q^{n+k}}d_qt$$

$$= [n-1]_q\sum_{k=0}^{\infty}\mathcal{P}_{n,k}^q(x)\begin{bmatrix}n+k-1\\k\end{bmatrix}_q q^{\frac{k^2}{2}}\frac{B_q(k+3,n-3)}{K(A,k+3)}$$

$$= \sum_{k=0}^{\infty}\mathcal{P}_{n,k}^q(x)q^{\frac{-5k}{2}-3}\frac{[k+2]_q[k+1]_q}{[n-2]_q[n-3]_q}.$$

Using $[k+2]_q = [k]_q + q^k[2]$ and $[k+1]_q = [k]_q + q^k$, we have

$$\mathcal{D}_n^q(t^2,x) = \sum_{k=0}^{\infty}\begin{bmatrix}n+k-1\\k\end{bmatrix}_q\frac{x^k}{(1+x)_q^{n+k}}q^{\frac{k^2-5k}{2}-3}\frac{\left([k]_q+q^k[2]_q\right)\left([k]_q+q^k\right)}{[n-2]_q[n-3]_q}$$

$$= \sum_{k=0}^{\infty}\begin{bmatrix}n+k-1\\k\end{bmatrix}_q\frac{x^k}{(1+x)_q^{n+k}}q^{\frac{k^2-5k}{2}-3}\frac{[k]_q^2}{[n-2]_q[n-3]_q}$$

$$+ \sum_{k=0}^{\infty}\begin{bmatrix}n+k-1\\k\end{bmatrix}_q\frac{x^k}{(1+x)_q^{n+k}}q^{\frac{k^2-3k}{2}-3}\frac{\left(1+[2]_q\right)[k]_q}{[n-2]_q[n-3]_q}$$

$$+ \sum_{k=0}^{\infty}\begin{bmatrix}n+k-1\\k\end{bmatrix}_q\frac{x^k}{(1+x)_q^{n+k}}q^{\frac{k^2-k}{2}-3}\frac{[2]_q}{[n-2]_q[n-3]_q}.$$

Again using (5.1) and Lemma 5.1, we have

$$\mathcal{D}_n^q(t^2,x) = \frac{q^{-5}[n]_q^2}{[n-2]_q[n-3]_q}\mathcal{B}_n^q(t^2,x) + \frac{q^{-4}\left(1+[2]_q\right)[n]_q}{[n-2]_q[n-3]_q}\mathcal{B}_n^q(t,x)$$

$$+ \frac{q^{-3}[2]}{[n-2]_q[n-3]_q} B_n^q(1,x)$$

$$= \frac{q[n]_q^2 + [n]_q}{q^6[n-2]_q[n-3]_q} x^2 + \frac{[n]_q + q(1+[2]_q)[n]_q}{q^5[n-2]_q[n-3]_q} x + \frac{[2]_q}{q^3[n-2]_q[n-3]_q}.$$

Since $[n]_q = [3]_q + q^3[n-3]_q$ and $[n]_q = [2]_q + q^2[n-2]_q$, we have the desired result. ∎

Remark 5.1. If we put $q = 1$, we get the moments of Baskakov–Durrmeyer operators as

$$D_n^1(t-x,x) = \frac{1+2x}{n-2}, n > 2$$

$$D_n^1(t,x) = \frac{1+nx}{n-2}, \; n > 2$$

and

$$D_n^1((t-x)^2,x) = \frac{2[(n+3)x^2 + (n+3)x + 1]}{(n-2)(n-3)}, \; n > 3$$

$$D_n^1(t^2,x) = \frac{(n^2+n)x^2 + 4nx + 2}{(n-2)(n-3)}, \; n > 3$$

Lemma 5.3. *Let $n > 3$ be a given number. For every $q \in (0,1)$ we have*

$$\mathcal{D}_n^q\left((t-x)^2,x\right) \leq \frac{15}{q^6[n-2]_q}\left(\varphi^2(x) + \frac{1}{[n-3]_q}\right),$$

where $\varphi^2(x) = x(1+x), x \in [0,\infty)$.

Proof. By Lemma 5.2, we have

$$\mathcal{D}_n^q\left((t-x)^2,x\right) = \left(\frac{[3]_q}{q^3[n-3]_q} - \frac{[2]_q}{q^2[n-2]_q} + \frac{q[2]_q[3]_q + [n]_q}{q^6[n-2]_q[n-3]_q}\right)x^2$$

$$+ \left(\frac{[n]_q + q(1+[2]_q)[n]_q}{q^5[n-2]_q[n-3]_q} - \frac{2}{q[n-2]_q}\right)x + \frac{[2]_q}{q^3[n-2]_q[n-3]_q}$$

$$= \left(\frac{q^3[3]_q[n-2]_q - q^4[n-3]_q[2]_q + q[2]_q[3]_q + [n]_q}{q^6[n-2]_q[n-3]_q}\right)x^2$$

$$+ \left(\frac{q[n]_q + q^2(1+[2]_q)[n]_q - 2q^5[n-3]_q}{q^6[n-2]_q[n-3]_q}\right)x + \frac{[2]_q}{q^3[n-2]_q[n-3]_q}$$

$$= x(1+x) \left(\frac{q^3 [3]_q [n-2]_q - q^4 [n-3]_q [2]_q + q [2]_q [3]_q + [n]_q}{q^6 [n-2]_q [n-3]_q} \right)$$

$$+ x \left(\frac{q[n]_q + q^2(1+[2]_q)[n]_q - 2q^5[n-3]_q - q^3 [3]_q [n-2]_q}{q^6 [n-2]_q [n-3]_q} \right.$$

$$\left. + \frac{q^4 [n-3]_q [2]_q - q [2]_q [3]_q - [n]_q}{q^6 [n-2]_q [n-3]_q} \right) + \frac{[2]_q}{q^3 [n-2]_q [n-3]_q}.$$

By direct computation, for $n > 3$, we have
$$q^3 [3]_q [n-2] - q^4 [n-3]_q [2]_q + q [2]_q [3]_q + [n]_q$$

$$= \left(q^3 + q^4 + q^5 \right) \left([n-3]_q + q^{n-3} \right) - \left(q^4 + q^5 \right) [n-3]_q + (q+q^2)(1+q+q^2)$$

$$+ \left([n-3]_q + q^{n-3} + q^{n-2} + q^{n-1} \right)$$

$$= [n-3]_q \left(q^3 + q^4 + q^5 - \left(q^4 + q^5 \right) + 1 \right) + q^n + q^{n+1} + q^{n+2}$$

$$+ (q+q^2)(1+q+q^2) + q^{n-3} + q^{n-2} + q^{n-1}$$

$$= [n-3]_q \left(q^3 + 1 \right) + q^n + q^{n+1} + q^{n+2} + (q+q^2)(1+q+q^2) + q^{n-3} + q^{n-2} + q^{n-1} > 0$$

for every $q \in (0,1)$. Furthermore
$$q[n]_q + q^2(1+[2]_q)[n]_q - 2q^5[n-3]_q$$

$$= q(1+2q+q^2)(1+q+\ldots+q^{n-1}) - 2q^5(1+q+\ldots+q^{n-4})$$

$$= q(1+q^2)(1+q+\ldots+q^{n-1}) + 2\left[(q^2+q^3+\ldots+q^{n+1}) - \left(q^5+q^6+\ldots+q^{n+1} \right) \right]$$

$$= (q+q^3)(1+q+\ldots+q^{n-1}) + 2[q^2 + q^3 + q^4]$$

and
$$[q+q^2(1+[2]_q)] [n]_q - 2q^5[n-3]_q - \left[q^3 [3]_q [n-2]_q - q^4 [n-3]_q [2]_q + q [2]_q [3]_q + [n]_q \right]$$

$$= (q+q^3)[n]_q + 2[q^2+q^3+q^4] - (q+q^2+q^3)\left([n]_q - (1+q) \right)$$

$$+ (q+q^2)\left([n]_q - (1+q+q^2) \right) - q(1+q)(1+q+q^2) - [n]_q$$

$$= (q-1)[n]_q + q^4 - q < 0$$

for every $q \in (0,1)$.

Thus we have

$$\mathcal{D}_n^q\left((t-x)^2,x\right) \le x(1+x)\left(\frac{q^3\,[3]_q\,[n-2]_q - q^4\,[n-3]_q\,[2]_q + q\,[2]_q\,[3]_q + [n]_q}{q^6\,[n-2]_q\,[n-3]_q}\right)$$

$$+\frac{[2]_q}{q^3\,[n-2]_q\,[n-3]_q}$$

$$\le \frac{15\,[n-3]_q}{q^6\,[n-2]_q\,[n-3]_q}\,\varphi^2(x) + \frac{2}{q^3\,[n-2]_q\,[n-3]_q}$$

$$\le \frac{15}{q^6\,[n-2]_q}\left(\varphi^2(x) + \frac{1}{[n-3]_q}\right)$$

for every $q \in (0,1)$ and $x \in [0,\infty)$. Thus the result holds. ∎

5.1.2 Local Approximation

In this section we establish direct and local approximation theorems in connection with the operators $\mathcal{D}_n^q$. Let $C_B[0,\infty)$ be the space of all real-valued continuous and bounded functions f on $[0,\infty)$ endowed with the norm $\|f\| = \sup\{|f(x)| : x \in [0,\infty)\}$. Further let us consider the following K-functional:

$$K_2(f,\delta) = \inf_{g\in W^2}\left\{\,\|f-g\| + \delta\,\|g''\|\,\right\},$$

where $\delta > 0$ and $W^2 = \{g \in C_B[0,\infty) : g',g'' \in C_B[0,\infty)\}$. By [50, p. 177, Theorem 2.4] there exists an absolute constant $C > 0$ such that

$$K_2(f,\delta) \le C\,\omega_2\left(f,\sqrt{\delta}\right), \tag{5.3}$$

where

$$\omega_2\left(f,\sqrt{\delta}\right) = \sup_{0<h\le\sqrt{\delta}}\ \sup_{x\in[0,\infty)} |f(x+2h) - 2f(x+h) + f(x)|$$

is the second-order modulus of smoothness of $f \in C_B[0,\infty)$. By

$$\omega(f,\delta) = \sup_{0<h\le\delta}\ \sup_{x\in[0,\infty)} |f(x+h) - f(x)|$$

we denote the usual modulus of continuity of $f \in C_B[0,\infty)$. In what follows we shall use the notations $\varphi(x) = \sqrt{x(1+x)}$ and $\delta_n^2(x) = \varphi^2(x) + \frac{1}{[n-3]_q}$, where $x \in [0,\infty)$ and $n \ge 4$.

Our first result is a direct local approximation theorem for the operators $\mathcal{D}_n^q$.

Theorem 5.1. *Let $q \in (0,1)$ and $n \geq 4$. We have*

$$|\mathcal{D}_n^q(f,x) - f(x)| \leq C\omega_2\left(f, \frac{\delta_n(x)}{\sqrt{q^6[n-2]_q}}\right) + \omega\left(f, \frac{q^{-2}[2]_q x + q^{-1}}{[n-2]_q}\right),$$

for every $x \in [0,\infty)$ and $f \in C_B[0,\infty)$, where C is a positive constant.

Proof. Let us introduce the auxiliary operators $\overline{\mathcal{D}}_n^q$ defined by

$$\overline{\mathcal{D}}_n^q(f,x) = \mathcal{D}_n^q(f,x) - f\left(x + \frac{q^{-2}[2]_q x + q^{-1}}{[n-2]_q}\right) + f(x), \qquad (5.4)$$

$x \in [0,\infty)$. The operators $\overline{\mathcal{D}}_n^q$ are linear and preserve the linear functions:

$$\overline{\mathcal{D}}_n^q(t - x, x) = 0 \qquad (5.5)$$

(see Lemma 5.2).

Let $g \in W^2$. From Taylor's expansion

$$g(t) = g(x) + g'(x)(t - x) + \int_x^t (t - u)\, g''(u)\, du, \qquad t \in [0,\infty)$$

and (5.5), we get

$$\overline{\mathcal{D}}_n^q(g,x) = g(x) + \overline{\mathcal{D}}_n^q\left(\int_x^t (t - u)\, g''(u)\, du, x\right).$$

Hence, by (5.4) one has
$|\overline{\mathcal{D}}_n^q(g,x) - g(x)| \leq$

$$\leq \left|\mathcal{D}_n^q\left(\int_x^t (t-u)\, g''(u)\, du, x\right)\right| + \left|\int_x^{x + \frac{q^{-2}[2]_q x + q^{-1}}{[n-2]_q}} \left(x + \frac{q^{-2}[2]_q x + q^{-1}}{[n-2]_q} - u\right) g''(u)\, du\right|$$

$$\leq \mathcal{D}_n^q\left(\left|\int_x^t |t-u|\,|g''(u)|\, du\right|, x\right) + \int_x^{x + \frac{q^{-2}[2]_q x + q^{-1}}{[n-2]_q}} \left|x + \frac{q^{-2}[2]_q x + q^{-1}}{[n-2]_q} - u\right| |g''(u)| du$$

$$\leq \left[\mathcal{D}_n^q\left((t-x)^2, x\right) + \left(\frac{q^{-2}[2]_q x + q^{-1}}{[n-2]_q}\right)^2\right] \|g''\|. \qquad (5.6)$$

Using Lemma 5.3 and $n \geq 4$, we obtain
$$\mathcal{D}_n^q\left((t-x)^2, x\right) + \left(\frac{q^{-2}[2]_q x + q^{-1}}{[n-2]_q}\right)^2 \leq$$

$$\leq \frac{15}{q^6[n-2]_q}\left(\varphi^2(x) + \frac{1}{[n-3]_q}\right) + \left(\frac{q^{-2}[2]_q x + q^{-1}}{[n-2]_q}\right)^2.$$

Since

$$\left(\frac{q^{-2}[2]_q x + q^{-1}}{[n-2]_q}\right)^2 \cdot \delta_n^{-2}(x)$$

$$= \frac{(1+q)^2 x^2 + 2q(1+q)x + q^2}{q^4[n-2]_q^2} \cdot \frac{[n-3]_q}{[n-3]_q x(x+1)+1}$$

$$\leq \frac{1}{q^4[n-2]_q} \cdot \frac{[n-3]_q}{[n-2]_q} \cdot \frac{4x^2 + 4x + 1}{[n-3]_q x(x+1)+1},$$

we have

$$D_n^q\left((t-x)^2, x\right) + \left(\frac{q^{-2}[2]_q x + q^{-1}}{[n-2]_q}\right)^2 \leq \frac{15}{q^6[n-2]_q}\delta_n^2(x).$$

Then, by (5.6), we get

$$|\overline{\mathcal{D}}_n^q(g,x) - g(x)| \leq \frac{15}{q^6[n-2]_q}\delta_n^2(x)\|g''\|. \tag{5.7}$$

On the other hand, by (5.4) and (5.2) and Lemma 5.2, we have

$$|\overline{\mathcal{D}}_n^q(f,x)| \leq |\mathcal{D}_n^q(f,x)| + 2\|f\| \leq \|f\|\mathcal{D}_n^q(1,x) + 2\|f\| \leq 3\|f\|. \tag{5.8}$$

Now (5.4), (5.7), and (5.8) imply

$$|\mathcal{D}_n^q(f,x) - f(x)| \leq |\overline{\mathcal{D}}_n^q(f-g,x) - (f-g)(x)|$$

$$+ |\overline{\mathcal{D}}_n^q(g,x) - g(x)| + \left| f\left(x + \frac{q^{-2}[2]_q x + q^{-1}}{[n-2]_q}\right) - f(x) \right|$$

$$\leq 4\|f-g\| + \frac{15}{q^6[n-2]_q}\delta_n^2(x)\|g''\|$$

$$+ \left| f\left(x + \frac{q^{-2}[2]_q x + q^{-1}}{[n-2]_q}\right) - f(x) \right|.$$

Hence taking infimum on the right-hand side over all $g \in W^2$, we get
$$|\mathcal{D}_n^q(f,x) - f(x)| \leq$$

$$\leq 15K_2\left(f, \frac{1}{q^6[n-2]_q}\delta_n^2(x)\right) + \omega\left(f, \frac{q^{-2}[2]_q x + q^{-1}}{[n-2]_q}\right).$$

In view of (5.3), for every $q \in (0,1)$ we get

$$|\mathcal{D}_n^q(f,x) - f(x)| \le C\omega_2\left(f, \frac{\delta_n(x)}{\sqrt{q^6[n-2]_q}}\right) + \omega\left(f, \frac{q^{-2}[2]_q x + q^{-1}}{[n-2]_q}\right).$$

This completes the proof of the theorem.

5.1.3   Rate of Convergence

Let $B_{x^2}[0, \infty)$ be the set of all functions f defined on $[0, \infty)$ satisfying the condition $|f(x)| \le M_f(1+x^2)$, where M_f is a constant depending only on f. By $C_{x^2}[0, \infty)$, we denote the subspace of all continuous functions belonging to $B_{x^2}[0, \infty)$. Also, let $C^*_{x^2}[0, \infty)$ be the subspace of all functions $f \in C_{x^2}[0, \infty)$, for which $\lim_{x\to\infty} \frac{f(x)}{1+x^2}$ is finite. The norm on $C^*_{x^2}[0, \infty)$ is $\|f\|_{x^2} = \sup_{x\in[0, \infty)} \frac{|f(x)|}{1+x^2}$. For any positive a, by

$$\omega_a(f, \delta) = \sup_{|t-x|\le\delta} \sup_{x,t\in[0, a]} |f(t) - f(x)|$$

we denote the usual modulus of continuity of f on the closed interval $[0, a]$. We know that for a function $f \in C_{x^2}[0, \infty)$, the modulus of continuity $\omega_a(f, \delta)$ tends to zero.

Now we give a rate of convergence theorem for the operator $\mathcal{D}_n^q$.

Theorem 5.2. *Let* $f \in C_{x^2}[0, \infty)$, $q = q_n \in (0, 1)$ *such that* $q_n \to 1$ *as* $n \to \infty$ *and* $\omega_{a+1}(f, \delta)$ *be its modulus of continuity on the finite interval* $[0, a+1] \subset [0, \infty)$, *where* $a > 0$. *Then for every* $n > 3$,

$$\|\mathcal{D}_n^q(f) - f\|_{C[0, a]} \le \frac{K}{q^6[n-3]_q} + 2\omega_{a+1}\left(f, \sqrt{\frac{K}{q^6[n-3]_q}}\right),$$

where $K = 90M_f(1+a^2)(1+a+a^2)$.

Proof. For $x \in [0, a]$ and $t > a+1$, since $t - x > 1$, we have

$$|f(t) - f(x)| \le M_f(2 + x^2 + t^2)$$
$$\le M_f(2 + 3x^2 + 2(t-x)^2)$$
$$\le 6M_f(1+a^2)(t-x)^2. \tag{5.9}$$

For $x \in [0, a]$ and $t \le a+1$, we have

$$|f(t) - f(x)| \le \omega_{a+1}(f, |t-x|) \le \left(1 + \frac{|t-x|}{\delta}\right)\omega_{a+1}(f, \delta) \tag{5.10}$$

with $\delta > 0$.

From (5.9) and (5.10) we can write

$$|f(t) - f(x)| \leq 6M_f \left(1 + a^2\right)(t-x)^2 + \left(1 + \frac{|t-x|}{\delta}\right)\omega_{a+1}(f, \delta) \qquad (5.11)$$

for $x \in [0,\ a]$ and $t \geq 0$. Thus

$$|\mathcal{D}_n^q(f,x) - f(x)| \leq \mathcal{D}_n^q(|f(t) - f(x)|, x)$$

$$\leq 6M_f \left(1 + a^2\right)\mathcal{D}_n^q\left((t-x)^2, x\right)$$

$$+ \omega_{a+1}(f,\ \delta)\left(1 + \frac{1}{\delta}\mathcal{D}_n^q\left((t-x)^2, x\right)^{\frac{1}{2}}\right).$$

Hence, by Schwarz's inequality and Lemma 5.3, for every $q \in (0,1)$ and $x \in [0,\ a]$

$$|\mathcal{D}_n^q(f,x) - f(x)| \leq \frac{90M_f\left(1 + a^2\right)}{q^6[n-2]_q}\left(\varphi^2(x) + \frac{1}{[n-3]_q}\right)$$

$$+ \omega_{a+1}(f,\ \delta)\left(1 + \frac{1}{\delta}\sqrt{\frac{15}{q^6[n-2]_q}\left(\varphi^2(x) + \frac{1}{[n-3]_q}\right)}\right)$$

$$\leq \frac{K}{q^6[n-3]_q} + \omega_{a+1}(f,\ \delta)\left(1 + \frac{1}{\delta}\sqrt{\frac{K}{q^6[n-3]_q}}\right).$$

By taking $\delta = \sqrt{\frac{K}{q^6[n-3]_q}}$, we get the assertion of our theorem. ■

Corollary 5.1. *If $f \in Lip_M\alpha$ on $[0,\ a+1]$, then for $n > 3$*

$$\|\mathcal{D}_n^q(f) - f\|_{C[0,\ a]} \leq (1 + 2M)\sqrt{\frac{K}{q^6[n-3]_q}}.$$

Proof. For a sufficiently large n,

$$\frac{K}{q^6[n-3]_q} \leq \sqrt{\frac{K}{q^6[n-3]_q}},$$

because of $\lim_{n\to\infty}[n-3]_q = \infty$. Hence, by $f \in Lip_M\alpha$, we obtain the assertion of the corollary. ■

5.1.4 Weighted Approximation

Now we shall discuss the weighted approximation theorem, where the approximation formula holds true on the interval $[0,\ \infty)$.

Theorem 5.3. *Let $q = q_n$ satisfies $0 < q_n < 1$ and let $q_n \to 1$ as $n \to \infty$. For each $f \in C_{x^2}^* [0, \infty)$, we have*

$$\lim_{n\to\infty} \left\| \mathcal{D}_n^{q_n} (f) - f \right\|_{x^2} = 0.$$

Proof. Using the theorem in [65] we see that it is sufficient to verify the following three conditions

$$\lim_{n\to\infty} \left\| \mathcal{D}_n^{q_n} (t^v, x) - x^v \right\|_{x^2} = 0, \quad v = 0, 1, 2. \tag{5.12}$$

Since $D_n^{q_n}(1, x) = 1$, the first condition of (5.12) is fulfilled for $v = 0$.

By Lemma 5.2 we have for $n > 2$

$$
\begin{aligned}
\left\| \mathcal{D}_n^{q_n} (t, x) - x \right\|_{x^2} &= \sup_{x \in [0, \infty)} \frac{\left| D_n^{q_n}(t, x) - x \right|}{1 + x^2} \\
&\leq \frac{[2]_{q_n}}{q_n^2 [n-2]_{q_n}} \sup_{x \in [0, \infty)} \frac{x}{1 + x^2} + \frac{1}{q_n [n-2]_{q_n}} \\
&\leq \frac{[2]_{q_n}}{q_n^2 [n-2]_{q_n}} + \frac{1}{q_n [n-2]_{q_n}},
\end{aligned}
$$

and the second condition of (5.12) holds for $v = 1$ as $n \to \infty$.

Similarly we can write for $n > 3$

$$
\begin{aligned}
\left\| \mathcal{D}_n^{q_n} (t^2, x) - x^2 \right\|_{x^2} &\leq \left(\frac{[3]_{q_n}}{q_n^3 [n-3]_{q_n}} + \frac{[2]_{q_n}}{q_n^2 [n-2]_{q_n}} + \frac{q_n [2]_{q_n} [3]_{q_n} + [n]_{q_n}}{q_n^6 [n-2]_{q_n} [n-3]_{q_n}} \right) \sup_{x \in [0, \infty)} \frac{x^2}{1 + x^2} \\
&\quad + \left(\frac{[n]_{q_n} + q_n \left(1 + [2]_{q_n} \right) [n]_{q_n}}{q_n^5 [n-2]_{q_n} [n-3]_{q_n}} \right) \sup_{x \in [0, \infty)} \frac{x}{1 + x^2} + \frac{[2]}{q_n^3 [n-2]_{q_n} [n-3]_{q_n}} \\
&\leq \frac{[3]_{q_n}}{q_n^3 [n-3]_{q_n}} + \frac{[2]_{q_n}}{q_n^2 [n-2]_{q_n}} + \frac{q_n [2]_{q_n} [3]_{q_n} + [n]_{q_n}}{q_n^6 [n-2]_{q_n} [n-3]_{q_n}} \\
&\quad + \frac{[n]_{q_n} + q_n \left(1 + [2]_{q_n} \right) [n]_{q_n}}{q_n^5 [n-2]_{q_n} [n-3]_{q_n}} + \frac{[2]_{q_n}}{q_n^3 [n-2]_{q_n} [n-3]_{q_n}},
\end{aligned}
$$

which implies that

$$\lim_{n\to\infty} \left\| \mathcal{D}_n^{q_n} (t^2, x) - x^2 \right\|_{x^2} = 0.$$

Thus the proof is completed.   ∎

We give the following theorem to approximate all functions in $C_{x^2} [0, \infty)$. This type of results are given in [71] for locally integrable functions.

Theorem 5.4. *Let* $q = q_n$ *satisfies* $0 < q_n < 1$ *and let* $q_n \to 1$ *as* $n \to \infty$. *For each* $f \in C_{x^2}[0, \infty)$ *and* $\alpha > 0$, *we have*

$$\lim_{n \to \infty} \sup_{x \in [0, \infty)} \frac{\left| \mathcal{D}_n^{q_n}(f,x) - f(x) \right|}{(1+x^2)^{1+\alpha}} = 0.$$

Proof. For any fixed $x_0 > 0$,

$$\sup_{x \in [0, \infty)} \frac{\left| \mathcal{D}_n^{q_n}(f,x) - f(x) \right|}{(1+x^2)^{1+\alpha}} \leq \sup_{x \leq x_0} \frac{\left| \mathcal{D}_n^{q_n}(f,x) - f(x) \right|}{(1+x^2)^{1+\alpha}} + \sup_{x \geq x_0} \frac{\left| \mathcal{D}_n^{q_n}(f,x) - f(x) \right|}{(1+x^2)^{1+\alpha}}$$

$$\leq \left\| \mathcal{D}_n^q(f) - f \right\|_{C[0, x_0]} + \|f\|_{x^2} \sup_{x \geq x_0} \frac{\left| \mathcal{D}_n^{q_n}(1+t^2,x) \right|}{(1+x^2)^{1+\alpha}}$$

$$+ \sup_{x \geq x_0} \frac{|f(x)|}{(1+x^2)^{1+\alpha}}.$$

The first term of the above inequality tends to zero from Theorem 5.2. By Lemma 5.2 for any fixed $x_0 > 0$, it is easily seen that $\sup\limits_{x \geq x_0} \dfrac{\left| \mathcal{D}_n^{q_n}(1+t^2,x) \right|}{(1+x^2)^{1+\alpha}}$ tends to zero as $n \to \infty$. We can choose $x_0 > 0$ so large that the last part of above inequality can be made small enough.

Thus the proof is completed.   ■

5.1.5   Recurrence Relation and Asymptotic Formula

The q-Baskakov–Durrmeyer operators $\mathcal{D}_n^q(f,x)$ can be defined in alternate form as

$$\mathcal{D}_n^q(f(t),x) := [n-1]_q \sum_{k=0}^{\infty} p_{n,k}^q(x) \int_0^{\infty/A} q^k p_{n,k}^q(t) f(t) \, d_q t, \qquad (5.13)$$

where

$$p_{n,k}^q(x) := \begin{bmatrix} n+k-1 \\ k \end{bmatrix}_q q^{\frac{k^2-k}{2}} \frac{x^k}{(1+x)_q^{n+k}}$$

for $x \in [0, \infty)$ and for every real-valued continuous and bounded function f on $[0, \infty)$ (see [88]). Also

$$(1+x)_q^n = (-x;q)_n := \begin{cases} (1+x)(1+qx) \dots (1+q^{n-1}x), & n = 1, 2, \dots \\ 1, & n = 0. \end{cases}$$

Lemma 5.4. *If we define the central moments as*

$$T_{n,m}(x) = \mathcal{D}_n^q \left((t-x)_q^m, x \right) = [n-1]_q \sum_{k=0}^{\infty} p_{n,k}^q(x) \int_0^{\infty/A} q^k p_{n,k}^q(t)(t-x)_q^m d_q t,$$

then

$$T_{n,0}(x) = 1, T_{n,1}(x) = \frac{[2]_q}{q^2[n-2]}x + \frac{1}{q[n-2]},$$

and for $n > m+2$, we have the following recurrence relation:

$$\left([n]_q - [m+2]_q \right) T_{n,m+1}(qx) = qx(1+x)\left[D_q T_{n,m}(x) + [m]_q T_{n,m-1}(qx) \right]$$

$$+ \left([3]_q q^m x + q - x \right) [m+1]_q T_{n,m}(qx)$$

$$+ \left[[2]_q q^m x \left([3]_q q^m x + q - x \right) - [3]_q q^{2m+1} x^2 - qx \right] [m]_q T_{n,m-1}(qx)$$

$$+ \left[q^m x \left\{ [2]_q q^m x \left([3]_q q^m x + q - x \right) - [3]_q q^{2m+1} x^2 - qx \right\} \right.$$

$$\left. + q^{2m+1} x^2 \left\{ q^2 x - [3]_q q^m x - q + x \right\} \right] [m-1]_q T_{n,m-2}(qx)$$

$$+ x(1 - q^{m+1})[n]_q T_{n,m}(qx) + qx(1 - q^{m-1})[n]_q T_{n,m}(qx)$$

$$- qx^2(1 - q^{m-1})(1 - q^m)[n]_q T_{n,m-1}(qx),$$

and we consider $T_{n,-ve}(x) = 0$.

Proof. Using the identity

$$qx(1+x)D_q[p_{n,k}^q(x)] = \left(\frac{[k]_q}{q^{k-1}[n]_q} - qx \right) [n]_q p_{n,k}^q(qx)$$

and q-derivatives of product rule, we have

$$qx(1+x)D_q[T_{n,m}(x)] = [n-1]_q \sum_{k=0}^{\infty} qx(1+x)D_q[p_{n,k}^q(x)] \int_0^{\infty/A} q^k p_{n,k}^q(t)(t-x)_q^m d_q t$$

$$- [m]_q[n-1]_q \sum_{k=0}^{\infty} qx(1+x)p_{n,k}^q(qx) \int_0^{\infty/A} q^k p_{n,k}^q(t)(t-qx)_q^{m-1} d_q t.$$

Thus

$$E := qx(1+x)\left[D_q[T_{n,m}(x)] + [m]_q T_{n,m-1}(qx)\right]$$

$$= [n]_q[n-1]_q \sum_{k=0}^{\infty} p_{n,k}^q(qx)\left(\frac{[k]_q}{q^{k-1}[n]_q} - qx\right)\int_0^{\infty/A} q^k p_{n,k}^q(t)(t-x)_q^m d_q t$$

$$= [n]_q[n-1]_q \sum_{k=0}^{\infty} p_{n,k}^q(qx)\int_0^{\infty/A} q^k\left(\frac{[k]_q}{q^{k-1}[n]_q} - t + t - q^m x - qx + q^m x\right) p_{n,k}^q(t)(t-x)_q^m d_q t$$

$$= [n-1]_q \sum_{k=0}^{\infty} p_{n,k}^q(qx)\int_0^{\infty/A} q^k q^2\left[\frac{t}{q}\left(1+\frac{t}{q}\right)\right] D_q\left[p_{n,k}^q\left(\frac{t}{q}\right)\right] (t-x)_q^m d_q t$$

$$+ [n]_q[n-1]_q \sum_{k=0}^{\infty} p_{n,k}^q(qx)\int_0^{\infty/A} q^k p_{n,k}^q(t)(t-x)_q^{m+1} d_q t$$

$$+ [n]_q[n-1]_q qx(q^{m-1}-1)\sum_{k=0}^{\infty} p_{n,k}^q(qx)\int_0^{\infty/A} q^k p_{n,k}^q(t)(t-x)_q^m d_q t$$

$$= [n-1]_q \sum_{k=0}^{\infty} p_{n,k}^q(qx)\int_0^{\infty/A} q^k(tq+t^2)D_q\left[p_{n,k}^q\left(\frac{t}{q}\right)\right] (t-x)_q^m d_q t$$

$$+ [n]_q[n-1]_q \sum_{k=0}^{\infty} p_{n,k}^q(qx)\int_0^{\infty/A} q^k p_{n,k}^q(t)(t-x)_q^{m+1} d_q t$$

$$+ [n]_q[n-1]_q qx(q^{m-1}-1)\sum_{k=0}^{\infty} p_{n,k}^q(qx)\int_0^{\infty/A} q^k p_{n,k}^q(t)(t-x)_q^m d_q t.$$

Using the identities

$$(t-q^m x)(t-q^{m+1}x) = t^2 - [2]_q q^m xt + q^{2m+1}x^2$$

and

$$(t-q^m x)(t-q^{m+1}x)(t-q^{m+2}x) = t^3 - [3]_q q^m xt^2 + [3]_q q^{2m+1}x^2 t - q^{3m+3}x^3,$$

we obtain the following identity after simple computation:

$$(qt+t^2)(t-x)_q^m = (qt+t^2)(t-x)(t-qx)_q^{m-1} = \left[t^3 + (q-x)t^2 - qxt\right](t-qx)_q^{m-1}$$

$$= (t-qx)_q^{m+2} + \left([3]_q q^m x + q - x\right)(t-qx)_q^{m+1}$$

$$+ \left[[2]_q q^m x\left\{[3]_q q^m x + q - x\right\} - [3]_q q^{2m+1} x^2 - qx)\right](t-qx)_q^m$$

$$+ \left[q^m x\left\{[2]_q q^m x\left([3]_q q^m x + q - x\right) - [3]_q q^{2m+1} x^2 - qx\right\}\right.$$

$$\left. + q^{2m+1} x^2\left\{q^2 x - [3]_q q^m x - q + x\right\}\right](t-qx)_q^{m-1}.$$

Using the above identity and q-integral by parts

$$\int_a^b u(t)D_q(v(t))d_qt = [u(t)v(t)]_a^b - \int_a^b v(qt)D_q[u(t)]d_qt,$$

we have

$$E = -[m+2]_q T_{n,m+1}(qx) - \left([3]_q q^m x + q - x\right)[m+1]_q T_{n,m}(qx)$$

$$- \left[[2]_q q^m x\left([3]_q q^m x + q - x\right) - [3]_q q^{2m+1} x^2 - qx\right][m]_q T_{n,m-1}(qx)$$

$$- \left[q^m x\left\{[2]_q q^m x\left([3]_q q^m x + q - x\right) - [3]_q q^{2m+1} x^2 - qx\right\}\right.$$

$$\left. + q^{2m+1} x^2\left\{q^2 x - [3]_q q^m x - q + x\right\}\right][m-1]_q T_{n,m-2}(qx)$$

$$+ [n]_q[n-1]_q \sum_{k=0}^{\infty} p_{n,k}^q(qx) \int_0^{\infty/A} q^k p_{n,k}^q(t)(t-x)_q^{m+1} d_qt$$

$$- [n]_q[n-1]_q qx(1-q^{m-1}) \sum_{k=0}^{\infty} p_{n,k}^q(qx) \int_0^{\infty/A} q^k p_{n,k}^q(t)(t-x)_q^m d_qt.$$

Finally using

$$(t-x)_q^{m+1} = (t-x)(t-qx)_q^m = (t-qx)_q^{m+1} - x(1-q^{m+1})(t-qx)_q^m$$

and

$$(t-x)_q^m = (t-x)(t-qx)_q^{m-1} = (t-qx)_q^m - x(1-q^m)(t-qx)_q^{m-1},$$

we get

$$E = -[m+2]_q T_{n,m+1}(qx) - \left([3]_q q^m x + q - x\right)[m+1]_q T_{n,m}(qx)$$

$$-\left[[2]_q q^m x\left([3]_q q^m x + q - x\right) - [3]_q q^{2m+1}x^2 - qx\right][m]_q T_{n,m-1}(qx)$$

$$-\left[q^m x\left\{[2]_q q^m x\left([3]_q q^m x + q - x\right) - [3]_q q^{2m+1}x^2 - qx\right\}\right.$$

$$\left.+q^{2m+1}x^2\left\{q^2 x - [3]_q q^m x - q + x\right\}\right][m-1]_q T_{n,m-2}(qx)$$

$$+[n]_q T_{n,m+1}(qx) - x(1-q^{m+1})[n]T_{n,m}(qx)$$

$$-qx(1-q^{m-1})[n]_q T_{n,m}(qx) + qx^2(1-q^{m-1})(1-q^m)[n]_q T_{n,m-1}(qx).$$

Thus, we have

$$\left([n]_q - [m+2]_q\right)T_{n,m+1}(qx) = qx(1+x)\left[D_q T_{n,m}(x) + [m]_q T_{n,m-1}(qx)\right]$$

$$+\left([3]_q q^m x + q - x\right)[m+1]_q T_{n,m}(qx)$$

$$+\left[[2]_q q^m x\left([3]q^m x + q - x\right) - [3]_q q^{2m+1}x^2 - qx\right][m]_q T_{n,m-1}(qx)$$

$$+\left[q^m x\left\{[2]_q q^m x\left([3]_q q^m x + q - x\right) - [3]_q q^{2m+1}x^2 - qx\right\}\right.$$

$$\left.+q^{2m+1}x^2\left\{q^2 x - [3]_q q^m x - q + x\right\}\right][m-1]_q T_{n,m-2}(qx)$$

$$+x(1-q^{m+1})[n]_q T_{n,m}(qx) + qx(1-q^{m-1})[n]_q T_{n,m}(qx)$$

$$-qx^2(1-q^{m-1})(1-q^m)[n]_q T_{n,m-1}(qx).$$

This completes the proof of recurrence relation.   ∎

Theorem 5.5 ([88]). *Let* $f \in C[0,\infty)$ *be a bounded function and* (q_n) *denote a sequence such that* $0 < q_n < 1$ *and* $q_n \to 1$ *as* $n \to \infty$*. Then we have for a point* $x \in (0,\infty)$

$$\lim_{n\to\infty} [n]_{q_n} \left(D_n^{q_n}(f,x) - f(x) \right) = (2x+1) \lim_{n\to\infty} D_{q_n} f(x) + x(1+x) \lim_{n\to\infty} D_{q_n}^2 f(x).$$

Proof. By q-Taylor formula [49] for f we have

$$f(t) = f(x) + D_q f(x)(t-x) + \frac{1}{[2]_q} D_q^2 f(x)(t-x)_q^2 + \Phi_q(x;t)(t-x)_q^2$$

for $0 < q < 1$ where

$$\Phi_q(x;t) = \begin{cases} \dfrac{f(t) - f(x) - D_q f(x)(t-x) - \frac{1}{[2]_q} D_q^2 f(x)(t-x)_q^2}{(t-x)_q^2}, & \text{if } x \neq y \\ 0, & \text{if } x = y. \end{cases} \qquad (5.14)$$

We know that for n large enough

$$\lim_{t\to x} \Phi_{q_n}(x;t) = 0. \qquad (5.15)$$

That is, for any $\varepsilon > 0, A > 0$, there exists a $\delta > 0$ such that

$$\left| \Phi_{q_n}(x;t) \right| < \varepsilon \qquad (5.16)$$

for $|t - x| < \delta$ and n sufficiently large. Using (5.14) we can write

$$D_n^{q_n}(f,x) - f(x) = D_{q_n} f(x) T_{n,1}(x) + \frac{D_{q_n}^2 f(x)}{[2]_{q_n}} T_{n,2}(x) + E_n^{q_n}(x),$$

where

$$E_n^q(x) = [n-1]_q \sum_{k=0}^{\infty} p_{n,k}^q(x) \int_0^{\infty/A} q^k p_{n,k}^q(t) \Phi_q(x;t)(t-x)_q^2 \, d_q t.$$

We can easily see that

$$\lim_{n\to\infty} [n]_{q_n} T_{n,1}(x) = 2x+1 \quad \text{and} \quad \lim_{n\to\infty} [n]_{q_n} T_{n,2}(x) = 2x(1+x).$$

In order to complete the proof of the theorem, it is sufficient to show that $\lim_{n\to\infty} [n]_{q_n} E_n^{q_n}(x) = 0$. We proceed as follows:
Let

$$R_{n,1}^{q_n}(x) = [n]_{q_n} [n-1]_{q_n} \sum_{k=0}^{\infty} p_{n,k}^{q_n}(x) \int_0^{\infty/A} q_n^k p_{n,k}^{q_n}(t) \Phi_{q_n}(x;t)(t-x)_{q_n}^2 \chi_x(t) \, d_q t$$

and

$$R^{q_n}_{n,2}(x) = [n]_{q_n} [n-1]_{q_n} \sum_{k=0}^{\infty} p^{q_n}_{n,k}(x) \int_0^{\infty/A} q_n^k p^{q_n}_{n,k}(t) \Phi_{q_n}(x;t)(t-x)^2_{q_n}(1-\chi_x(t)) d_q t,$$

so that

$$[n]_{q_n} E_n^{q_n}(x) = R^{q_n}_{n,1}(x) + R^{q_n}_{n,2}(x),$$

where $\chi_x(t)$ is the characteristic function of the interval $\{t : |t-x| < \delta\}$.

It follows from (5.14)

$$\left| R^{q_n}_{n,1}(x) \right| < \varepsilon 2x(x+1) \quad \text{as } n \to \infty.$$

If $|t-x| \geq \delta$, then $\left| \Phi_{q_n}(x;t) \right| \leq \frac{M}{\delta^2}(t-x)^2$, where $M > 0$ is a constant. Since

$$(t-x)^2 = (t-q^2x+q^2x-x)(t-q^3x+q^3x-x)$$
$$= (t-q^2x)(t-q^3x) + x(q^3-1)(t-q^2x) + x(q^2-1)(t-q^2x)$$
$$+ x^2(q^2-1)(q^2-q^3) + x^2(q^2-1)(q^3-1),$$

we have

$$\left| R^{q_n}_{n,2}(x) \right| \leq \frac{M}{\delta^2}[n]_{q_n}[n-1]_{q_n} \sum_{k=0}^{\infty} p^{q_n}_{n,k}(x) \int_0^{\infty/A} q_n^k p^{q_n}_{n,k}(t)(t-x)^4_{q_n} d_q t$$

$$+ \frac{M}{\delta^2} x \left(|(q_n^3-1) + (q_n^2-1)| \right) [n]_{q_n}[n-1]_{q_n} \sum_{k=0}^{\infty} p^{q_n}_{n,k}(x) \int_0^{\infty/A} q_n^k p^{q_n}_{n,k}(t)(t-x)^3_{q_n} d_q t$$

$$+ \frac{M}{\delta^2} x^2 (q_n^2-1)^2 [n]_{q_n}[n-1]_{q_n} \sum_{k=0}^{\infty} p^{q_n}_{n,k}(x) \int_0^{\infty/A} q_n^k p^{q_n}_{n,k}(t)(t-x)^2_{q_n} d_q t$$

and

$$\left| R^{q_n}_{n,2}(x) \right| \leq \frac{M}{\delta^2} \left\{ [n]_{q_n} T_{n,4}(x) + x(2-q_n^2-q_n^3)[n]_{q_n} T_{n,3}(x) + x^2 (q_n^2-1)^2 [n]_{q_n} T_{n,2}(x) \right\}.$$

Using Lemma 5.4, we have

$$T_{n,4}(x) \leq \frac{C_m}{[n]^2_{q_n}}, \quad T_{n,3}(x) \leq \frac{C_m}{[n]^2_{q_n}} \quad \text{and } T_{n,4}(x) \leq \frac{C_m}{[n]_{q_n}}.$$

We have the desired result. ∎

Corollary 5.2. *Let $f \in C[0,\infty)$ be a bounded function and (q_n) denote a sequence such that $0 < q_n < 1$ and $q_n \to 1$ as $n \to \infty$. Suppose that the first and second derivatives $f'(x)$ and $f''(x)$ exist at a point $x \in (0,\infty)$, we have*

$$\lim_{n \to \infty} [n]_{q_n} (D_n^{q_n}(f,x) - f(x)) = f'(x)(2x+1) + x(1+x)f''(x).$$

5.2   q-Szász-Beta Operators

Very recently Radu [136] established the approximation properties of certain q-operators. She also proposed the q-analogue of well-known Szász–Mirakian operators, different from [29]. After the Durrmeyer variants of well-known exponential-type operators, namely, Bernstein, Baskakov, and Szász–Mirakian operators, several researchers proposed the hybrid operators. In this direction Gupta and Noor [90] introduced certain Szász-beta operators, which reproduce constant as well as linear functions. In approximation theory because of this property, the convergence becomes faster. Very recently Song et al. [143] observed that signals are often of random characters and random signals play an important role in signal processing, especially in the study of sampling results. For this purpose, one usually uses stochastic processes which are stationary in the wide sense as a model [141]. A wide-sense stationary process is only a kind of second-order moment processes. They obtained a Korovkin-type approximation theorem and mentioned the operators such as Bernstein, Baskakov, and Szász operators and their Kantorovich variants as applications. Here we extend the study and consider more complex operators by dealing with the q-summation–integral operators. In the present study, as an application of q-beta functions [49], we introduce the q- analogue of the Szász-beta operators and obtain its moments up to second order to study their convergence behaviors.

Radu [136] proposed q-generalization of the Szász operators as

$$S_{n,q}(f,x) = \sum_{k=0}^{\infty} s_{n,k}^q(x) q^{k(k-1)} f\left(\frac{[k]_q}{q^{k-1}[n]_q}\right),\qquad(5.17)$$

where

$$s_{n,k}^q(x) = \frac{([n]_q x)^k}{[k]_q!} E_q\left(-[n]_q q^k x\right).$$

Lemma 5.5 ([136]). *We have the following:*

$$S_{n,q}(1,x) = 1.$$
$$S_{n,q}(t,x) = x.$$
$$S_{n,q}(t^2,x) = x^2 + \frac{x}{[n]_q}.$$

5.2.1   Construction of Operators

For every $n \in \mathbb{N}$, $q \in (0, 1)$, the linear positive operators $\mathcal{D}_n^q$ are defined by

$$\mathcal{D}_n^q\left(f\left(t\right),x\right) := \sum_{k=1}^{\infty} q^{\frac{3k^2-3k}{2}} s_{n,k}^q\left(x\right) \int_0^{\infty/A} p_{n,k}^q\left(t\right) f\left(qt\right) d_q t + E_q\left(-[n]_q x\right) f\left(0\right) \quad (5.18)$$

where

$$p_{n,k}^q\left(t\right) := \frac{1}{B_q\left(n+1,k\right)} \frac{t^{k-1}}{\left(1+t\right)_q^{n+k+1}}$$

and

$$s_{n,k}^q\left(x\right) = \frac{\left([n]_q x\right)^k}{[k]_q!} E_q\left(-[n]_q q^k x\right)$$

for $x \in [0, \infty)$ and for every real-valued continuous and bounded function f on $[0, \infty)$ (see [87]). In case $q = 1$ the above operators reduce to the Szász-beta operators discussed in [90] .

Lemma 5.6 ([87]). *The following equalities hold:*

(i) $\mathcal{D}_n^q\left(1,x\right) = 1$.

(ii) $\mathcal{D}_n^q\left(t,x\right) = x$.

(iii) $\mathcal{D}_n^q\left(t^2,x\right) = \frac{[n]_q x^2 + [2]_q x}{q[n-1]_q}$, *for $n > 1$.*

Proof. For $x \in [0, \infty)$ by (5.18), we have

$$\mathcal{D}_n^q\left(1,x\right) = \sum_{k=1}^{\infty} s_{n,k}^q\left(x\right) \frac{q^{\frac{3k^2-3k}{2}}}{B_q\left(n+1,k\right)} \int_0^{\infty/A} \frac{t^{k-1}}{\left(1+t\right)_q^{n+k+1}} d_q t + E_q\left(-[n]_q x\right)$$

$$= \sum_{k=1}^{\infty} s_{n,k}^q\left(x\right) \frac{q^{\frac{3k^2-3k}{2}}}{B_q\left(n+1,k\right)} \int_0^{\infty/A} \frac{t^{k-1}}{\left(1+t\right)_q^{n+k+1}} d_q t + E_q\left(-[n]_q x\right).$$

Using (1.15) and (1.17), we can write

$$\mathcal{D}_n^q\left(1,x\right) = \sum_{k=1}^{\infty} s_{n,k}^q\left(x\right) \frac{q^{\frac{3k^2-3k}{2}}}{B_q\left(n+1,k\right)} \frac{B_q\left(n+1,k\right)}{K\left(A,k\right)} + E_q\left(-[n]_q x\right)$$

$$= \sum_{k=1}^{\infty} s_{n,k}^q\left(x\right) q^{\frac{3k^2-3k}{2}} \frac{1}{q^{\frac{k(k-1)}{2}}} + E_q\left(-[n]_q x\right)$$

$$= \sum_{k=1}^{\infty} s_{n,k}^q\left(x\right) q^{k(k-1)} + E_q\left(-[n]_q x\right)$$

$$= \sum_{k=0}^{\infty} \frac{\left([n]_q x\right)^k}{[k]_q!} q^{k(k-1)} E_q\left(-[n]_q q^k x\right) = S_{n,q}(1,x) = 1,$$

where $\mathcal{S}_n^q\left(f,x\right)$ is the q-Szász operator defined by (5.17).

Similarly

$$\mathcal{D}_n^q(t,x) = \sum_{k=1}^{\infty} s_{n,k}^q(x) \int_0^{\infty/A} p_{n,k}^q(t) \, qt d_q t$$

$$= \sum_{k=1}^{\infty} s_{n,k}^q(x) \frac{q^{\frac{3k^2-3k}{2}}}{B_q(n+1,k)} \int_0^{\infty/A} \frac{qt^k}{(1+t)_q^{n+k+1}} d_q t$$

$$= \sum_{k=1}^{\infty} s_{n,k}^q(x) \frac{q^{\frac{3k^2-3k}{2}}}{B_q(n+1,k)} \frac{qB_q(n,k+1)}{K(A,k+1)}$$

$$= \sum_{k=1}^{\infty} s_{n,k}^q(x) q^{\frac{3k^2-3k+2}{2}} \frac{[n+k]_q!}{[n]_q![k-1]_q!} \frac{[k]_q![n-1]_q!}{[n+k]_q! q^{\frac{k(k+1)}{2}}}$$

$$= \sum_{k=0}^{\infty} \frac{[k]_q}{[n]_q} s_{n,k}^q(x) q^{k^2-2k+1} = \sum_{k=0}^{\infty} \frac{[k]_q}{q^{k-1}[n]_q} s_{n,k}^q(x) q^{k^2-k} = \mathcal{S}_{n,q}(t,x) = x.$$

Finally for $n > 1$, we have

$$\mathcal{D}_n^q(t^2,x) = \sum_{k=1}^{\infty} q^{\frac{3k^2-3k}{2}} s_{n,k}^q(x) \int_0^{\infty/A} p_{n,k}^q(t) q^2 t^2 d_q t$$

$$= \sum_{k=1}^{\infty} s_{n,k}^q(x) \frac{q^{\frac{3k^2-3k}{2}}}{B_q(n+1,k)} \int_0^{\infty/A} \frac{q^2 t^{k+1}}{(1+t)_q^{n+k+1}} d_q t$$

$$= \sum_{k=1}^{\infty} s_{n,k}^q(x) \frac{q^{\frac{3k^2-3k}{2}}}{B_q(n+1,k)} \frac{q^2 B_q(n-1,k+2)}{K(A,k+2) q^{\frac{k^2+3k+2}{2}}}$$

$$= \sum_{k=1}^{\infty} s_{n,k}^q(x) q^{\frac{3k^2-3k}{2}} \frac{[k+1]_q [k]_q}{[n]_q [n-1]_q} \frac{q^2}{q^{\frac{k^2+3k+2}{2}}}.$$

Using $[k+1]_q = [k]_q + q^k$, we have

$$\mathcal{D}_n^q(t^2,x) = \sum_{k=0}^{\infty} s_{n,k}^q(x) q^{k^2-3k+1} \frac{\left([k]_q + q^k\right)[k]_q}{[n]_q[n-1]_q}$$

$$= \frac{[n]_q}{q[n-1]_q} \mathcal{S}_{n,q}(t^2,x) + \frac{1}{[n-1]_q} \mathcal{S}_{n,q}(t,x)$$

$$= \frac{[n]_q}{q[n-1]_q} \left(x^2 + \frac{x}{[n]_q}\right) + \frac{x}{[n-1]_q}$$

$$= \frac{[n]_q x^2 + [2]_q x}{q[n-1]_q}.$$

∎

Remark 5.2. Let $n > 1$ and $x \in [0, \infty)$, and then for every $q \in (0, 1)$, we have

$$\mathcal{D}_n^q((t - x), x) = 0$$

and

$$\mathcal{D}_n^q((t - x)^2, x) = \frac{x^2 + [2]_q x}{q[n - 1]_q}.$$

5.2.2   Direct Theorem

By $C_B[0, \infty)$, we denote the space of real-valued continuous and bounded functions f defined on the interval $[0, \infty)$. The norm-$||.||$ on the space $C_B[0, \infty)$ is given by

$$||f|| = \sup_{0 \leq x < \infty} |f(x)|.$$

The Peetre K-functional is defined as

$$K_2(f, \delta) = \inf\{||f - g|| + \delta||g''|| : g \in W_\infty^2\},$$

where $W_\infty^2 = \{g \in C_B[0, \infty) : g', g'' \in C_B[0, \infty)\}$. For $f \in C_B[0, \infty)$ the modulus of continuity of second order is defined by

$$\omega_2(f, \sqrt{\delta}) = \sup_{0 < h \leq \sqrt{\delta}} \sup_{0 \leq x < \infty} |f(x + 2h) - 2f(x + h) + f(x)|.$$

By [50], there exists a positive constant $C > 0$ such that

$$K_2(f, \delta) \leq C\omega_2(f, \delta^{1/2}), \delta > 0.$$

Theorem 5.6. *Let $f \in C_B[0, \infty)$ and $0 < q < 1$. Then for all $x \in [0, \infty)$ and $n > 1$, there exists an absolute constant $C > 0$ such that*

$$|\mathcal{D}_n^q(f, x) - f(x)| \leq C\omega_2\left(f, \sqrt{\frac{x^2 + [2]_q x}{2q[n - 1]_q}}\right).$$

Proof. Let $g \in W_\infty^2$ and $x, t \in [0, \infty)$. By Taylor's expansion, we have

$$g(t) = g(x) + g'(x)(t - x) + \int_x^t (t - u)g''(u)du$$

Applying Remark 5.2, we obtain

$$\mathcal{D}_n^q(g, x) - g(x) = \mathcal{D}_n^q\left(\int_x^t (t - u)g''(u)du, x\right).$$

Obviously, we have $|\int_x^t (t-u)g''(u)du| \le (t-x)^2 ||g''||$. Therefore

$$|\mathcal{D}_n^q(g,x) - g(x)| \le \mathcal{D}_n^q((t-x)^2,x)||g''|| = \frac{x^2 + [2]_q x}{q[n-1]_q}||g''||.$$

Using Lemma 5.6, we have

$$|\mathcal{D}_n^q(f,x)| \le \sum_{k=1}^{\infty} q^{\frac{3k^2-3k}{2}} s_{n,k}^q(x) \int_0^{\infty/A} p_{n,k}^q(t)|f(qt)|d_q t + E_q(-[n]_q x)|f(0)| \le ||f||.$$

Thus

$$|\mathcal{D}_n^q(f,x) - f(x)| \le |\mathcal{D}_n^q(f-g,x) - (f-g)(x)| + |\mathcal{D}_n^q(g,x) - g(x)|$$

$$\le 2||f-g|| + \frac{x^2 + [2]_q x}{q[n-1]_q}||g''||.$$

Finally taking the infimum over all $g \in W_\infty^2$ and using the inequality $K_2(f,\delta) \le C\omega_2(f,\delta^{1/2}), \delta > 0$, we get the required result. This completes the proof of Theorem 5.6. ∎

We consider the following class of functions.

Let $H_{x^2}[0,\infty)$ be the set of all functions f defined on $[0,\infty)$ satisfying the condition $|f(x)| \le M_f(1+x^2)$, where M_f is a constant depending only on f. By $C_{x^2}[0,\infty)$, we denote the subspace of all continuous functions belonging to $H_{x^2}[0,\infty)$. Also, let $C_{x^2}^*[0,\infty)$ be the subspace of all functions $f \in C_{x^2}[0,\infty)$, for which $\lim_{|x|\to\infty} \frac{f(x)}{1+x^2}$ is finite. The norm on $C_{x^2}^*[0,\infty)$ is $||f||_{x^2} = \sup_{x\in[0,\infty)} \frac{|f(x)|}{1+x^2}$.

We denote the modulus of continuity of f on closed interval $[0,a], a > 0$ as by

$$\omega_a(f,\delta) = \sup_{|t-x|\le\delta} \sup_{x,t\in[0,a]} |f(t) - f(x)|.$$

We observe that for function $f \in C_{x^2}[0,\infty)$, the modulus of continuity $\omega_a(f,\delta)$ tends to zero.

Theorem 5.7. *Let $f \in C_{x^2}[0,\infty)$, $q \in (0,1)$ and $\omega_{a+1}(f,\delta)$ be its modulus of continuity on the finite interval $[0,a+1] \subset [0,\infty)$, where $a > 0$. Then for every $n > 1$,*

$$||\mathcal{D}_n^q(f) - f||_{C[0,a]} \le \frac{6M_f a(1+a^2)(2+a)}{q[n-1]_q} + 2\omega\left(f, \sqrt{\frac{a(2+a)}{q[n-1]_q}}\right).$$

Proof. For $x \in [0,a]$ and $t > a+1$, since $t - x > 1$, we have

$$|f(t) - f(x)| \le M_f \left(2 + x^2 + t^2\right)$$

$$\le M_f \left(2 + 3x^2 + 2(t-x)^2\right)$$

$$\le 6M_f \left(1 + a^2\right)(t-x)^2. \tag{5.19}$$

For $x \in [0, a]$ and $t \le a + 1$, we have

$$|f(t) - f(x)| \le \omega_{a+1}(f, |t-x|) \le \left(1 + \frac{|t-x|}{\delta}\right) \omega_{a+1}(f, \delta) \tag{5.20}$$

with $\delta > 0$.

From (5.19) and (5.20), we can write

$$|f(t) - f(x)| \le 6M_f \left(1 + a^2\right)(t-x)^2 + \left(1 + \frac{|t-x|}{\delta}\right) \omega_{a+1}(f, \delta) \tag{5.21}$$

for $x \in [0, a]$ and $t \ge 0$. Thus

$$|\mathcal{D}_n^q(f,x) - f(x)| \le \mathcal{D}_n^q(|f(t) - f(x)|, x)$$

$$\le 6M_f \left(1 + a^2\right) \mathcal{D}_n^q\left((t-x)^2, x\right)$$

$$+ \omega_{a+1}(f, \delta) \left(1 + \frac{1}{\delta} \mathcal{D}_n^q\left((t-x)^2, x\right)\right)^{\frac{1}{2}}.$$

Hence, by using Schwarz inequality and Remark 5.2, for every $q \in (0,1)$ and $x \in [0, a]$,

$$|\mathcal{D}_n^q(f,x) - f(x)| \le \frac{6M_f \left(1 + a^2\right)(x^2 + [2]_q x)}{q[n-1]_q}$$

$$+ \omega_{a+1}(f, \delta) \left(1 + \frac{1}{\delta} \sqrt{\frac{x^2 + [2]_q x}{q[n-1]_q}}\right)$$

$$\le \frac{6M_f a(1 + a^2)(2 + a)}{q[n-1]_q} + \omega_{a+1}(f, \delta) \left(1 + \frac{1}{\delta} \sqrt{\frac{a(2+a)}{q[n-1]_q}}\right).$$

By taking $\delta = \sqrt{\frac{a(2+a)}{q[n-1]_q}}$, we get the assertion of our theorem. This completes the proof of the theorem. ∎

Remark 5.3. It is observed that under the assumptions of Theorem 5.7, the point-wise convergence rate of the operators (5.18) to f is $\frac{1}{\sqrt{q_n[n-1]_{q_n}}}$ for $0 < q_n < 1$ and $q_n \to 1$ as $n \to \infty$. Also this convergence rate can be made better depending on the choice of q_n and is at least as fast as than $\frac{1}{\sqrt{n-1}}$.

5.2.3 *Weighted Approximation*

Now, we shall discuss the weighted approximation theorem as follows:

Theorem 5.8. *Let* $q = q_n$ *satisfies* $0 < q_n < 1$ *and let* $q_n \to 1$ *as* $n \to \infty$. *For each* $f \in C^*_{x^2}[0, \infty)$, *we have*

$$\lim_{n \to \infty} \left\| \mathcal{D}^{q_n}_n (f) - f \right\|_{x^2} = 0.$$

Proof. Using Korovkin's theorem (see [65]), it is sufficient to verify the following three conditions:

$$\lim_{n \to \infty} \left\| \mathcal{D}^{q_n}_n (t^v, x) - x^v \right\|_{x^2} = 0, \quad v = 0, 1, 2. \tag{5.22}$$

Since $\mathcal{D}^{q_n}_n (1, x) = 1$ and $\mathcal{D}^{q_n}_n (t, x) = x$, (5.22) holds for $v = 0$ and $v = 1$.
  Next for $n > 1$, we have

$$\left\| \mathcal{D}^{q_n}_n (t^2, x) - x^2 \right\|_{x^2} \leq \left(\frac{[n]_q}{q_n [n-1]_{q_n}} - 1 \right) \sup_{x \in [0, \infty)} \frac{x^2}{1 + x^2} + \frac{[2]_q}{q_n [n-1]_{q_n}} \sup_{x \in [0, \infty)} \frac{x}{1 + x^2}$$

$$\leq \frac{1}{q_n [n-1]_{q_n}} + \frac{[2]_q}{q_n [n-1]_{q_n}}$$

which implies that

$$\lim_{n \to \infty} \left\| \mathcal{D}^{q_n}_n (t^2, x) - x^2 \right\|_{x^2} = 0.$$

Thus the proof is completed. ∎

  Next we give the following theorem to approximate all functions in $C_{x^2}[0, \infty)$. This type of result is given in [70] for locally integrable functions.

Theorem 5.9. *Let* $q = q_n$ *satisfies* $0 < q_n < 1$ *and let* $q_n \to 1$ *as* $n \to \infty$. *For each* $f \in C_{x^2}[0, \infty)$ *and* $\alpha > 0$, *we have*

$$\lim_{n \to \infty} \sup_{x \in [0, \infty)} \frac{\left| \mathcal{D}^{q_n}_n (f, x) - f(x) \right|}{(1 + x^2)^{1 + \alpha}} = 0.$$

Proof. For any fixed $x_0 > 0$,

$$\sup_{x \in [0, \infty)} \frac{\left| \mathcal{D}^{q_n}_n (f, x) - f(x) \right|}{(1 + x^2)^{1 + \alpha}} \leq \sup_{x \leq x_0} \frac{\left| \mathcal{D}^{q_n}_n (f, x) - f(x) \right|}{(1 + x^2)^{1 + \alpha}} + \sup_{x \geq x_0} \frac{\left| \mathcal{D}^{q_n}_n (f, x) - f(x) \right|}{(1 + x^2)^{1 + \alpha}}$$

$$\leq \left\| \mathcal{D}^q_n (f) - f \right\|_{C[0, x_0]} + \left\| f \right\|_{x^2} \sup_{x \geq x_0} \frac{\left| \mathcal{D}^{q_n}_n (1 + t^2, x) \right|}{(1 + x^2)^{1 + \alpha}}$$

$$+ \sup_{x \geq x_0} \frac{|f(x)|}{(1 + x^2)^{1 + \alpha}}.$$

Obviously, the first term of the above inequality tends to zero, which is evident from Theorem 5.6. By Lemma 5.6 for any fixed $x_0 > 0$, it is easily seen that $\sup\limits_{x \geq x_0} \dfrac{\left| \mathcal{D}_n^{qn} \left(1 + t^2, x\right) \right|}{\left(1 + x^2\right)^{1+\alpha}}$ tends to zero as $n \to \infty$. Finally, we can choose $x_0 > 0$ so large that the last part of above inequality can be made small enough.   ∎

5.3 q-Szász–Durrmeyer Operators

In this section we present direct approximation result in weighted function space with the help of a weighted Korovkin-type theorem for new q-Szász–Durrmeyer operators (see [33]). Then we give the weighted approximation error of these operators in terms of weighted modulus of continuity. Finally, we establish an asymptotic formula.

Recently for $0 < q < 1$, Aral [25] (also see [29]) defined the q-Szász–Mirakian operators as

$$S_n^q(f,x) = E_q\left(-[n]_q \frac{x}{b_n}\right) \sum_{k=0}^{\infty} f\left(\frac{[k]_q b_n}{[n]_q}\right) \frac{\left([n]_q x\right)^k}{[k]_q!(b_n)^k}, \tag{5.23}$$

where $0 \leq x < \alpha_q(n), \alpha_q(n) := \frac{b_n}{(1-q)[n]_q}, f \in C([0,\infty))$ and (b_n) is a sequence of positive numbers such that $\lim_{n \to \infty} b_n = \infty$. Some approximation properties of these operators are studied in [29].

Based on this, we now propose the q-Szász–Durrmeyer operators for $0 < q < 1$ as

$$\mathcal{Z}_n^q\left(f\left(t\right), x\right) = \frac{[n]_q}{b_n} \sum_{k=0}^{\infty} s_{n,k}^q(x) \int_0^{\frac{qb_n}{1-q^n}} s_{n,k}^q(t) f(t) \, d_q t , \tag{5.24}$$

where

$$s_{n,k}^q(x) = \frac{\left([n]_q x\right)^k}{q^{\frac{k+1}{2}} [k]_q! (b_n)^k} E_q\left(-[n]_q \frac{x}{b_n}\right).$$

Remark 5.4. Note that the q-Szász–Durrmeyer operators can be rewritten via an improper integral by using Definition (1.13). We can easily see that $E_q\left(-\frac{q^n}{1-q}\right) = 0$ for $n \leq 0$. Thus for $0 < q < 1$ we can write

$$\mathcal{Z}_n^q\left(f\left(t\right), x\right) = \frac{[n]_q}{b_n} \sum_{k=0}^{\infty} s_{n,k}^q(x) \int_0^{\frac{\infty}{1-q^n}/qb_n} s_{n,k}^q(t) f(t) \, d_q t.$$

By [29], we have

$$S_n^q(1,x) = 1, \qquad S_n^q(t,x) = x, \qquad S_n^q(t^2,x) = qx^2 + \frac{b_n}{[n]_q}x,$$

$$S_n^q(t^3,x) = q^3x^3 + \left([2]_q + 1\right)q\frac{b_n}{[n]_q}x^2 + \frac{b_n}{[n]_q}x,$$

$$S_n^q(t^4,x) = q^6x^4 + a_1(q)S_n^q(t^3,x) + a_2(q)S_n^q(t^2,x) + [2]_q[3]_q\left(\frac{b_n}{[n]_q}\right)^2 x,$$

where

$$a_1(q) = \left(1 + [2]_q + [3]_q\right)\frac{b_n}{[n]_q} \text{ and } a_2(q) = -\left([2]_q[3]_q + [2]_q + [3]_q\right)\left(\frac{b_n}{[n]_q}\right)^2.$$

5.3.1   Auxiliary Results

In the sequel, we shall need the following auxiliary results.

Lemma 5.7. *We have*

$$\mathcal{Z}_n^q(1,x) = 1, \qquad \mathcal{Z}_n^q(t,x) = q^2x + \frac{qb_n}{[n]_q},$$

$$\mathcal{Z}_n^q(t^2,x) = q^6x^2 + \left(q^5 + 2q^4 + q^3\right)\frac{b_n}{[n]_q}x + q^2(1+q)\left(\frac{b_n}{[n]_q}\right)^2,$$

$$\mathcal{Z}_n^q(t^3,x) = q^9S_n^q(t^3,x) + d_1(q)S_n^q(t^2,x) + d_2(q)S_n^q(t,x) + q^3\left(\frac{b_n}{[n]_q}\right)^3[2]_q[3]_q,$$

$$\mathcal{Z}_n^q(t^4,x) = q^{14}S_n^q(t^4,x) + d_3(q)S_n^q(t^3,x) + d_4(q)S_n^q(t^2,x) + d_5(q)S_n^q(t,x)$$

$$+ q^4[2]_q[3]_q[4]_q\left(\frac{b_n}{[n]_q}\right)^4,$$

where

$$d_1(q) = q^6\left([3]_q + q[2]_q + q^2\right)\frac{b_n}{[n]_q}, \quad d_2(q) = q^4\left([2]_q[3]_q + q[3]_q + q^2[2]_q\right)\left(\frac{b_n}{[n]_q}\right)^2,$$

$$d_3(q) = q^{10}\left[[4]_q + \left(q[3]_q + q^2[2]_q + q^3\right)\right]\frac{b_n}{[n]_q},$$

$$d_4(q) = q^7 \left[[4]_q \left([3]_q + q[2]_q + q^2 \right) + q^2 \left([2]_q[3]_q + q[3]_q + q^2[2]_q \right) \right] \left(\frac{b_n}{[n]_q} \right)^2, \; and$$

$$d_5(q) = q^4 \left(q^4[2]_q[3]_q + [4]_q \left(q[2]_q[3]_q + q^2[3]_q + q^3[2]_q \right) \right) \left(\frac{b_n}{[n]_q} \right)^3.$$

Proof. Using (5.24), we have

$$\mathcal{Z}_n^q(1,x) = \frac{[n]_q}{b_n} \sum_{k=0}^{\infty} s_{n,k}^q(x) \int_0^{\frac{qb_n}{1-q^n}} s_{n,k}^q(t)\, d_q t$$

$$= \frac{[n]_q}{b_n} \sum_{k=0}^{\infty} s_{n,k}^q(x) \frac{\left([n]_q \right)^k}{q^{\frac{k+1}{2}} [k]_q!(b_n)^k} \int_0^{\frac{qb_n}{1-q^n}} t^k E_q\left(-[n]_q \frac{t}{b_n} \right) d_q t.$$

Using (5.19) and change of variable formula for q-integral with $t = q\frac{b_n}{[n]_q} y$, then we have

$$\mathcal{Z}_n^q(1,x) = \frac{[n]_q}{b_n} \sum_{k=0}^{\infty} s_{n,k}^q(x) \frac{q^{k+1} b_n}{q^{\frac{k+1}{2}} [n]_q [k]_q!} \int_0^{\frac{1}{1-q}} y^k E_q(-qy)\, d_q y.$$

From (5.24) and (5.23), it follows that

$$\mathcal{Z}_n^q(1,x) = \frac{[n]_q}{b_n} \sum_{k=0}^{\infty} s_{n,k}^q(x) \frac{q^{\frac{k+1}{2}} b_n}{[n]_q [k]_q!} \Gamma_q(k+1)$$

$$= E_q\left(-[n]_q \frac{x}{b_n} \right) \sum_{k=0}^{\infty} \frac{\left([n]_q x \right)^k}{[k]_q!(b_n)^k}$$

$$= \mathcal{S}_n^q(1,x) = 1.$$

Also, using a similar technique, from (5.24) with $t = q\frac{b_n}{[n]_q} y$, we have

$$\mathcal{Z}_n^q(t,x) = \frac{[n]_q}{b_n} \sum_{k=0}^{\infty} s_{n,k}^q(x) \int_0^{\frac{qb_n}{1-q^n}} s_{n,k}^q(t)\, t\, d_q t$$

$$= \frac{[n]_q}{b_n} \sum_{k=0}^{\infty} s_{n,k}^q(x) \frac{\left([n]_q \right)^k}{q^{\frac{k+1}{2}} [k]_q!(b_n)^k} \int_0^{\frac{qb_n}{1-q^n}} t^{k+1} E_q\left(-[n]_q \frac{t}{b_n} \right) d_q t$$

$$= \frac{qb_n}{[n]_q} \sum_{k=0}^{\infty} s_{n,k}^q(x) \frac{q^{\frac{k+1}{2}}}{[k]_q!} \int_0^{\frac{1}{1-q}} y^{k+1} E_q(-qy)\, d_q y$$

$$= \frac{qb_n}{[n]_q} \sum_{k=0}^{\infty} s_{n,k}^q(x) \frac{q^{\frac{k+1}{2}}}{[k]_q!} \Gamma_q(k+2).$$

From (5.24) and (5.23), it follows that

$$\mathcal{Z}_n^q(t,x) = qE_q\left(-[n]_q\frac{x}{b_n}\right)\sum_{k=0}^{\infty}\frac{\left([n]_qx\right)^k}{[k]_q!(b_n)^k}b_n\frac{[k+1]_q}{[n]_q}$$

$$= qE_q\left(-[n]_q\frac{x}{b_n}\right)\sum_{k=0}^{\infty}\frac{\left([n]_qx\right)^k}{[k]_q!(b_n)^k}b_n\frac{\left(q[k]_q+1\right)}{[n]_q}$$

$$= q^2\mathcal{S}_n^q(t,x) + \frac{qb_n}{[n]_q}\mathcal{S}_n^q(1,x) = q^2x + \frac{qb_n}{[n]_q}.$$

On the other hand

$$\mathcal{Z}_n^q(t^2,x) = \frac{[n]_q}{b_n}\sum_{k=0}^{\infty}s_{n,k}^q(x)\frac{\left([n]_q\right)^k}{q^{\frac{k+1}{2}}[k]_q!(b_n)^k}\frac{q^{k+3}(b_n)^{k+3}}{\left([n]_q\right)^{k+3}}\int_0^{\frac{1}{1-q}}y^{k+2}E_q(-qy)d_qy$$

$$= \frac{[n]_q}{b_n}\sum_{k=0}^{\infty}s_{n,k}^q(x)\frac{\left([n]_q\right)^k}{q^{\frac{k+1}{2}}[k]_q!(b_n)^k}\frac{q^{k+3}(b_n)^{k+3}}{\left([n]_q\right)^{k+3}}\Gamma_q(k+3)$$

$$= q^2E_q\left(-[n]_q\frac{x}{b_n}\right)\sum_{k=0}^{\infty}\frac{\left([n]_qx\right)^k}{[k]_q!(b_n)^k}\frac{(b_n)^2[k+1]_q[k+2]_q}{\left([n]_q\right)^2}$$

$$= q^2E_q\left(-[n]_q\frac{x}{b_n}\right)\sum_{k=0}^{\infty}\frac{\left([n]_qx\right)^k}{[k]_q!(b_n)^k}\frac{(b_n)^2\left(q[k]_q+1\right)\left(q^2[k]_q+[2]_q\right)}{\left([n]_q\right)^2}$$

$$= q^5\mathcal{S}_n^q(t^2,x) + q^2\left(q[2]_q+q^2\right)\frac{b_n}{[n]_q}\mathcal{S}_n^q(t,x) + q^2\left(\frac{b_n}{[n]_q}\right)^2[2]_q\mathcal{S}_n^q(1,x)$$

$$= q^6x^2 + \left(q^5+2q^4+q^3\right)\frac{b_n}{[n]_q}x + q^2(1+q)\left(\frac{b_n}{[n]_q}\right)^2.$$

Other moments can be calculated similarly.

Lemma 5.8. *We have the following:*

1. $\mathcal{Z}_n^q(t-x,x) = (q^2-1)x + q\frac{b_n}{[n]_q}.$

2. $\mathcal{Z}_n^q\left((t-x)^2,x\right) = \left((q^6-2q^2+1)x^2 + q^2(q^3+2q^2+q-2)\frac{b_n}{[n]_q}x + q^2(1+q)\left(\frac{b_n}{[n]_q}\right)^2\right).$

3. $\mathcal{Z}_n^q\left((t-x)^4,x\right) = x^4\left(q^{20}-4q^{12}+6q^6-4q^2+1\right)$

$$+ \left(\begin{array}{c} q^{17}a_1\left(q\right)+d_3\left(q\right)q^3-4\left(\left[2\right]_q+1\right)\frac{b_n}{\left[n\right]_q}q^{10}-4d_1\left(q\right) \\ +6\left(q^5+2q^4+q^3\right)\frac{b_n}{\left[n\right]_q}-4\frac{qb_n}{\left[n\right]_q} \end{array} \right) x^3$$

$$+ \left(\begin{array}{c} q^{15}a_1\left(q\right)\left(\left[2\right]_q+1\right)\frac{b_n}{\left[n\right]_q}+q^{15}a_2\left(q\right)+d_3\left(q\right)q\left(\left[2\right]_q+1\right)\frac{b_n}{\left[n\right]_q} \\ +d_4\left(q\right)q-4q^9\left(\frac{b_n}{\left[n\right]_q}\right)^2-4d_2\left(q\right)-4d_1\left(q\right)\frac{b_n}{\left[n\right]_q}+6\left(1+q\right)q^2\left(\frac{b_n}{\left[n\right]_q}\right)^2 \end{array} \right) x^2$$

$$+ \left(\begin{array}{c} q^{14}a_1\left(q\right)\left(\frac{b_n}{\left[n\right]_q}\right)^2+q^{14}a_2\left(q\right)\frac{b_n}{\left[n\right]_q}+d_3\left(q\right)\left(\frac{b_n}{\left[n\right]_q}\right)^2+d_4\left(q\right)\frac{b_n}{\left[n\right]_q} \\ -4q^3\left(\frac{b_n}{\left[n\right]_q}\right)^3\left[2\right]_q\left[3\right]_q \end{array} \right) x$$

$$+ q^{14}\left(\frac{b_n}{\left[n\right]_q}\right)^3\left[2\right]_q\left[3\right]_q+d_5\left(q\right)+q^4\left[2\right]_q\left[3\right]_q\left[4\right]_q\left(\frac{b_n}{\left[n\right]_q}\right)^2.$$

5.3.2 Approximation Properties

Let B_2 be the set of all functions f defined on $[0, \infty)$ satisfying the condition $|f(x)| \leq M_f(1+x^2)$, where M_f is a constant depending only f. C_2 denotes the subspace of all continuous function in B_2, and C_2^* denotes the subspace of all functions $f \in C_2$ for which $\lim\limits_{x \to \infty} \frac{|f(x)|}{1+x^2}$ exists finitely.

Let (α_n) be a sequence of positive numbers, such that $\lim\limits_{n \to \infty} a_n = \infty$ and

$$\|f\|_{2,[0,a_n]} = \sup_{0 \leq x \leq \alpha_n} \frac{|f(x)|}{1+x^2},$$

for $f \in B_2$. These type functions are mentioned in [71].

Theorem 5.10. *Let $f \in C_2^*$ and $q = q_n$ satisfies $0 < q_n < 1$ such that $q_n \to 1$ as $n \to \infty$. If $\lim_{n \to \infty} \frac{b_n}{[n]_q} = 0$, we have*

$$\lim_{n \to \infty} \|\mathcal{Z}_n^q(f) - f\|_{2,[0,\alpha_{q_n}(n)]} = 0.$$

Proof. On account of Theorem 1 in [71], it is enough to show the validity of the following:

$$\lim_{n \to \infty} \|\mathcal{Z}_n^{q_n}(t^v, x) - x^v\|_{2,[0,\alpha_{q_n}(n)]} = 0, \quad v = 0, 1, 2. \tag{5.25}$$

Since, $\mathcal{Z}_n^{q_n}(1, x) = 1$, it obvious that

$$\lim_{n \to \infty} \|\mathcal{Z}_n^{q_n}(1, x) - 1\|_{2,[0,\alpha_{q_n}(n)]} = 0.$$

Using Lemma 5.7, we obtain

$$\lim_{n\to\infty} \left\| \mathcal{Z}_n^{q_n}(t,x) - x \right\|_{2,[0,\alpha_{q_n}(n)]} \leq \left(1 - q_n^2\right) \sup_{0 \leq x \leq \alpha_{q_n}(n)} \frac{x}{1+x^2} + \frac{q_n b_n}{[n]_{q_n}}$$

$$\leq \left(1 - q_n^2\right) + \frac{q_n b_n}{[n]_{q_n}}$$

and

$$\lim_{n\to\infty} \left\| \mathcal{Z}_n^{q_n}(t^2,x) - x^2 \right\|_{2,[0,\alpha_{q_n}(n)]}$$

$$= \lim_{n\to\infty} \sup_{0 \leq x \leq \alpha_{q_n}(n)} \frac{\left| \mathcal{Z}_n^{q_n}(t^2,x) - x^2 \right|}{1+x^2}$$

$$\leq \left(1 - q_n^6\right) \sup_{0 \leq x \leq \alpha_{q_n}(n)} \frac{x^2}{1+x^2} + \left(q_n^5 + 2q_n^4 + q_n^3\right) \frac{b_n}{[n]_{q_n}} \sup_{0 \leq x \leq \alpha_{q_n}(n)} \frac{x}{1+x^2}$$

$$+ q_n^2(1+q_n)\left(\frac{b_n}{[n]_{q_n}}\right)^2 \sup_{0 \leq x \leq \alpha_{q_n}(n)} \frac{1}{1+x^2}$$

$$\leq \left(1 - q_n^6\right) + + \left(q_n^5 + 2q_n^4 + q_n^3\right)\frac{b_n}{[n]_{q_n}} + \left(\frac{b_n}{[n]_{q_n}}\right)^2 q_n^2(1+q_n).$$

Since $\lim_{n\to\infty} \frac{b_n}{[n]_{q_n}} = 0$ and $\lim_{n\to\infty} q_n = 1$, we have $\lim_{n\to\infty} \left\| \mathcal{Z}_n^{q_n}(t,x) - x \right\|_{2,[0,\alpha_{q_n}(n)]} = 0$ and $\lim_{n\to\infty} \left\| \mathcal{D}_n^{q_n}(t^2,x) - x^2 \right\|_{2,[0,\alpha_{q_n}(n)]} = 0$. Hence the conditions of (5.25) are fulfilled and we get $\lim_{n\to\infty} \left\| \mathcal{Z}_n^q(f) - f \right\|_{2,[0,\alpha_{q_n}(n)]} = 0$ for every $f \in C_2^*$. ■

Now, we find the order of approximation of the functions $f \in C_2^*$ by the operators $\mathcal{Z}_n^q$ with the help of following weighted modulus of continuity (see [153]).

Let

$$\Omega_2(f;\delta) = \sup_{0<h<\delta, x\in[0,a(n)]} \frac{|f(x+h) - f(x)|}{1+(x+h)^2}, \quad \text{for each } f \in C_2^*.$$

The weighted modulus of continuity has the following properties which are similar to usual first modulus of continuity.

Lemma 5.9. *Let $f \in C_2^*$. Then, we have the following:*

(i) $\Omega_2(f;\delta)$ is a monotone increasing function of δ.
(ii) For each $f \in C_2^*$, $\lim_{\delta\to0^+} \Omega_2(f;\delta)$.
(iii) For each $\lambda > 0$, $\Omega_2(f;\lambda\delta) \leq (1+\lambda)\Omega_2(f;\delta)$.

Now we give the main theorem of this section.

Theorem 5.11. *Let $f \in C_2^*$ and $q = q_n$ satisfies $0 < q_n < 1$ such that $q_n \to 1$ as $n \to \infty$. If $\lim\limits_{n \to \infty} \frac{b_n}{[n]_{q_n}} = 0$, then there exists a positive constant A such that the inequality*

$$\sup_{x \in [0, \alpha_{q_n}(n)]} \frac{\left| \mathcal{Z}_n^q(f, x) - f(x) \right|}{(1 + x^2)^{\frac{3}{2}}} \leq A \Omega_2 \left(f; \sqrt{a_q(n)} \right)$$

holds, where $a_q(n) = \max \left\{ 1 - q^3, \frac{b_n}{[n]_q} \right\}$ and A is a positive constant.

Proof. For $t \geq 0$, $x \in [0, \alpha_{q_n}(n)]$ and $\delta > 0$, using the definition of $\Omega_2(f; \delta)$ and Lemma 5.9 (*iii*), we get

$$|f(t) - f(x)| \leq \left(1 + (x + |t - x|)^2 \right) \left(1 + \frac{|t - x|}{\delta} \right) \Omega_2(f; \delta)$$

$$\leq 2 \left(1 + x^2 \right) \left(1 + (t - x)^2 \right) \left(1 + \frac{|t - x|}{\delta} \right) \Omega_2(f; \delta).$$

Since $\mathcal{Z}_n^q$ is linear and positive, we have

$$|\mathcal{Z}_n^q(f, x) - f(x)|$$
$$\leq 2 \left(1 + x^2 \right) \Omega_2(f; \delta)$$
$$\times \left\{ 1 + \mathcal{Z}_n^q((t - x)^2, x) + \mathcal{Z}_n^q((1 + (t - x)^2) \frac{|t - x|}{\delta}, x) \right\}. \tag{5.26}$$

To estimate the first term of above inequality, using Lemma 5.7, we have

$$\mathcal{Z}_n^q \left((t - x)^2, x \right) = \mathcal{Z}_n^q(t^2, x) - 2x \mathcal{Z}_n^q(t, x) + x^2 \mathcal{Z}_n^q(1, x)$$

$$= \left(q^6 - 2q^2 + 1 \right) x^2 + \left(q^5 + 2q^4 + q^3 - 2q \right) \frac{b_n}{[n]_q} x$$

$$+ q^2 (1 + q) \left(\frac{b_n}{[n]_q} \right)^2$$

$$\leq \left(q^6 - 2q^2 + 1 \right) x^2 + \frac{2b_n}{[n]_q} x + 2 \left(\frac{b_n}{[n]_q} \right)^2$$

$$\leq (1 - q^3)^2 x^2 + \frac{2b_n}{[n]_q} x + 2 \left(\frac{b_n}{[n]_q} \right)^2$$

$$\leq A_1 \mathcal{O}(a_q(n)) \left(1 + x + x^2 \right), \tag{5.27}$$

where $A_1 > 0$ and $a_q(n) = \max\left\{1 - q^3, \frac{b_n}{[n]_q}\right\}$. Since $\lim\limits_{n\to\infty} \frac{b_n}{[n]_{q_n}} = 0$ and $\lim\limits_{n\to\infty} q_n = 1$, there exists a positive constant A_2 such that

$$\mathcal{Z}_n^q\left((t-x)^2, x\right) \leq A_2\left(1 + x^2\right).$$

To estimate the second term of (5.26), applying the Cauchy–Schwarz inequality, we have

$$\mathcal{Z}_n^q\left(\left(1+(t-x)^2\right)\frac{|t-x|}{\delta}, x\right) \leq 2\left(\mathcal{Z}_n^q\left(1+(t-x)^4, x\right)\right)^{\frac{1}{2}}\left(\mathcal{Z}_n^q\left(\frac{(t-x)^2}{\delta^2}, x\right)\right)^{\frac{1}{2}}.$$

Using (5.27) and Lemma 5.8, by direct computation we get

$$\left(\mathcal{Z}_n^q\left(1+(t-x)^4, x\right)\right)^{\frac{1}{2}} \leq A_3\left(1 + x + x^2\right)$$

and

$$\left(\mathcal{Z}_n^q\left(\frac{(t-x)^2}{\delta^2}, x\right)\right)^{\frac{1}{2}} \leq \frac{A_4}{\delta}\mathcal{O}\left(a_q(n)\right)^{\frac{1}{2}}\left(1 + x + x^2\right)^{\frac{1}{2}}$$

for $A_3 > 0$ and $A_4 > 0$. If we take $\delta = a_q(n)^{\frac{1}{2}}$, $A = 2\left(1 + A_2 + 2A_3A_4\right)$ and combine above estimates, we have the inequality of the theorem. ∎

Now we give an asymptotic formula with respect to weighted norm. The symbol UC_2^2 will stand for the space of all twice-differentiable functions on $[0, \infty)$ with uniformly continuous and bounded second derivative.

Theorem 5.12. *Let $f \in UC_2^2$, $q = q_n$ satisfies $0 < q_n < 1$ such that $q_n \to 1$ as $n \to \infty$ and $\lim\limits_{n\to\infty} \frac{b_n}{[n]_{q_n}} = 0$ then*

$$\lim_{n\to\infty} \frac{1}{(1+x^2)^2}\left\{\frac{b_n}{[n]_{q_n}}\left(\mathcal{Z}_n^{q_n}(f, x) - f(x) - \left(xf' + f\right)\right)\right\} = 0$$

uniformly on $[0, \alpha_{q_n}(n)]$. Particularly

$$\lim_{n\to\infty} \frac{b_n}{[n]_{q_n}}\left(\mathcal{Z}_n^{q_n}(f, x) - f(x) - \left(xf' + f\right)\right) = 0$$

uniformly on compact subsets of $\alpha_{q_n}(n)$.

Proof. On account of Theorem 1 in [12], we need the show that:

1. $\lim_{n\to\infty} \frac{1}{(1+x^2)^2}\left(\frac{[n]_{q_n}}{b_n}\mathcal{Z}_n^{q_n}\left((t-x)^2, x\right) - 2x\right) = 0$.

2. $\lim_{n\to\infty} \frac{x^k}{(1+x^2)^2}\left(\frac{[n]_{q_n}}{b_n}\mathcal{Z}_n^{q_n}\left((t-x), x\right) - 1\right) = 0$, for $k = 0, 1$.

3. $\lim_{n\to\infty} \frac{1}{\left(1+x^2\right)^2} \frac{[n]_{q_n}}{b_n} \mathcal{Z}_n^{q_n}\left((t-x)^4, x\right) = 0.$

4. $\sup_{x\in[0,\alpha_n(q))} \sup_{n\geq 1} \frac{1}{\left(1+x^2\right)^2} \frac{[n]_{q_n}}{b_n} \mathcal{Z}_n^{q_n}\left((t-x)^2, x\right) < \infty.$

Since $\lim_{n\to\infty} \frac{[n]_{q_n}}{b_n}\left(q_n^6 - 2q_n^2 + 1\right) = \lim_{n\to\infty} \frac{1-q_n^n}{b_n}\left(\frac{q_n^6 - 2q_n^2 + 1}{1-q_n}\right) = 0$, we have

$$\lim_{n\to\infty} \frac{1}{\left(1+x^2\right)^2}\left(\frac{[n]_{q_n}}{b_n}\mathcal{Z}_n^{q_n}\left((t-x)^2, x\right) - 2x\right)$$

$$= \frac{x^2}{\left(1+x^2\right)^2} \lim_{n\to\infty} \frac{[n]_{q_n}}{b_n}\left(q_n^6 - 2q_n^2 + 1\right)$$

$$+ \frac{x}{\left(1+x^2\right)^2} \lim_{n\to\infty}\left(q_n^5 + 2q_n^4 + q_n^3 - 2q_n - 2\right)$$

$$+ \frac{1}{\left(1+x^2\right)^2} \lim_{n\to\infty} q_n^2(1+q_n)\frac{b_n}{[n]_{q_n}} = 0$$

uniformly on $[0, \alpha_{q_n}(n)]$. Also, for every $x \in [0, \alpha_{q_n}(n)]$

$$\lim_{n\to\infty} \frac{x^k}{\left(1+x^2\right)^2}\left(\frac{[n]_{q_n}}{b_n}\mathcal{Z}_n^{q_n}\left((t-x), x\right) - 1\right)$$

$$= \frac{x^k}{\left(1+x^2\right)^2} \lim_{n\to\infty} \frac{[n]_{q_n}}{b_n}\left(q_n^2 - 1\right)x + q_n - 1$$

$$= \frac{x^k}{\left(1+x^2\right)^2} \lim_{n\to\infty} \frac{1-q_n^n}{b_n}\left(q_n - 1\right)x + q_n - 1$$

$$= 0$$

for $k = 0, 1$ uniformly on $[0, \alpha_{q_n}(n)]$. Since

$$\lim_{n\to\infty} \frac{[n]_{q_n}}{b_n}\left(q_n^{20} - 4q_n^{12} + 6q_n^{6n} - 4q_n^2 + 1\right) = \lim_{n\to\infty} \frac{1-q_n^n}{b_n}\left(\frac{q_n^{20} - 4q_n^{12} + 6q_n^6 - 4q_n^2 + 1}{1-q_n}\right) = 0,$$

we have

$$\lim_{n\to\infty} \frac{1}{\left(1+x^2\right)^2} \frac{[n]_{q_n}}{b_n} \mathcal{Z}_n^{q_n}\left((t-x)^4, x\right) = 0.$$

Finally

$$\sup_{x\in[0,\alpha_n(q))}\sup_{n\geq1}\frac{1}{(1+x^2)^2}\frac{[n]_{q_n}}{b_n}\mathcal{Z}_n^{q_n}\left((t-x)^2,x\right)$$

$$=\sup_{x\in[0,\alpha_n(q))}\sup_{n\geq1}\frac{x^2}{(1+x^2)^2}\left[\frac{[n]_{q_n}}{b_n}\left(q_n^6-2q_n^2+1\right)\right.$$

$$\left.+\left(q_n^5+2q_n^4+q_n^3-2q_n\right)+q_n^2(1+q_n)\frac{b_n}{[n]_{q_n}}\right]$$

$$\leq\sup_{n\geq1}\left[\frac{[n]_{q_n}}{b_n}\left(q_n^6-2q_n^2+1\right)\right.$$

$$\left.+\left(q_n^5+2q_n^4+q_n^3-2q_n\right)+q_n^2(1+q_n)\frac{b_n}{[n]_{q_n}}\right]<\infty,$$

and hence the result follows.   ∎

5.4 q-Phillips Operators

Phillips [135] defined the well-known linear positive operators

$$P_n(f;x)=n\sum_{k=1}^{\infty}e^{-nx}\frac{n^kx^k}{n!}\int_0^{\infty}e^{-nt}\frac{n^{k-1}t^{k-1}}{n!}f(t)dt+e^{-nx}f(0),$$

where $x\in[0,\infty)$. Some approximation properties of these operators were studied by Gupta and Srivastava [93] and by May [123]. Bézier variant of these Phillips operators was proposed and studied by Gupta [85], where the rate of convergence for the Bézier variant of the Phillips operators for bounded variation functions was discussed. Very recently, Mahmudov in [119] introduced the following q-Szász–Mirakian operator

$$S_{n,q}(f;x)=\frac{1}{\prod_{j=0}^{\infty}\left(1+(1-q)q^j[n]_qx\right)}\sum_{k=0}^{\infty}f\left(\frac{[k]_q}{q^{k-2}[n]_q}\right)q^{\frac{k(k-1)}{2}}\frac{[n]_q^kx^k}{[k]_q!},$$

where $x\in[0,\infty)$, $0<q<1$, $f\in C[0,\infty)$, and investigated their approximation properties.

Definition 5.1 ([118]). For $f\in R^{[0,\infty)}$, we define the following q-parametric Phillips operators

$$\mathcal{P}_{n,q}(f;x)=[n]_q \sum_{k=1}^{\infty} q^{k-1} S_{n,k}(q;qx) \int_0^{\infty/(1-q)} S_{n,k-1}(q;t)f(t)d_q t + e_q\left(-[n]_q qx\right) f(0),$$

$$(5.28)$$

where $x \in [0,\infty)$ and $S_{n,k}(q;x) = e_q(-[n]x)q^{\frac{k(k-1)}{2}} \frac{[n]_q^k x^k}{[k]_q!}$.

These operators generalize the sequence of classical Phillips operators.

In this section we present the approximation properties of the q-Phillips operators defined by (5.28), establish some local approximation result for continuous functions in terms of modulus of continuity, and obtain inequalities for the weighted approximation error of q-Phillips operators. Furthermore, we study Voronovskaja-type asymptotic formula for the q-Phillips operators.

5.4.1 Moments

There are two q-analogues of the exponential function e^z; see [104]:

$$e_q(z) = \sum_{k=0}^{\infty} \frac{z^k}{[k]_q!} = \frac{1}{(1-(1-q)z)_q^\infty}, \quad |z| < \frac{1}{1-q}, \quad |q| < 1,$$

and

$$E_q(z) = \prod_{j=0}^{\infty} \left(1+(1-q)q^j z\right) = \sum_{k=0}^{\infty} q^{k(k-1)/2} \frac{z^k}{[k]_q!} = (1+(1-q)z)_q^\infty, \quad |q| < 1,$$

$$(5.29)$$

where $(1-x)_q^\infty = \prod_{j=0}^{\infty}\left(1-q^j x\right)$.

We set

$$s_{n,k}(q;x) = \frac{1}{E_q\left([n]_q x\right)} q^{\frac{k(k-1)}{2}} \frac{[n]_q^k x^k}{[k]_q!}$$

$$= e_q\left(-[n]_q x\right) q^{\frac{k(k-1)}{2}} \frac{[n]_q^k x^k}{[k]_q!}, \quad n = 1,2,\dots. \quad (5.30)$$

It is clear that $s_{n,k}(q;x) \geq 0$ for all $q \in (0,1)$ and $x \in [0,\infty)$ and moreover

$$\sum_{k=0}^{\infty} s_{n,k}(q;x) = e_q\left(-[n]_q x\right) \sum_{k=0}^{\infty} q^{\frac{k(k-1)}{2}} \frac{\left([n]_q x\right)^k}{[k]_q!} = 1.$$

The two q-gamma functions are defined as

$$\Gamma_q(x) = \int_0^{\frac{1}{1-q}} t^{x-1} E_q(-qt)\,d_qt, \quad \gamma_q^A(x) = \int_0^{\infty/A(1-q)} t^{x-1} e_q(-qt)\,d_qt.$$

For every $A, x > 0$ one has

$$\Gamma_q(x) = K(A;x)\,\gamma_q^A(x), 0$$

where $K(A;x) = \frac{1}{1+A}A^x\left(1+\frac{1}{A}\right)_q^x (1+A)_q^{1-x}$. In particular for any positive integer n

$$K(A;n) = q^{\frac{n(n-1)}{2}} \quad \text{and} \quad \Gamma_q(n) = q^{\frac{n(n-1)}{2}} \gamma_q^A(n);$$

see [49].

In this section, we will calculate $\mathcal{P}_{n,q}(t^i;x)$ for $i = 0,1,2$. By the definition of q-gamma function γ_q^1, we have

$$\int_0^{\infty/(1-q)} t^s S_{n,k}(q;t)\,d_qt = \int_0^{\infty/(1-q)} t^s e_q(-[n]_q t) q^{\frac{k(k-1)}{2}} \frac{[n]_q^k t^k}{[k]_q!}\,d_qt$$

$$= \frac{1}{[n]_q^{s+1}} \frac{1}{[k]_q!} q^{\frac{k(k-1)}{2}} \int_0^{\infty/(1-q)} \left([n]_q t\right)^{k+s} e_q(-[n]_q t)\,[n]_q\,d_qt$$

$$= \frac{1}{[n]_q^{s+1}} \frac{1}{[k]_q!} q^{\frac{k(k-1)}{2}} \int_0^{\infty/(1-q)} (u)^{k+s} e_q(-u)\,d_qu$$

$$= \frac{1}{[n]_q^{s+1}} \frac{1}{[k]_q!} q^{\frac{k(k-1)}{2}} \gamma_q^1(k+s+1)$$

$$= \frac{1}{[n]_q^{s+1}} \frac{1}{[k]_q!} q^{\frac{k(k-1)}{2}} \frac{[k+s]_q!}{q^{(k+s+1)(k+s)/2}}$$

$$= \frac{1}{[n]_q^{s+1}} \frac{[k+s]_q!}{[k]_q!} \frac{1}{q^{(2k+s)(s+1)/2}}.$$

Lemma 5.10. *We have*

$$\mathcal{P}_{n,q}(1;x) = 1, \quad \mathcal{P}_{n,q}(t;x) = x,$$

$$\mathcal{P}_{n,q}(t^2;x) = \frac{1}{q^2}x^2 + \frac{(1+q)}{q^2[n]_q}x,$$

$$\mathcal{P}_{n,q}((t-x)^2;x) = \left(\frac{1}{q^2} - 1\right)x^2 + \frac{(1+q)}{q^2[n]_q}x.$$

Proof. For $f(t) = 1$,

$$\mathcal{P}_{n,q}(1;x) = [n]_q \sum_{k=1}^{\infty} q^{k-1} S_{n,k}(q;qx) \int_0^{\infty/(1-q)} S_{n,k-1}(q;t) d_q t + e_q\left(-[n]_q qx\right)$$

$$= [n] \sum_{k=1}^{\infty} q^{k-1} S_{n,k}(q;qx) \frac{1}{[n]_q} \frac{1}{q^{k-1}} + e_q\left(-[n]_q qx\right)$$

$$= \sum_{k=1}^{\infty} S_{n,k}(q;qx) + e_q\left(-[n]_q qx\right) = \sum_{k=0}^{\infty} S_{n,k}(q;qx) = 1.$$

For $f(t) = t$

$$\mathcal{P}_{n,q}(t;x) = [n] \sum_{k=1}^{\infty} q^{k-1} S_{n,k}(q;qx) \int_0^{\infty/(1-q)} t S_{n,k-1}(q;t) d_q t$$

$$= [n] \sum_{k=1}^{\infty} q^k S_{n,k}(q;qx) \frac{[k]}{[n]^2} \frac{1}{q^{2k-1}} = \sum_{k=0}^{\infty} S_{n,k}(q;qx) \frac{[k]}{[n]} \frac{1}{q^k}$$

$$= \frac{1}{q^2} \sum_{k=0}^{\infty} S_{n,k}(q;qx) \frac{[k]}{[n]} \frac{1}{q^{k-2}} = \frac{1}{q^2} q^2 x = x.$$

For $f(t) = t^2$

$$\mathcal{P}_{n,q}(t^2;x) = [n]_q \sum_{k=1}^{\infty} q^{k-1} S_{n,k}(q;qx) \int_0^{\infty/(1-q)} t^2 S_{n,k-1}(q;t) d_q t$$

$$= \sum_{k=1}^{\infty} S_{n,k}(q;qx) \frac{[k+1]_q [k]_q}{[n]_q^2} \frac{1}{q^{2k+1}} = \sum_{k=0}^{\infty} S_{n,k}(q;qx) \frac{[k+1][k]}{[n]^2} \frac{1}{q^{2k+1}}$$

$$= \sum_{k=0}^{\infty} S_{n,k}(q;qx) \frac{\left([k]_q + q^k\right)[k]_q}{[n]_q^2} \frac{1}{q^{2k+1}} = \sum_{k=0}^{\infty} S_{n,k}(q;qx) \frac{[k]_q^2}{[n]_q^2} \frac{1}{q^{2k+1}}$$

$$+ \sum_{k=0}^{\infty} S_{n,k}(q;qx) \frac{q^k [k]_q}{[n]_q^2} \frac{1}{q^{2k+1}}$$

$$= \frac{1}{q^5} \sum_{k=0}^{\infty} S_{n,k}(q;qx) \frac{[k]_q^2}{[n]_q^2} \frac{1}{q^{2k-4}} + \frac{1}{[n]_q q^3} \sum_{k=0}^{\infty} S_{n,k}(q;qx) \frac{[k]_q}{[n]_q} \frac{1}{q^{k-2}}$$

$$= \frac{1}{q^5} \left(q^3 x^2 + \frac{q^3}{[n]_q} x\right) + \frac{1}{[n]_q q} x = \frac{1}{q^2} x^2 + \frac{1}{q^2 [n]_q} x + \frac{1}{[n]_q q} x$$

$$= \frac{1}{q^2} x^2 + \frac{(1+q)}{q^2 [n]_q} x.$$

∎

Lemma 5.11. *For all $0 < q < 1$ the following identity holds:*

$$P_{n,q}(t^m;x) = \frac{1}{[n]_q^m q^{(m^2-m)/2}} \sum_{s=1}^{m} C_s(m)\,[n]_q^s \sum_{k=0}^{\infty} \frac{[k]_q^s}{[n]_q^s}\frac{1}{q^{k(m+1)}} S_{n,k}(q;qx).$$

Proof. We have

$$P_{n,q}(t^m;x) = [n]_q \sum_{k=1}^{\infty} q^{k-1} S_{n,k}(q;qx) \int_0^{\infty/(1-q)} t^m S_{n,k-1}(q;t)\,d_qt$$

$$= [n]_q \sum_{k=1}^{\infty} q^{k-1} S_{n,k}(q;qx) \frac{1}{[n]_q^{m+1}} \frac{1}{[k-1]_q!} q^{\frac{(k-1)(k-2)}{2}} \frac{[k-1+m]_q!}{q^{(k+m)(k-1+m)/2}}$$

$$= \sum_{k=1}^{\infty} \frac{[k-1+m]_q \cdots [k]_q}{[n]_q^m} \frac{1}{q^{(m^2+2mk+2k-m)/2}} S_{n,k}(q;qx)$$

$$= \sum_{k=0}^{\infty} \frac{[k-1+m]_q \cdots [k]_q}{[n]_q^m\, q^{(m^2+2mk+2k-m)/2}} S_{n,k}(q;qx).$$

Using $[k+s]_q = [s]_q + q^s [k]_q$, we obtain

$$[k]_q\,[k+1]_q \cdots [k+m-1]_q = \prod_{s=0}^{m-1}\left([s]_q + q^s [k]_q\right) = \sum_{s=1}^{m} C_s(m)\,[k]_q^s$$

where $C_s(m) > 0$, $s = 1,2,\ldots,m$ are the constants independent of k. Hence

$$P_{n,q}(t^m;x) = \frac{1}{[n]_q^m q^{(m^2-m)/2}} \sum_{k=0}^{\infty} \frac{1}{q^{k(m+1)}} \sum_{s=1}^{m} C_s(m)\,[k]_q^s S_{n,k}(q;qx)$$

$$= \frac{1}{[n]_q^m q^{(m^2-m)/2}} \sum_{k=0}^{\infty}\sum_{s=1}^{m} C_s(m)\,[k]_q^s \frac{1}{q^{k(m+1)}} S_{n,k}(q;qx)$$

$$= \frac{1}{[n]_q^m q^{(m^2-m)/2}} \sum_{s=1}^{m} C_s(m)\,[n]_q^s \sum_{k=0}^{\infty} \left(\frac{[k]_q}{[n]_q}\right)^s \frac{1}{q^{k(m+1)}} S_{n,k}(q;qx). \quad\blacksquare$$

5.4.2 Direct Results

Let $C_B[0,\infty)$ be the space of all real-valued continuous and bounded functions f on $[0,\infty)$, endowed with the norm $\|f\| = \sup\limits_{x\in[0,\infty)} |f(x)|$. The Peetre K-functional is defined by

$$K_2(f;\delta) = \inf_{g\in C^2[0,\infty)} \left\{\|f-g\| + \delta\,\|g''\|\right\},$$

where $C_B^2[0,\infty) := \{g \in C_B[0,\infty) : g',g'' \in C_B[0,\infty)\}$. By [50, Theorem 2.4] there exists an absolute constant $M > 0$ such that

$$K_2(f,\delta) \leq M\omega_2(f;\sqrt{\delta}), \tag{5.31}$$

where $\delta > 0$ and the second-order modulus of smoothness is defined as

$$\omega_2(f;\sqrt{\delta}) = \sup_{0<h\leq\delta} \sup_{x\in[0,\infty)} |f(x+2h) - 2f(x+h) + f(x)|,$$

where $f \in C_B[0,\infty)$ and $\delta > 0$. Also, we let

$$\omega(f;\delta) = \sup_{0<h\leq\delta} \sup_{x\in[0,\infty)} |f(x+h) - f(x)|.$$

Lemma 5.12. *Let $f \in C_B[0,\infty)$. Then, for all $f \in C_B^2[0,\infty)$, we have*

$$|\mathcal{P}_{n,q}(f;x) - f(x)| \leq \left\{ \left(\frac{1-q^2}{q^2}\right)x^2 + \frac{(1+q)}{q^2\,[n]_q}x \right\} \|f''\|. \tag{5.32}$$

Proof. Let $x \in [0,\infty)$ and $f \in C_B^2[0,\infty)$. Using Taylor's formula

$$f(t) - f(x) = (t-x)f'(x) + \int_x^t (t-u)f''(u)du,$$

we can write

$$\mathcal{P}_{n,q}(f;x) - f(x) = \mathcal{P}_{n,q}((t-x)f'(x);x) + \mathcal{P}_{n,q}\left(\int_x^t (t-u)f''(u)du;x\right)$$

$$= f'(x)\mathcal{P}_{n,q}((t-x);x) + \mathcal{P}_{n,q}\left(\int_x^t (t-u)f''(u)du;x\right) - \int_x^x (x-u)f''(u)du$$

$$= \mathcal{P}_{n,q}\left(\int_x^t (t-u)f''(u)du;x\right).$$

On the other hand, since

$$\left|\int_x^t (t-u)f''(u)du\right| \leq \int_x^t |t-u|\,|f''(u)|\,du \leq \|f''\|\int_x^t |t-u|\,du \leq (t-x)^2\,\|f''\|.$$

we conclude that

$$\left| P_{n,q}(f;x) - f(x) \right| = \left| P_{n,q}\left(\int_x^t (t-u)g''(u)du; x \right) \right|$$

$$\leq P_{n,q}((t-x)^2 \|f''\|; x)$$

$$= \left\{ \left(\frac{1-q^2}{q^2} \right) x^2 + \frac{(1+q)}{q^2 [n]_q} x \right\} \|f''\| . \qquad \blacksquare$$

Lemma 5.13. *For* $f \in C[0,\infty)$, *we have*

$$\|P_{n,q}f\| \leq \|f\| .$$

Theorem 5.13. *Let* $f \in C_B[0,\infty)$. *Then, for every* $x \in [0,\infty)$, *there exists a constant* $M > 0$ *such that*

$$\left| P_{n,q}(f;x) - f(x) \right| \leq M\omega_2(f; \sqrt{\delta_n(x)}),$$

where

$$\delta_n(x) = \left(\frac{1-q^2}{q^2} \right) x^2 + \frac{(1+q)}{q^2 [n]_q} x.$$

Proof. Now, taking into account boundedness of $P_{n,q}$, we get

$$\left| P_{n,q}(f;x) - f(x) \right| = \left| P_{n,q}(f;x) - P_{n,q}(g,x) - f(x) + g(x) + P_{n,q}(g,x) - g(x) \right|$$

$$\leq \left| P_{n,q}(f-g;x) - (f-g)(x) \right| + \left| P_{n,q}(g;x) - g(x) \right|$$

$$\leq . \left| P_{n,q}(f-g;x) + (f-g)(x) \right| + \left| P_{n,q}(g;x) - g(x) \right|$$

$$\leq 2\|f-g\| + \left\{ \left(\frac{1-q^2}{q^2} \right) x^2 + \frac{(1+q)}{q^2 [n]_q} x \right\} \|g''\|$$

$$\leq 2 \left(\|f-g\| + \delta_n(x) \|g''\| \right).$$

Now, taking infimum on the right-hand side over all $g \in C_B^2[0,\infty)$ and using (5.31), we get the following result

$$\left| P_{n,q}(f;x) - f(x) \right| \leq 2K_2(f; \delta_n(x)) \leq 2M\omega_2(f; \sqrt{\delta_n(x)}). \qquad \blacksquare$$

Theorem 5.14. *Let* $0 < \alpha \leq 1$ *and* E *be any subset of the interval* $[0,\infty)$. *Then, if* $f \in C_B[0,\infty)$ *is locally* $Lip(\alpha)$, *i.e., the condition*

$$|f(y) - f(x)| \leq L|y-x|^\alpha, \ y \in E \ and \ x \in [0,\infty) \qquad (5.33)$$

holds, then, for each $x \in [0,\infty)$, *we have*

$$\left|\mathcal{P}_{n,q}(f;x)-f(x)\right|\leq L\left\{\delta_n^{\frac{\alpha}{2}}(x)+2\left(d\left(x,E\right)\right)^{\alpha}\right\},$$

where L is a constant depending on α and f ; and d (x,E) is the distance between x and E defined as

$$d\left(x,E\right)=\inf\left\{|t-x|:t\in E\right\}.$$

Proof. Let $\overline{E}$ denote the closure of E in $[0,\infty)$. Then, there exists a point $x_0 \in \overline{E}$ such that $|x - x_0| = d(x,E)$. Using the triangle inequality

$$|f(t)-f(x)|\leq|f(t)-f(x_0)|+|f(x)-f(x_0)|$$

we get, by (5.33)

$$
\begin{aligned}
\left|\mathcal{P}_{n,q}(f;x)-f(x)\right| &\leq \mathcal{P}_{n,q}(|f(y)-f(x_0)|;x)+\mathcal{P}_{n,q}(|f(x)-f(x_0)|;x)\\
&\leq L\left\{\mathcal{P}_{n,q}(|t-x_0|^{\alpha};x)+|x-x_0|^{\alpha}\right\}\\
&\leq L\left\{\mathcal{P}_{n,q}(|t-x|^{\alpha}+|x-x_0|^{\alpha};x)+|x-x_0|^{\alpha}\right\}\\
&= L\left\{\mathcal{P}_{n,q}(|t-x|^{\alpha};x)+2|x-x_0|^{\alpha}\right\}.
\end{aligned}
$$

Using the Hölder inequality with $p=\frac{2}{\alpha},q=\frac{2}{2-\alpha}$, we find that

$$
\begin{aligned}
\left|\mathcal{P}_{n,q}(f;x)-f(x)\right| &\leq L\left\{\left[\mathcal{P}_{n,q}(|t-x|^{\alpha p};x)\right]^{\frac{1}{p}}\left[\mathcal{P}_{n,q}(1^q;x)\right]^{\frac{1}{q}}+2\left(d\left(x,E\right)\right)^{\alpha}\right\}\\
&= M\left\{\left[\mathcal{P}_{n,q}(|t-x|^2;x)\right]^{\frac{\alpha}{2}}+2\left(d\left(x,E\right)\right)^{\alpha}\right\}\\
&\leq M\left\{\left[\left(\frac{1-q^2}{q^2}\right)x^2+\frac{(1+q)}{q^2[n]_q}x\right]^{\frac{\alpha}{2}}+2\left(d\left(x,E\right)\right)^{\alpha}\right\}\\
&= M\left\{\delta_n^{\frac{\alpha}{2}}(x)+2\left(d\left(x,E\right)\right)^{\alpha}\right\}. \qquad\blacksquare
\end{aligned}
$$

We consider the following classes of functions:

$$C_m[0,\infty):=\left\{f\in C[0,\infty):\exists M_f>0\ |f(x)|<M_f(1+x^m)\ \text{and}\ \|f\|_m:=\sup_{x\in[0,\infty)}\frac{|f(x)|}{1+x^m}\right\}.$$

$$C_m^*[0,\infty):=\left\{f\in C_m[0,\infty):\lim_{x\to\infty}\frac{|f(x)|}{1+x^m}<\infty\right\},\quad m\in\mathbb{N}.$$

Next, we obtain a direct approximation theorem in $C_1^*[0,\infty)$ and an estimation in terms of the weighted modulus of continuity. It is known that if f is not uniformly continuous on the interval $[0,\infty)$, then the usual first modulus of continuity $\omega(f,\delta)$

does not tend to zero, as $\delta \to 0$. For every $f \in C_m^* [0, \infty)$ the weighted modulus of continuity is defined as follows

$$\Omega_m (f, \delta) = \sup_{x \geq 0,\ 0 < h \leq \delta} \frac{|f(x+h) - f(x)|}{1 + (x+h)^m}.$$

See [112].

Lemma 5.14 ([112]). *Let $f \in C_m^* [0, \infty)$, $m \in \mathbb{N}$. Then, we have the following:*

1. *$\Omega_m (f, \delta)$ is a monotone increasing function of δ.*
2. *$\lim_{\delta \to 0^+} \Omega_m (f, \delta) = 0$.*
3. *For any $\alpha \in [0, \infty)$, $\Omega_m (f, \alpha\delta) \leq (1 + \alpha) \Omega_m (f, \delta)$.*

In the next theorem we give an expression of the approximation error with the operators $S_{n,q}$ by means of Ω_1.

Theorem 5.15. *If $f \in C_1^* [0, \infty)$, then the inequality*

$$\left\| \mathcal{P}_{n,q} (f) - f \right\|_2 \leq k(q) \Omega_1 \left(f; \frac{1}{\sqrt{[n]_q}} \right),$$

where k is a constant independent of f and n.

Proof. From the definition of $\Omega_1 (f, \delta)$ and Lemma 5.14, we may write

$$|f(t) - f(x)| \leq (1 + x + |t - x|) \left(\frac{|t - x|}{\delta} + 1 \right) \Omega_1 (f, \delta)$$

$$\leq (1 + 2x + t) \left(\frac{|t - x|}{\delta} + 1 \right) \Omega_1 (f, \delta).$$

Then

$$\left| \mathcal{P}_{n,q} (f; x) - f(x) \right| \leq \mathcal{P}_{n,q} (|f(t) - f(x)|; x) \leq \Omega_1 (f, \delta) \left(\mathcal{P}_{n,q} ((1 + 2x + t); x) \right.$$

$$\left. + \mathcal{P}_{n,q} \left((1 + 2x + t) \frac{|t - x|}{\delta}; x \right) \right).$$

Applying the Cauchy–Schwarz inequality to the second term, we get

$$\mathcal{P}_{n,q} \left((1 + 2x + t) \frac{|t - x|}{\delta}; x \right) \leq \left(\mathcal{P}_{n,q} \left((1 + 2x + t)^2; x \right) \right)^{1/2} \left(\mathcal{P}_{n,q} \left(\frac{|t - x|^2}{\delta^2}; x \right) \right)^{1/2}.$$

Consequently

$$\left|\mathcal{P}_{n,q}\left(f;x\right)-f\left(x\right)\right| \leq \Omega_m\left(f,\delta\right)\left(\mathcal{P}_{n,q}\left(\left(1+2x+t\right);x\right)\right.$$

$$\left.+\left(\mathcal{P}_{n,q}\left(\left(1+2x+t\right)^2;x\right)\right)^{1/2}\left(\mathcal{P}_{n,q}\left(\frac{\left|t-x\right|^2}{\delta^2};x\right)\right)^{1/2}\right).$$

$$(5.34)$$

On the other hand, there is a positive constant $K\left(q\right)$ such that

$$\mathcal{P}_{n,q}\left(\left(1+2x+t\right);x\right) = 1+3x \leq 3\left(1+x\right),$$

$$\left(\mathcal{P}_{n,q}\left(\left(1+2x+t\right)^2;x\right)\right)^{1/2} = \left(\left(\left(1+2x\right)^2+\left(1+2x\right)x+\frac{1}{q^2}x^2+\frac{\left(1+q\right)}{q^2\left[n\right]_q}x;x\right)\right)^{1/2}$$

$$\leq K\left(q\right)\left(1+x\right),\qquad (5.35)$$

and

$$\left(\mathcal{P}_{n,q}\left(\frac{\left|t-x\right|^2}{\delta^2};x\right)\right)^{1/2} = \frac{1}{\delta q}\sqrt{\left(1-q^2\right)x^2+\frac{\left(1+q\right)}{\left[n\right]_q}x} \leq \frac{1}{\delta q}\sqrt{\frac{\left(1-q^n\right)}{\left[n\right]_q}x^2+\frac{2x}{\left[n\right]_q}}$$

$$\leq \frac{2}{\delta q\sqrt{\left[n\right]_q}}\sqrt{x^2+x} \leq \frac{2}{\delta q\sqrt{\left[n\right]_q}}\left(1+x\right). \qquad (5.36)$$

Now from (5.34)–(5.36), we have

$$\left|\mathcal{P}_{n,q}\left(f;x\right)-f\left(x\right)\right| \leq \Omega_1\left(f,\delta\right)\left(3\left(1+x\right)+K\left(q\right)\frac{2\left(1+x\right)^2}{q\delta\sqrt{\left[n\right]_q}}\right)$$

$$\leq \left(1+x^2\right)\Omega_1\left(f,\delta\right)\left(3K_1+K\left(q\right)\frac{4}{q\delta\sqrt{\left[n\right]_q}}\right),$$

where

$$K_1 = \sup_{x\geq 0}\frac{1+x^m+x+x^{m+1}}{1+x^{m+1}}.$$

If we take $\delta=\left[n\right]^{q-\frac{1}{2}}$, then from the above inequality we obtain the desired result.   ∎

5.4.3   Voronovskaja-Type Theorem

In this section, we proceed to state and prove a Voronovskaja-type theorem for the q-Phillips operators. We first prove the following lemma:

Lemma 5.15. *Let* $0 < q < 1$. *We have*

$$P_{n,q}(t^3;x) = \frac{1}{q^6}x^3 + \frac{[2]_q[3]_q}{[n]_q q^6}x^2 + \frac{[2][3]}{[n]^2 q^5}x$$

$$P_{n,q}(t^4;x) = \frac{1}{q^{12}}x^4 + \frac{[2]_q[3]_q(1+q^2)}{[n]_q q^{12}}x^3 + \frac{[2]_q[3]_q^2(1+q^2)}{[n]_q^2 q^{11}}x^2 + \frac{[2]_q^2[3]_q(1+q^2)}{[n]_q^3 q^9}x.$$

Proof. Simple calculations show that

$$P_{n,q}(t^3;x) = \frac{1}{[n]_q^3 q^3}\sum_{k=0}^{\infty}\frac{[k+2]_q[k+1]_q[k]_q}{q^{3k}}S_{n,k}(q;qx)$$

$$= \frac{1}{[n]_q^3 q^3}\sum_{k=0}^{\infty}\frac{[k]_q^3+q^k(2+q)[k]_q^2+q^{2k}(1+q)[k]_q}{q^{3k}}S_{n,k}(q;qx)$$

$$= \frac{1}{[n]_q^3 q^3}\left\{\sum_{k=0}^{\infty}\frac{[k]_q^3}{q^{3k}}S_{n,k}(q;qx) + \sum_{k=0}^{\infty}\frac{(2+q)[k]_q^2}{q^{2k}}S_{n,k}(q;qx)\right.$$

$$\left. + \sum_{k=0}^{\infty}\frac{(1+q)[k]_q}{q^k}S_{n,k}(q;qx)\right\}$$

$$= \frac{1}{q^3}\sum_{k=0}^{\infty}\frac{[k]_q^3}{[n]_q^3 q^{3k}}S_{n,k}(q;qx) + \frac{(2+q)}{[n]_q q^3}\sum_{k=0}^{\infty}\frac{[k]_q^2}{[n]_q^2 q^{2k}}S_{n,k}(q;qx)$$

$$+ \frac{(1+q)}{[n]_q^2 q^3}\sum_{k=0}^{\infty}\frac{[k]_q}{[n]_q q^k}S_{n,k}(q;qx)$$

$$= \frac{1}{q^9}\sum_{k=0}^{\infty}\frac{[k]_q^3}{[n]_q^3 q^{3k-6}}S_{n,k}(q;qx) + \frac{(2+q)}{[n]_q q^7}\sum_{k=0}^{\infty}\frac{[k]_q^2}{[n]_q^2 q^{2k-4}}S_{n,k}(q;qx)$$

$$+ \frac{(1+q)}{[n]_q^2 q^5}\sum_{k=0}^{\infty}\frac{[k]_q}{[n]_q q^{k-2}}S_{n,k}(q;qx)$$

$$= \frac{1}{q^9}\sum_{k=0}^{\infty}\left(\frac{[k]_q}{[n]_q q^{k-2}}\right)^3 S_{n,k}(q;qx) + \frac{(2+q)}{[n]_q q^7}\sum_{k=0}^{\infty}\left(\frac{[k]_q}{[n]_q q^{k-2}}\right)^2 S_{n,k}(q;qx)$$

$$+ \frac{(1+q)}{[n]_q^2 q^5} \sum_{k=0}^{\infty} \frac{[k]_q}{[n]_q q^{k-2}} S_{n,k}(q;qx)$$

$$= \frac{1}{q^9} \left(\frac{q^4}{[n]_q^2} x + (2q^4 + q^3) \frac{x^2}{[n]_q} + q^3 x^3 \right) + \frac{(2+q)}{[n]_q q^7} \left(q^3 x^2 + \frac{q^3}{[n]_q} x \right) + \frac{(1+q)q^2}{[n]_q^2 q^5} x$$

$$= \frac{1}{q^5 [n]_q^2} x + \frac{(2q+1)}{q^6 [n]_q} x^2 + \frac{1}{q^6} x^3 + \frac{(2+q)}{[n]_q q^4} x^2 + \frac{(2+q)}{[n]_q^2 q^4} x + \frac{(1+q)}{[n]_q^2 q^3} x$$

$$= \frac{1}{q^6} x^3 + \frac{(1+2q+2q^2+q^3)}{q^6 [n]_q} x^2 + \frac{(1+2q+2q^2+q^3)}{q^5 [n]_q^2} x$$

$$= \frac{1}{q^6} x^3 + \frac{(1+q)(1+q+q^2)}{[n]_q q^6} x^2 + \frac{(1+q)(1+q+q^2)}{[n]_q^2 q^5} x$$

$$\mathcal{P}_{n,q}(t^4;x)$$

$$= \frac{1}{[n]_q^4 q^6} \sum_{k=0}^{\infty} \frac{[k+3]_q [k+2]_q [k+1]_q [k]_q}{q^{4k}} S_{n,k}(q;qx)$$

$$= \frac{1}{q^{14}} \sum_{k=0}^{\infty} \left(\frac{[k]_q}{[n]_q q^{k-2}} \right)^4 S_{n,k}(q;qx) + \frac{(3+2q+q^2)}{[n]_q q^{12}} \sum_{k=0}^{\infty} \left(\frac{[k]_q}{[n]_q q^{k-2}} \right)^3 S_{n,k}(q;qx)$$

$$+ \frac{(3+4q+3q^2+q^3)}{[n]_q^2 q^{10}} \sum_{k=0}^{\infty} \left(\frac{[k]_q}{[n]_q q^{k-2}} \right)^2 S_{n,k}(q;qx)$$

$$+ \frac{(1+2q+2q^2+q^3)}{[n]_q^3 q^8} \sum_{k=0}^{\infty} \frac{[k]_q}{[n]_q q^{k-2}} S_{n,k}(q;qx)$$

$$= \frac{1}{q^{14}} \left(\frac{q^5}{[n]_q^3} x + (3q^3 + 3q^2 + q) \frac{q^2}{[n]_q^2} x^2 + \left(3q + 2 + \frac{1}{q} \right) \frac{q^3}{[n]_q} x^3 + q^2 x^4 \right)$$

$$+ \frac{(3+2q+q^2)}{[n]_q q^{12}} \left(\frac{q^4}{[n]_q^2} x + (2q^2 + q) \frac{q^2}{[n]_q} x^2 + q^3 x^3 \right)$$

$$+ \frac{(3+4q+3q^2+q^3)}{[n]_q^2 q^{10}} \left(q^3 x^2 + \frac{q^3}{[n]_q} x \right) + \frac{(1+2q+2q^2+q^3)q^2}{[n]_q^3 q^8} x$$

$$= \frac{1}{q^{12}} x^4 + \frac{1+2q+3q^2+(3+2q+q^2)q^3}{[n]_q q^{12}} x^3$$

$$+\frac{1+3q+3q^2+(3+2q+q^2)(2q+1)q^2+(3+4q+3q^2+q^3)q^4}{[n]_q^2 q^{11}}x^2$$

$$+\frac{1+(3+2q+q^2)q+(3+4q+3q^2+q^3)q^2+(1+2q+2q^2+q^3)q^3}{[n]_q^3 q^9}x$$

$$=\frac{1}{q^{12}}x^4+\frac{(1+q)(1+q^2)(1+q+q^2)}{[n]_q q^{12}}x^3+\frac{(1+q)(1+q^2)(1+q+q^2)^2}{[n]_q^2 q^{11}}x^2$$

$$+\frac{(1+q)^2(1+q^2)(1+q+q^2)}{[n]_q^3 q^9}x. \qquad\blacksquare$$

Theorem 5.16. *Let $q_n \in (0,1)$. Then the sequence $\{\mathcal{P}_{n,q_n}(f)\}$ converges to f uniformly on $[0,A]$ for each $f \in C_2^*[0,\infty)$ if and only if $\lim\limits_{n\to\infty} q_n = 1$.*

Proof. The proof is similar to that of Theorem 2 [86]. $\qquad\blacksquare$

Lemma 5.16. *Assume that $q_n \in (0,1)$, $q_n \to 1$, and $q_n^n \to a$ as $n \to \infty$. For every $x \in [0,\infty)$ there hold*

$$\lim_{n\to\infty} [n]_{q_n} \mathcal{P}_{n,q_n}((t-x)^2;x) = 2(1-a)x^2+2x,$$

$$\lim_{n\to\infty} [n]_{q_n}^2 \mathcal{P}_{n,q_n}((t-x)^4;x) = 12x^2+24(1-a)x^3+12(1-a)^2x^4.$$

Proof. First, we have

$$\lim_{n\to\infty} [n]_{q_n} \mathcal{P}_{n,q_n}((t-x)^2;x) = \lim_{n\to\infty} [n]_{q_n}\left\{\left(\frac{1}{q_n^2}-1\right)x^2+\frac{(1+q_n)}{q_n^2[n]_{q_n}}x\right\}$$

$$= \lim_{n\to\infty}\left(\frac{(1-q_n^n)(1+q_n)}{q_n^2}x^2+\frac{(1+q_n)}{q_n^2}x\right)$$

$$= 2(1-a)x^2+2x.$$

In order to calculate the second limit, we need expression for $\mathcal{P}_{n,q_n}((t-x)^4;x)$:

$$\mathcal{P}_{n,q_n}((t-x)^4;x)$$

$$= \mathcal{P}_{n,q_n}(t^4;x) - 4x\mathcal{P}_{n,q_n}(t^3;x) + 6x^2\mathcal{P}_{n,q_n}(t^2;x) - 4x^3\mathcal{P}_{n,q_n}(t;x) + x^4$$

$$= \frac{1}{q_n^{12}}x^4 + \frac{[2]_{q_n}[3]_{q_n}(1+q_n^2)}{[n]_{q_n}q_n^{12}}x^3 + \frac{[2]_{q_n}[3]_{q_n}^2(1+q_n^2)}{[n]_{q_n}^2 q_n^{11}}x^2 + \frac{[2]_{q_n}^2[3]_{q_n}(1+q_n^2)}{[n]_{q_n}^3 q_n^9}x$$

$$-4x\left\{\frac{1}{q_n^6}x^3+\frac{[2]_{q_n}[3]_{q_n}}{[n]_{q_n}q_n^6}x^2+\frac{[2]_{q_n}[3]_{q_n}}{[n]_{q_n}^2q_n^5}x\right\}+6x^2\left\{\frac{1}{q_n^2}x^2+\frac{[2]_{q_n}}{q_n^2[n]_{q_n}}x\right\}-3x^4$$

$$=\frac{(1-4q_n^6+6q_n^{10}-3q_n^{12})}{q_n^{12}}x^4+\left\{\frac{[2]_{q_n}[3]_{q_n}(1+q_n^2)-4[2]_{q_n}[3]_{q_n}q_n^6+6q_n^{10}[2]_{q_n}}{q_n^{12}[n]_{q_n}}\right\}x^3$$

$$+\left\{\frac{[2]_{q_n}[3]_{q_n}^2(1+q_n^2)-4q_n^6[2]_{q_n}[3]_{q_n}}{q_n^{11}[n]_{q_n}^2}\right\}x^2$$

$$+\frac{[2]_{q_n}^2[3]_{q_n}(1+q_n^2)}{[n]_{q_n}^3q_n^9}x$$

$$=\frac{(1+2q_n^2+3q_n^4-3q_n^8)(1-q_n^n)^2(q_n+1)^2}{q_n^{12}[n]_{q_n}^2}x^4$$

$$+\left\{\frac{(q_n^n-1)(q_n+1)(2q_n^7-4q_n^2-5q_n^3-6q_n^4-6q_n^5-2q_n^6-2q_n+6q_n^8+6q_n^9-1)}{q_n^{12}[n]_{q_n}^2}\right\}x^3$$

$$+\left\{\frac{[2]_{q_n}[3]_{q_n}^2(1+q_n^2)-4q_n^6[2]_{q_n}[3]_{q_n}}{q_n^{11}[n]_{q_n}^2}\right\}x^2+\frac{[2]_{q_n}^2[3]_{q_n}(1+q_n^2)}{[n]_{q_n}^3q_n^9}x.$$

Thus

$$\lim_{n\to\infty}[n]_{q_n}^2\mathcal{P}_{n,q_n}((t-x)^4;x)$$

$$=\lim_{n\to\infty}\frac{(1-q_n^n)^2}{(1-q_n)^2}\left\{\frac{(2q_n^2+3q_n^4-3q_n^8+1)(q_n-1)^2(q_n+1)^2}{q_n^{12}}x^4\right.$$

$$+\left(\frac{(q_n-1)(q_n+1)(2q_n^7-4q_n^2-5q_n^3-6q_n^4-6q_n^5-2q_n^6-2q_n+6q_n^8+6q_n^9-1)}{q_n^{12}[n]_{q_n}}\right)x^3$$

$$+\left(\frac{(q_n+1)(q_n+2q_n^2+q_n^3+q_n^4-4q_n^6+1)(q_n+q_n^2+1)}{q_n^{11}[n]_{q_n}^2}\right)x^2$$

$$+\left.\left(\frac{(1+q_n)^2(1+q_n^2)(1+q_n+q_n^2)}{q_n^9[n]_{q_n}^3}\right)x\right\}$$

$$=12(1-a)^2x^4+24(1-a)x^3+12x^2.\qquad\blacksquare$$

Theorem 5.17. *Assume that $q_n \in (0,1)$, $q_n \to 1$, and $q_n^n \to a$ as $n \to \infty$. For any $f \in C_2^*[0,\infty)$ such that $f', f'' \in C_2^*[0,\infty)$, the following equality holds*

$$\lim_{n\to\infty} [n]_{q_n} (\mathcal{P}_{n,q_n}(f;x) - f(x)) = ((1-a)x^2 + x) f''(x)$$

uniformly on any $[0,A]$, $A > 0$.

Proof. Let $f, f', f'' \in C_2^*[0,\infty)$ and $x \in [0,\infty)$ be fixed. By the Taylor formula we may write

$$f(t) = f(x) + f'(x)(t-x) + \frac{1}{2}f''(x)(t-x)^2 + r(t;x)(t-x)^2, \qquad (5.37)$$

where $r(t;x)$ is the Peano form of the remainder, $r(.;x) \in C_2^*[0,\infty)$, and $\lim_{t\to x} r(t;x) = 0$. Applying $\mathcal{P}_{n,q_n}$ to (5.37) we obtain

$$[n]_{q_n} (\mathcal{P}_{n,q_n}(f;x) - f(x)) = \frac{1}{2}f''(x) [n]_{q_n} \mathcal{P}_{n,q_n} \left((t-x)^2;x\right)$$

$$+ [n]_{q_n} \mathcal{P}_{n,q_n} \left(r(t;x)(t-x)^2;x\right).$$

By the Cauchy–Schwarz inequality, we have

$$\mathcal{P}_{n,q_n} \left(r(t;x)(t-x)^2;x\right) \leq \sqrt{\mathcal{P}_{n,q_n}(r^2(t;x);x)} \sqrt{\mathcal{P}_{n,q_n}\left((t-x)^4;x\right)}. \qquad (5.38)$$

Observe that $r^2(x;x) = 0$ and $r^2(.;x) \in C_2^*[0,\infty)$. Then it follows from Theorem 5.16 that

$$\lim_{n\to\infty} \mathcal{P}_{n,q_n}\left(r^2(t;x);x\right) = r^2(x;x) = 0 \qquad (5.39)$$

uniformly with respect to $x \in [0,A]$. Now from (5.38) and (5.39) and Lemma 5.16, we get immediately

$$\lim_{n\to\infty} [n]_{q_n} \mathcal{P}_{n,q_n} \left(r(t;x)(t-x)^2;x\right) = 0.$$

Then we get the following

$$\lim_{n\to\infty} [n]_{q_n} (\mathcal{P}_{n,q_n}(f;x) - f(x))$$

$$= \lim_{n\to\infty} \left(\frac{1}{2}f''(x) [n]_{q_n} \mathcal{P}_{n,q_n}\left((t-x)^2;x\right) + [n]_{q_n} \mathcal{P}_{n,q_n}\left(r(t;x)(t-x)^2;x\right)\right)$$

$$= ((1-a)x^2 + x) f''(x). \qquad \blacksquare$$

Chapter 6
Statistical Convergence of q-Operators

One of the most recently studied subject in approximation theory is the approximation of function by linear positive operators using A-statistical convergence or a matrix summability method. In approximation theory by linear positive operators, the statistical convergence has been examined for the first time by Gadjiev and Orhan [69].

First of all, we recall the concept of statistical convergence.

Let us recall the concept of statistical convergence. The density of a set $K \subset \mathbb{N}$ is defined by

$$\delta(K) = \lim_{n \to \infty} \frac{1}{n} \sum_{k=1}^{n} \chi_K(k)$$

provided the limit exists, where χ_K is the characteristic function of K. Clearly, the sum of the right-hand side represents the cardinality of the set $\{k \leq n : k \in K\}$. Following [69], a sequence $x = (x_k)_{k \geq 1}$ is statistically convergent to a real number L if, for every $\varepsilon > 0$,

$$\delta(\{k \in \mathbb{N} : |x_k - L| \geq \varepsilon\}) = 0.$$

In this case we write $st - \lim_n x_n = L$. It is known that any convergent sequence is statistically convergent, but not conversely. Closely related to this notion is A-statistical convergence, where $A = (a_{n,k})$ is an infinite summability matrix. For a given sequence $x = (x_k)_{k \geq 1}$, the A-transform of x denoted by $(Ax)_n$ $n \in \mathbb{N}$ is defined by

$$(Ax)_n = \sum_{k=1}^{\infty} a_{n,k} x_k, \quad n \in \mathbb{N},$$

provided the series converges for each n. Suppose that A is nonnegative regular summability matrix. Then x is A-statistically convergent to the real number L if, for every $\varepsilon > 0$, one has

$$\lim_{n \to \infty} \sum_{k \in I(\varepsilon)} a_{n,k} = 0$$

where $I(\varepsilon) = \{k \in \mathbb{N} : |x_k - L| \geq \varepsilon\}$. We write $st_A - \lim_n x = L$, see, e.g., [64, 69].

A. Aral et al., *Applications of q-Calculus in Operator Theory*,
DOI 10.1007/978-1-4614-6946-9_6, © Springer Science+Business Media New York 2013

6.1 General Class of Positive Linear Operators

In this section, a general class of positive linear operators of discrete type based on q-calculus is presented, and their weighted statistical approximation properties are investigated by using a Bohman–Korovkin-type theorem. We also mark out two particular cases of this general class of operators.

This section is based on [136].

6.1.1 Notations and Preliminary Results

We use the following notations:

$$(a+b)_q^n = \prod_{s=0}^{n-1} (a+q^s b), \quad n \in \mathbb{N}, \quad a,b \in \mathbb{R}, \tag{6.1}$$

$$(1+a)_q^\infty = \prod_{s=0}^{n-1} (1+q^s a) \qquad a \in \mathbb{R}, \tag{6.2}$$

$$(1+a)_q^t = \frac{(1+a)_q^\infty}{(1+q^t a)_q^\infty}, \quad a,t \in \mathbb{R}. \tag{6.3}$$

Note that the infinite product (6.2) is convergent if $q \in (0,1)$. Throughout the section we consider $q \in (0,1)$.

We recall the q-Taylor theorem as it is given in [59, pp. 103].

Theorem 6.1. *If the function $f(x)$ is capable of expansion as a convergent power series and q is not a root of unity, then*

$$f(x) = \sum_{n=0}^{\infty} \frac{(x-a)_q^n}{[n]_q!} D_q^n f(a),$$

where

$$(x-a)_q^n = \prod_{s=0}^{n-1} (x-q^s a) = \sum_{k=0}^{n} \begin{bmatrix} n \\ k \end{bmatrix}_q q^{\frac{k(k-1)}{2}} x^{n-k}(-a)^k. \tag{6.4}$$

6.1.2 Construction of the Operators

Letting $\mathbb{R}_+ = [0,\infty)$ and $\mathbb{N}_0 = \{0\} \cup \mathbb{N}$, by $C_B(\mathbb{R}_+)$, we denote the space of all continuous real-valued functions on $\mathbb{R}_+$ and bounded on the entire positive axis. Baskakov [37] introduced the operators $L_n : C_B(\mathbb{R}_+) \longrightarrow C(J), n \in \mathbb{N}$,

$$L_n f(x) = \sum_{k=0}^{\infty} \frac{(-x)^k}{k!} \varphi_n^{(k)}(x) f\left(\frac{k}{n}\right), \tag{6.5}$$

which are generated by a sequence of functions $(\varphi_n)_{n\geq 1}$, $\varphi_n : \mathbb{C} \longrightarrow \mathbb{C}$, having the following properties:

(1) $\varphi_n, n \in \mathbb{N}$ are analytic on a domain D containing the disc $\{z \in \mathbb{C} : |z - R| \leq R\}$ and $J = [0, R]$.
(2) $\varphi_n(0) = 1, n \in \mathbb{N}$.
(3) $\varphi_n, n \in \mathbb{N}$ are completely monotone on J, i.e., $(-1)^k \varphi_n(k) \geq 0$ for $x \in J, k \in \mathbb{N}_0, n \in \mathbb{N}$.
(4) There exists a positive integer $m(n)$ such that

$$\varphi_n^{(k)}(x) = -n\varphi_{m(n)}^{(k-1)}(x)(1 + \alpha_{k,n}(x)), \quad x \in J, \quad (n,k) \in \mathbb{N} \times \mathbb{N},$$

where $\alpha_{k,n}(x)$ converges to zero uniformly in k and x on J for n tending to infinity.
(5) $\lim_{n\to\infty} \frac{n}{m(n)} = 1$.

We set $e_i, e_i(x) = x^i, i \geq 0$.

Let $(\phi_n)_{n \in \mathbb{N}}$ be a sequence of real functions on $\mathbb{R}_+$ which are continuously infinitely q-differentiable on $\mathbb{R}_+$ satisfying the following conditions:

(P1)
$$\phi_n(0) = 1, \quad n \in \mathbb{N}. \tag{6.6}$$

(P2)
$$(-1^k) D_q^k \phi_n(x) \geq 0, \ n \in \mathbb{N}, \ k \in \mathbb{N}_0, x \geq 0. \tag{6.7}$$

(P3)   For all $(x,k) \in \mathbb{N} \times \mathbb{N}_0$, there exists a positive integer $i_k, 0 \leq i_k \leq k$, such that

$$D_q^{k+1} \phi_n(x) = (-1)^{i_k+1} D_q^{k-i} \phi_n(q^{i_k+1}x)\beta_{n,k,i_k,q}(x), \tag{6.8}$$

where

$$st - \lim_n \frac{\beta_{n,k,i_k,q}(0)}{[n]_q^{i_k+1} q^k} = 1. \tag{6.9}$$

Remark 6.1. Multiplying (6.8) by $(-1)^{k-2i_k+1}$, we get

$$(-1)^{k+1} D_q^{k+1} \phi_n(x) = (-1)^{k-i_k} D_q^{k-i_k} \phi_n(q^{i_k+1}x)\beta_{n,k,i_k,q}(x).$$

The last equality and (6.7) yield that

$$\beta_{n,k,i_k,q}(x) \geq 0, \tag{6.10}$$

for all $x \in \mathbb{R}_+, (n,k) \in \mathbb{N} \times \mathbb{N}_0, q \in (0,1)$.

We set

$$C_N(\mathbb{R}_+) = \left\{ f \in C(\mathbb{R}_+) : (\exists) \lim_{x \longrightarrow \infty} \frac{f(x)}{1+x^N} < \infty \right\}, \quad N \geq 2.$$

Endowed with the norm $\|.\|_N$, this space is Banach space, where

$$\|f\|_N = \sup_{x \geq 0} \frac{|f(x)|}{1+x^N}. \tag{6.11}$$

Inspired by the Baskakov operators (6.5), we introduce the announced q-operators as follows:

$$T_n(f;q;x) = \sum_{k=0}^{\infty} \frac{(-x)^k}{[k]_q!} q^{\frac{k(k-1)}{2}} D_q^k \phi_n(x) f\left(\frac{[k]_q}{[n]_q q^{k-1}} \right) \tag{6.12}$$

for all $f \in C_2(\mathbb{R}_+), x \in \mathbb{R}_+, q \in (0,1), n \in \mathbb{N}$, where $(\phi_n)_n$ is a sequence of functions satisfying $(P1)$–$(P3)$.-

It is obvious that $T_n, n \in \mathbb{N}$, are positive and linear operators.

Lemma 6.1. *For all $n \in \mathbb{N}, x \in \mathbb{R}_+$, and $0 < q < 1$, we have*

$$T_n(e_0;q;x) = 1, \tag{6.13}$$

$$T_n(e_1;q;x) = -x \frac{D_q\phi_n(0)}{[n]_q} \tag{6.14}$$

$$T_n(e_2;q;x) = x^2 \frac{D_q^2\phi_n(0)}{q[n]_q^2} - \frac{D_q\phi_n(0)}{[n]_q^2} \tag{6.15}$$

Proof. For a fixed $x \in \mathbb{R}_+$, by Theorem 6.1, we obtain

$$\phi_n(t) = \sum_{k=0}^{\infty} \frac{(t-x)_q^k}{[k]_q!} D_q^k \phi_n(x). \tag{6.16}$$

Choosing $t = 0$ in the above relation and taking into account (6.6), (6.1) and $(-x)_q^k = (-x)^k q^{\frac{k(k-1)}{2}}$, we get

$$\sum_{k=0}^{\infty} \frac{(-x)^k}{[k]_q!} q^{\frac{k(k-1)}{2}} D_q^k \phi_n(x) = \phi_n(0) = 1,$$

and (6.13) is proved.

Using (6.16) we can write the q-derivative of ϕ_n with respect to t as

$$D_q\phi_n(t) = \sum_{k=1}^{\infty} \frac{[k]_q}{[k]_q!} (t-x)_q^{k-1} D_q^k \phi_n(x). \tag{6.17}$$

For getting the above identity we used the formula $D_q(t+a)^k = [k]_q (t+a)^{k-1}$, see (1.4).

Multiplying (6.17) by $(-x)$ and choosing $t = 0$ we obtain

$$-xD_q\phi_n(0) = \sum_{k=1}^{\infty} \frac{[k]_q}{[k]_q!} (-x)^k q^{\frac{(k-1)(k-2)}{2}} D_q^k \phi_n(x), \qquad (6.18)$$

which yields (6.14).

We use a similar technique to get (6.15). Differentiating (6.17) with respect to t we get

$$D_q^2\phi_n(t) = \sum_{k=2}^{\infty} \frac{[k]_q [k-1]_q}{[k]_q!} (t-x)_q^{k-2} D_q^k \phi_n(x).$$

Now choosing again $t = 0$ one has

$$D_q^2\phi_n(0) = \sum_{k=2}^{\infty} \frac{[k]_q [k-1]_q}{[k]_q!} (-x)^{k-2} q^{\frac{(k-2)(k-3)}{2}} D_q^k \phi_n(x). \qquad (6.19)$$

From (6.18) and (6.19) we have

$$x^2 D_q^2\phi_n(0) - xqD_q\phi_n(0) = \sum_{k=1}^{\infty} \frac{[k]_q [k-1]_q}{[k]_q!} (-x)^k q^{\frac{(k-2)(k-3)}{2}} D_q^k \phi_n(x)$$

$$+ q\sum_{k=1}^{\infty} \frac{[k]_q}{[k]_q!} (-x)^k q^{\frac{(k-1)(k-2)}{2}} D_q^k \phi_n(x)$$

$$= \sum_{k=1}^{\infty} \frac{[k]_q}{[k]_q!} (-x)^k q^{\frac{(k-2)(k-3)}{2}} D_q^k \phi_n(x) \left([k-1]_q + q^{k-1}\right)$$

$$= \sum_{k=1}^{\infty} \frac{[k]_q^2}{[k]_q!} (-x)^k q^{\frac{(k-2)(k-3)}{2}} D_q^k \phi_n(x),$$

where we used the fact that $[k-1]_q + q^{k-1} = [k]_q$ for all $k \in \mathbb{N}$.

On the other hand

$$T_n(e_2;q;x) = \frac{1}{q [n]_q^2} \sum_{k=1}^{\infty} \frac{[k]_q^2}{[k]_q!} (-x)^k q^{\frac{(k-2)(k-3)}{2}} D_q^k \phi_n(x)$$

$$= x^2 \frac{D_q^2\phi_n(0)}{q [n]_q^2} - x \frac{D_q\phi_n(0)}{[n]_q^2},$$

and (6.15) follows. The proof is complete. ∎

Remark 6.2. Since any linear and positive operator is monotone, relation (6.15) guarantees that $T_n f \in C_2(\mathbb{R}_+)$.

6.1.3   Statistical Approximation Properties in Weighted Space

In this section, by using a Bohman–Korovkin-type theorem proved in [57], we present the statistical approximation properties of the operator T_n given by (6.12).

Let $\mathbb{R}$ denote the set of real numbers. A real function ρ is called a weight function if it is continuous on $\mathbb{R}$ and $\lim_{|x| \to \infty} \rho(x) = \infty, \rho(x) \geq 1$ for all $x \in \mathbb{R}$

Let us denote by $B_\rho(\mathbb{R})$ the weighted space of real-valued functions f defined on $\mathbb{R}$ with the property $|f(x)| \leq M_f \rho(x)$ for all $x \in \mathbb{R}$, where M_f is a constant depending on the function f. We also consider the weighted subspace $C_\rho(\mathbb{R})$ of $B_\rho(\mathbb{R})$ given by

$$C_\rho(\mathbb{R}) = \{ f \in B_\rho(\mathbb{R}) : f \text{ continous on } \mathbb{R} \}.$$

Endowed with the norm $\|.\|_\rho$, where $\|f\|_\rho = \sup_{x \in \mathbb{R}} \frac{f(x)}{\rho(x)}$, $B_\rho(\mathbb{R})$, and $C_\rho(\mathbb{R})$ are Banach spaces.

Using A-statistical convergence Duman and Orhan proved the following Bohman–Korovkin-type theorem [57, Theorem 3].

Theorem 6.2. *Let $A = (a_{jn})_{j,n}$ be a non negative regular summability matrix and let $(L_n)_n$ be a sequence of positive linear operators from $C_{\rho_1}(\mathbb{R})$ into $B_{\rho_2}(\mathbb{R})$, where ρ_1 and ρ_2 satisfy*

$$\lim_{|x| \to \infty} \frac{\rho_1}{\rho_2} = 0. \tag{6.20}$$

Then

$$st_A - \lim_n \|L_n f - f\|_{\rho_2} = 0 \text{ for all } f \in C_{\rho_1}(\mathbb{R})$$

if and only if

$$st_A - \lim_n \|L_n F_v - F_v\|_{\rho_1} = 0, \ v = 0, 1, 2$$

where $F_v(x) = \frac{x^v \rho_1(x)}{1 + x^2}, \ v = 0, 1, 2.$

Examining this result, clearly, replacing $\mathbb{R}$ by $\mathbb{R}_+$, the theorem holds true. Further on, we consider a sequence $(q_n)_n$, $q_n \in (0, 1)$, such that

$$st - \lim_n q_n = 1. \tag{6.21}$$

Theorem 6.3. *Let $(q_n)_n$ be a sequence satisfying (6.21). Then for all $f \in C_2(\mathbb{R}_+)$, we have*

$$st - \lim_n \|T_n(f; q_n; .) - f\|_{2\alpha} = 0, \ \alpha > 1.$$

Proof. It is clear that

$$st - \lim_n \|T_n(e_0; q_n; .) - e_0\|_2 = 0. \tag{6.22}$$

Based on (6.8) we have

$$D_{q_n} \phi_n(x) = -\phi_n(q_n x) \beta_{n,0,0,q_n}(x) \qquad x \in \mathbb{R}_+, \; n \in \mathbb{N},$$

where

$$st - \lim_n \frac{\beta_{n,0,0,q_n}(0)}{[n]_{q_n}} = 1. \tag{6.23}$$

Thus, by (6.14) and (6.6), we obtain

$$\frac{|T_n(e_1; q_n; x) - e_1(x)|}{1 + x^2} \leq \|e_1\|_2 \left| \frac{\beta_{n,0,0,q_n}(0)}{[n]_{q_n}} - 1 \right|.$$

Consequently,

$$\|T_n(e_1; q_n; .) - e_1\|_2 \leq \frac{1}{2} \left| \frac{\beta_{n,0,0,q_n}(0)}{[n]_{q_n}} - 1 \right|$$

and for any $\varepsilon > 0$ we have $\delta(A) \leq \delta(B) = 0$, where

$$A = \{n \in \mathbb{N} : \|T_n(e_1; q_n; .) - e_1\|_2 \geq \varepsilon\}$$

$$B = \left\{ n \in \mathbb{N} : \left| \frac{\beta_{n,0,0,q_n}(0)}{[n]_{q_n}} - 1 \right| \geq 2\varepsilon \right\}.$$

Hence, we get

$$st - \lim_n \|T_n(e_1; q_n; .) - e_1\|_2 = 0. \tag{6.24}$$

The condition (6.8) implies that for any $n \in \mathbb{N}$, we have

$$D_{q_n}^2 \phi_n(x) = -D_{q_n} \phi_n(q_n x) \beta_{n,1,0,q_n}(x) \tag{6.25}$$

or

$$D_{q_n}^2 \phi_n(x) = \phi_n q_n^2(x) \beta_{n,1,1,q_n}(x). \tag{6.26}$$

■

Case 6.1. If (6.25) holds true, then $D_{q_n}^2 \phi_n(0) = \beta_{n,0,0,q_n}(0)\beta_{n,1,0,q_n}(0)$. From (6.15) we get

$$|T_n(e_2; q_n; x) - e_2(x)| \leq \left| \frac{\beta_{n,0,0,q_n}(0)\beta_{n,1,0,q_n}(0)}{q_n [n]_{q_n}^2} - 1 \right| x^2 + \frac{\beta_{n,0,0,q_n}(0)}{[n]_{q_n}^2} x.$$

By using the elementary inequality

$$|XY - 1| \leq \max\left\{\left|X^2 - 1\right|, \left|Y^2 - 1\right|\right\}, \qquad X, Y \in \mathbb{R}, \ XY \geq 0,$$

we can write

$$\frac{|T_n(e_2; q_n; x) - e_2(x)|}{1 + x^2} \leq \|e_2\|_2 \max_{k=\overline{0,1}} \left\{\left|\left(\frac{\beta_{n,k,0,q_n}(0)}{q_n^k [n]_{q_n}}\right)^2 - 1\right|\right\}$$

$$+ \|e_1\|_2 \frac{\beta_{n,0,0,q_n}(0)}{[n]_{q_n}^2}$$

$$\leq \max_{k=\overline{0,1}} \left\{\left|\frac{\beta_{n,k,0,q_n}(0)}{q_n^k [n]_{q_n}} - 1\right| \left(2 + \left|\frac{\beta_{n,k,0,q_n}(0)}{q_n^k [n]_{q_n}} - 1\right|\right)\right\}$$

$$+ \frac{1}{2} \frac{\beta_{n,0,0,q_n}(0)}{[n]_{q_n}^2}.$$

Since $st - \lim_n q_n = 1$ we have

$$st - \lim_n \frac{1}{[n]_{q_n}} = 0. \tag{6.27}$$

From (6.9) and (6.27) we obtain

$$st - \lim_n \|T_n(e_2; q_n; .) - e_2\|_2 = 0. \tag{6.28}$$

Case 6.2. If (6.26) holds true, then $D_{q_n}^2 \phi_n(0) = \beta_{n,1,1,q_n}(0)$. By using (6.15), we get

$$|T_n(e_2; q_n; x) - e_2(x)| \leq \left|\frac{\beta_{n,1,1,q_n}(0)}{q_n [n]_{q_n}^2} - 1\right| x^2 + \frac{\beta_{n,0,0,q_n}(0)}{[n]_{q_n}^2} x$$

and

$$\|T_n(e_2; q_n; .) - e_2\|_2 \leq \left|\frac{\beta_{n,1,1,q_n}(0)}{q_n [n]_{q_n}^2} - 1\right| + \frac{1}{2} \frac{\beta_{n,0,0,q_n}(0)}{[n]_{q_n}^2}.$$

Taking into account (6.9) and (6.27), the last inequality implies

$$st - \lim_n \|T_n(e_2; q_n; .) - e_2\|_2 = 0. \tag{6.29}$$

Finally, using (6.22), (6.24), and (6.28) or (6.29), the proof follows from Theorem 6.2 by choosing $A = C_1$, the Cesáro matrix of order one and $\rho_1(x) = 1 + x^2, \rho_2(x) = 1 + x^{2\alpha}, x \in \mathbb{R}_+, \alpha > 1$.

6.1.4 Special Cases of T_n Operator

In this section we present two particular cases of operator T_n given by (6.12), which turn into the well-known Szász Mirakyan (also called Mirakjan) operator and classical Baskakov operator, respectively, in the case $q \longrightarrow 1^-$.

A q-Analogue of Szász–Mirakyan Operator

For some $q \in (0,1)$, let $(\alpha_n)_n$ be a sequence of positive real numbers satisfying $st - \lim_n \alpha_n = 1$ and let $\phi_n(x) = E_q\left(-[n]_q \alpha_n x\right)$, $x \in \mathbb{R}_+, n \in \mathbb{N}$. Then, for all $(n,k) \in \mathbb{N} \times \mathbb{N}$, we have $\phi_n(0) = 1$ and

$$D_q^k \phi_n(x) = (-1)^k [n]_q^k \alpha_n^k q^{\frac{k(k-1)}{2}} E_q\left(-[n]_q \alpha_n q^k x\right), \quad x \geq 0.$$

It is obvious that under the assumption made upon sequence $(\alpha_n)_n$, the condition (6.7) is fulfilled. Furthermore, for all $k \in \mathbb{N}$, we get

$$D_q^{k+1} \phi_n(x) = -D_q^k \phi_n(qx)\beta_{n,k,0,q}(x),$$

where $\beta_{n,k,0,q}(x) = [n]_q \alpha_n q^k$. Consequently,

$$\frac{\beta_{n,k,0,q}(x)}{[n]_q q^k} = \alpha_n, \quad n \in \mathbb{N},$$

and (6.8), (6.9) are also satisfied.

In this case the operator T_n turns into S_n^*, given as follows:

$$S_n^*(f;q;x) = \sum_{k=0}^{\infty} \frac{\left(\alpha_n [n]_q x\right)^k}{[k]_q!} q^{k(k-1)} E_q\left(-[n]_q q^k \alpha_n x\right) f\left(\frac{[k]_q}{[n]_q q^{k-1}}\right), \quad (6.30)$$

for $f \in C_2(\mathbb{R}_+)$, $x \in \mathbb{R}_+, n \in \mathbb{N}, q \in (0,1)$.

Remark 6.3. Choosing $\alpha_n = 1, n \in \mathbb{N}$, the operator S_n^* given by (6.30) reduces to the classical Szász–Mirakyan operator, when $q \longrightarrow 1^-$.

Based on Lemma 6.1, we have

$$S_n^*(e_0;q;x) = 1$$
$$S_n^*(e_1;q;x) = \alpha_n x,$$
$$S_n^*(e_2;q;x) = \alpha_n^2 x^2 + \frac{\alpha_n}{[n]_q} x,$$

for $x \in \mathbb{R}_+, n \in \mathbb{N}, q \in (0,1)$.

We point out that our q-generalization $S_n^*, n \in \mathbb{N}$, is different by the q-analogue of Szász–Mirakyan operator, recently introduced by Aral (see [25]) as follows:

$$S_n^q(f;x) = E_q\left(-[n]_q \frac{x}{b_n}\right) \sum_{k=0}^{\infty} f\left(\frac{[k]_q b_n}{[n]_q}\right) \frac{\left([n]_q x\right)^k}{[k]_q! (b_n)^k},$$

where $0 \leq x \leq \frac{b_n}{1-q^n}$, b_n is a sequence of positive numbers such that $lim_n b_n = \infty$.

The approximation function $S_n^*(f;q;.)$ is defined on $[0,\infty]$ for each $n \in \mathbb{N}$, while the domain of $S_n^q(f;.)$ depends on n. Moreover, in the case $\alpha_n = 1, n \in \mathbb{N}$, since $|S_n^*(e_2;q;x) - e_2(x)| = \frac{1}{[n]_q}x$ and $|S_n^*(e_2;q;x) - e_2(x)| = (1-q)x^2 + \frac{b_n}{[n]_q}x$, the behavior of $S_n^*(.;q;.)$ on e_2 is better than S_n^q on e_2.

A q-Analogue of Classical Baskakov Operator

In order to give second particular case of the operator T_n, we consider the next lemma. The proof follows immediately from (6.1)–(6.3) (see [104, pp. 106–107]).

Lemma 6.2. *Let $t, s, a \in \mathbb{R}$*

$$D_q(1+ax)_q^t = [t]_q a(1+aqx)_q^{t-1}, \tag{6.31}$$

$$(1+x)_q^{s+t} = (1+x)_q^s(1+q^s x)_q^t, \tag{6.32}$$

$$(1+x)_q^{-t} = \frac{1}{(1+q^{-t}x)_q^t}. \tag{6.33}$$

By using (6.31) and the identity $[-n]_q = \frac{[-n]_q}{q^n}, n \in \mathbb{N}$, it is easy to see that

$$D_q(1+ax)_q^{-n} = \frac{[-n]_q}{q^n}a(1+aqx)_q^{n-1}. \tag{6.34}$$

Let $\phi_n(x) = (1+q^{n-1}x)_q^{-n}, x \in \mathbb{R}_+, n \in \mathbb{N}$. Taking into account (6.33), (6.34), and definition of the high q-derivatives, we obtain

$$D_q^k \phi_n(x) = (-1)^k [n]_q \cdots [n+k-1]_q q^k (1+q^{n+k+1}x)_q^{-n-k}$$

$$= (-1)^k \frac{[n+k-1]_q!}{[n-1]_q!} q^k \frac{1}{(1+qx)_q^{n+k}},$$

for all $x \in \mathbb{R}_+, (n,k) \in \mathbb{N} \times \mathbb{N}, q \in (0,1)$.

Consequently, by using (6.32) we can write

$$
D_q^{k+1} \phi_n(x) = (-1)^{k+1} \frac{[n+k]_q!}{[n-1]_q!} q^{k+1} (1+q^{n+k+2}x)_q^{-n-k-1}
$$

$$
= (-1)^{k+1} \frac{[n+k]_q!}{[n-1]_q!} q^{k+1} (1+q^{n+k+2}x)_q^{-n-1} (1+q^{n+k+2}x)_q^{-k}
$$

$$
= (-1)^k D_q \phi_n(q^k x) \beta_{n,k,k-1,q}(x),
$$

where $\beta_{n,k,k-1,q}(x) = \frac{[n+k]_q!}{[n]_q!} q^k (1+q^{k+1}x)_q^{-k}$, and $\frac{\beta_{n,k,k-1,q}(0)}{[n]_q^k q^k} = \frac{[n+k]_q!}{[n]_q^k [n]_q!}$.

Since, for $0 < q < 1$, we have $\lim_n \frac{[n+k]_q}{[n]_q} = 1, (3.4)$ and (6.9) are also verified with $i_k = k-1$.

In this case the operator T_n turns into V_n^*, given as follows:

$$
V_n^*(f;q;x) = \sum_{k=0}^{\infty} \begin{bmatrix} n+k-1 \\ k \end{bmatrix}_q q^{\frac{k(k-1)}{2}} (qx)^k \frac{1}{(1+qx)_q^{n+k}} f\left(\frac{[k]_q}{[n]_q q^{k-1}} \right), \quad (6.35)
$$

for $f \in C_2(\mathbb{R}_+)$, $x \in \mathbb{R}_+, n \in \mathbb{N}, q \in (0,1)$.

Remark 6.4. The operator V_n^* given by (6.35) becomes the classical *n*th Baskakov operator in the case $q \longrightarrow 1^-$.

Based on Lemma 6.1 we have

$$
V_n^*(e_0;q;x) = 1,
$$

$$
V_n^*(e_1;q;x) = qx,
$$

$$
V_n^*(e_2;q;x) = \frac{[n+1]_q}{[n]_q} qx^2 + \frac{q}{[n]_q}x, \quad x \in \mathbb{R}_+, n \in \mathbb{N}.
$$

6.2 *q*-Szász–King-type Operators

6.2.1 *Notations and Preliminaries*

In order to introduce a *q*-variant for Szász–Mirakjan operators, right from the start, we present a construction due to Aral [25] and studied in deepness by Aral and Gupta [29]. Let $(b_n)_{n \geq 1}$ be a sequence of positive numbers such that $\lim_n b_n = \infty$. For each $n \in \mathbb{N}$, $q \in (0,1)$, and $f \in C(\mathbb{R}_+)$, the authors defined

$$(S_n^q f)(x) = E_q\left(-[n]_q \frac{x}{b_n}\right) \sum_{k=0}^{\infty} f\left(\frac{[k]_q b_n}{[n]_q}\right) \frac{([n]_q x)^k}{[k]_q! b_n^k} \tag{6.36}$$

where $0 \le x < \frac{b_n}{1-q^n}$. The following explicit expressions for $S_n^q e_k$, $k = 0, 1, 2$ have been established [29, (3.5)–(3.7)]:

$$S_n^q e_0 = e_0, \quad S_n^q e_1 = e_1, \quad S_n^q e_2 = q e_2 + \frac{b_n}{[n]_q} e_1$$

In [9] the classical Szász–Mirakjan operators have been modified in King's sense. Following a similar route, we transform the operators defined at (6.36) in order to preserve the quadratic function e_2. Defining the functions

$$v_{n,q}(x) = \frac{1}{2q[n]_q}\left(-b_n + \sqrt{b_n^2 + 4q[n]_q^2 x^2}\right), \quad x \ge 0, \tag{6.37}$$

we consider the linear and positive operators

$$(S_{n,q}^* f)(x) = E_q\left(-[n]_q \frac{v_{n,q}(x)}{b_n}\right) \sum_{k=0}^{\infty} f\left(\frac{[k]_q b_n}{[n]_q}\right) \frac{([n]_q v_{n,q}(x))^k}{[k]_q! b_n^k} \tag{6.38}$$

where $x \in J_n(q) := \left[0, \frac{b_n}{1-q^n}\right)$.

Lemma 6.3. *The operators defined at (6.38) verify for each $x \in J_n(q)$ the following identities*

$$\left(S_{n,q}^* e_0\right)(x) = 1, \quad \left(S_{n,q}^* e_1\right)(x) = v_{n,q}(x), \quad \left(S_{n,q}^* e_2\right)(x) = x^2 \tag{6.39}$$

$$\left(S_{n,q}^* \psi_x^2\right)(x) = 2x\left(x - v_{n,q}(x)\right) \tag{6.40}$$

where $\psi_x(t) = t - x$, $t \ge 0$.

Since the identities are easily obtained by direct computation, we omit the proof.

Examining relations (6.37), (6.39), and based on Bohman–Korovkin theorem, it is clear that $\left(S_{n,q}^*\right)_{n \ge 1}$ does not form an approximation process. The next step is to transform it for employing of this property. For each $n \in \mathbb{N}$, the constant q will be replaced by a number $q_n \in (0, 1)$ such that $\lim_n b_n = 1$. At this stage we also need a connection between the involved sequences $(b_n)_{n \ge 1}$, $(q_n)_{n \ge 1}$.

Theorem 6.4. *Let $(q_n)_{n \ge 1}$, $0 < q_n < 1$ be a sequence and let the operators S_{n,q_n}^*, $n \in \mathbb{N}$, be defined as in (6.38). If*

$$\lim_n q_n = 1, \quad \lim_n \frac{b_n}{1 - q_n^n} = \infty \quad \text{and} \quad \lim_n \frac{b_n}{[n]_q} = 0 \tag{6.41}$$

then for any compact $K \subset \mathbb{R}_+$ and for each $f \in C(\mathbb{R}_+)$, one has

$$\lim_n \left(S_{n,q}^* f\right)(x) = f(x), \quad \text{uniformly in } x \in K.$$

Proof. The second limit in (6.41) guarantees that $\bigcup_{n=1}^{\infty} J_n(q_n) = \mathbb{R}_+$. Consequently, the sequence of operators is properly defined, this meaning that it is suitable to approximate functions defined on $\mathbb{R}_+$. The third limit in (6.41) implies $\lim_n v_{n,q}(x) = x^2$ uniformly in $x \in K$. The result follows from Bohman–Korovkin criterion via Lemma 6.3. $\blacksquare$

Remark 6.5. For removing any doubt, we indicate pairs of sequences $(q_n)_{n \geq 1}$, $(b_n)_{n \geq 1}$ which verify the plurality of requirements imposed in Theorem 6.4.

1 $q_1 = \frac{1}{\sqrt{2}}$ and $q_n = \frac{1}{\sqrt[n]{n}}$ $(n \geq 2)$; $b_n = n^{\lambda}$ for any fixed $\lambda \in \left(0, \frac{1}{2}\right)$.

2 $q_1 = \frac{1}{2}$ and $q_n = 1 - \frac{1}{n}$ $(n \geq 2)$; $b_n = \left([n]_{q_n}\right)^{\lambda}$ for any fixed $\lambda \in (0, 1)$.

6.2.2 Weighted Statistical Approximation Property

A real-valued function ρ defined on $\mathbb{R}$ is usually called a weight function if it is continuous on the domain satisfying the conditions $\rho \geq e_0$ and $\lim_{|x| \to \infty} \rho(x) = \infty$. For example, the mapping $x \longmapsto 1 + x^{2+\lambda}$, λ a nonnegative parameter, is often used as weight function. Let us consider the spaces

$$B_{\rho}(\mathbb{R}) = \{f : \mathbb{R} \longrightarrow \mathbb{R} | \text{ a constant } M_f \text{ depending on } f \text{ exists such that } |f| \leq M_f \rho\}$$

$$C_{\rho}(\mathbb{R}) = \{f \in B_{\rho}(\mathbb{R}) | f \text{ continuous on } \mathbb{R}\}$$

endowed with the usual norm $\|\cdot\|_{\rho}$, this meaning

$$\|f\|_{\rho} = \sup_{x \in \mathbb{R}} \frac{|f(x)|}{\rho(x)}$$

Clearly, all the notations and results in Theorem 6.2 are still valid if we replace the domain $\mathbb{R}$ by $\mathbb{R}_+$. The main result of this section is based on Theorem 6.2. We choose the pair of weight functions (ρ_0, ρ_{λ}), where

$$\rho_0(x) = 1 + x^2, \quad \rho_{\lambda}(x) = 1 + x^{2+\lambda}, \quad x \in \mathbb{R}_+ \tag{6.42}$$

$\lambda > 0$ being a fixed parameter. Relation (6.20) is fulfilled and $B_{\rho_0}(\mathbb{R}_+) \subset B_{\rho_{\lambda}}(\mathbb{R}_+)$. Moreover, taking A the Cesàro matrix of first order, Theorem 6.2 implies

Corollary 6.1. *For any sequence* $(T_n)_{n\geq 1}$ *of linear positive operators acting from* $C_{\rho_0}(\mathbb{R}_+)$ *into* $C_{\rho_\lambda}(\mathbb{R}_+)$, $\lambda > 0$, *one has*

$$st - \lim_n \|T_n f - f\|_{\rho_\lambda} = 0, \quad f \in C_{\rho_0}.(\mathbb{R}_+) \tag{6.43}$$

if and only if

$$st - \lim_n \|T_n e_k - e_k\|_{\rho_0} = 0, \quad k = 0, 1, 2 \tag{6.44}$$

Next, we collect some elementary properties of the functions defined by (6.37).

Lemma 6.4. *Let* $v_{n,q}$, $n \in \mathbb{N}$, *be defined by (6.37), where* $q \in (0, 1)$ *and* $b_n > 0$, $n \in \mathbb{N}$. *The following statements are true:*

(i)

$$v_{n,q}(0) = 0, \quad v_{n,q}\left(\frac{b_n}{1 - q^n}\right) = \frac{b_n}{1 - q^n}$$

(ii)

$$0 \leq v_{n,q}(x) \leq x, \quad x \in \left[0, \frac{b_n}{1 - q^n}\right] \tag{6.45}$$

(iii)

$$x - v_{n,q}(x) \leq x_0 - v_{n,q}(x_0) \leq \frac{b_n}{2q[n]_q}, \quad x \geq 0 \tag{6.46}$$

where $x_0 = \frac{b_n}{2[n]_q\sqrt{1-q}}$.

Proof. Both *(i)* and *(ii)* are obtained by a straightforward calculation. Because $S_{n,q}^*$ is a positive operator, actually, the inequality $v_{n,q}(x) \leq x$ springs from (6.40).

For proving *(iii)* we can consider the function $h : [0, \infty) \longrightarrow \mathbb{R}$, $h(x) = x - v_{n,q}(x)$. The unique solution of the equation $\frac{d}{dx}h(x) = 0$ being x_0, the monotonicity of h implies $h(x) \leq h(x_0) = \frac{b_n(1-\sqrt{1-q})}{2q[n]_q}$, and (6.46) follows. ∎

The main result of this section will be read as follows.

Theorem 6.5. *Let the sequence* $(q_n)_{n\geq 1}$, $0 < q_n < 1$, *be given such that* $st - \lim_n q_n = 1$. *Let the operators* S_{n,q_n}^*, $n \in \mathbb{N}$, *be defined as in (6.38). If*

$$st - \lim_n \frac{b_n}{[n]_{q_n}} = 0 \tag{6.47}$$

then, for each function $f \in C_{\rho_0}(\mathbb{R}_+)$, *one has*

$$st - \lim_n \|S_{n,q_n}^* f - f\|_{\rho_\lambda} = 0$$

where $\lambda > 0$.

Proof. Each function $S^*_{n,q_n} f$, $f \in C_{\rho_0}(\mathbb{R}_+)$, is defined on $J_n(q_n)$. We extend it on $\mathbb{R}_+$ in the classical manner. Let $\tilde{S}^*_{n,q_n}$ be defined as follows

$$\left(\tilde{S}^*_{n,q_n} f \right)(x) = \begin{cases} \left(S^*_{n,q_n} f \right)(x) & , \ x \in J_n(q_n) \\ f(x) & , \ x \geq \frac{b_n}{1-q_n}. \end{cases}$$

For each $n \in \mathbb{N}$, the norm $\left\| \tilde{S}^*_{n,q_n} f - f \right\|_{\rho_\lambda}$ coincides with the norm of the element $\left(S^*_{n,q_n} f - f \right)$ in the space $B_{\rho_\lambda}(J_n(q_n))$, for any $\lambda \geq 0$. Applying Corollary 6.1 to the operators $T_n \equiv \tilde{S}^*_{n,q_n}$, the proof of Theorem 6.5 will be finished. In this respect, it is sufficient to prove that, under our hypothesis, the operators verify the conditions given at (6.44).

By the first and the third identity of relation (6.39), it is clear that

$$st - \lim_n \left\| \tilde{S}^*_{n,q_n} e_k - e_k \right\|_{\rho_0} = 0$$

for $k = 0$ and $k = 2$. The second identity of (6.39) and Lemma 6.4 allow us to write

$$\sup_{x \in J_n(q_n)} \frac{1}{\rho_0(x)} \left| \left(S^*_{n,q_n} e_1 \right)(x) - x \right| = \sup_{x \in J_n(q_n)} \frac{x - v_{n,q_n}(x)}{1+x^2}$$

$$\leq \sup_{x \geq 0} \frac{x - v_{n,q_n}(x)}{1+x^2}$$

$$\leq \frac{b_n}{2 q_n [n]_{q_n}}$$

Based on (6.47) we get $st - \lim_n \left\| \tilde{S}^*_{n,q_n} e_1 - e_1 \right\|_{\rho_0} = 0$, and the proof is completed. ∎

6.2.3   Rate of Weighted Approximation

The q-Stirling numbers of the second kind denoted by $S_q(m,k)$ $(m,k \in \mathbb{N}_0, \ m \geq k)$ are described by the recurrence formula

$$S_q(m,k) = S_q(m-1,k-1) + [k]_q S_q(m-1,k), \quad m \geq k \geq 1$$

with $S_q(0,0) = 1$ and $S_q(m,0) = 1$ for $m \in \mathbb{N}$. We agree $S_q(m,k) = 0$ for $k \geq m$. The closed form is the following

$$S_q(m,k) = \frac{1}{[k]_q! \, q^{\frac{k(k-1)}{2}}} \sum_{j=0}^{k} (-1)^j q^{\frac{j(j-1)}{2}} \begin{bmatrix} k \\ j \end{bmatrix}_q [k-j]_q^m, \quad 1 \leq k \leq m \qquad (6.48)$$

For $q \longrightarrow 1^-$, $S_1(m,k)$ represents the number of ways of partitioning a set of m elements into k nonempty subsets [4, pp. 824].

Lemma 6.5. *Let the sequence* $(q_n)_{n\geq1}$, $0 < q_n < 1$, *be given and let the operators* S^*_{n,q_n}, $n \in \mathbb{N}$, *be defined as in (6.38). One has*

$$\left(S^*_{n,q_n}e_m\right)(x) = q_n^{\frac{m(m-1)}{2}} v^m_{n,q_n}(x) + \sum_{k=1}^{m-1} \left(\frac{b_n}{[n]_{q_n}}\right)^{m-k} S_{q_n}(m,k) q_n^{\frac{k(k-1)}{2}} v^k_{n,q_n}(x)$$

$x \in J_n(q_n)$, *where* $S_{q_n}(m,k)$ *are* q_n-*Stirling numbers given by (6.48.) Here* e_m *stands for the monomial of m degree.*

Proof. Taking in view both (6.38) and (1.8) and using the Cauchy rule (or Mertens formula) for multiplication of two series, we can write

$$\left(S^*_{n,q_n}e_m\right)(x) = \sum_{k=0}^{\infty}\sum_{i=0}^{k}(-1)^i q_n^{\frac{i(i-1)}{2}} e_m\left(\frac{[k-i]_{q_n}b_n}{[n]_{q_n}}\right) \frac{\left([n]_{q_n}v_{n,q_n}(x)\right)^k}{[i]_{q_n}![k-i]_{q_n}!b_n^k}$$

$$= \sum_{k=0}^{\infty} S_{q_n}(m,k)[k]_{q_n}!q_n^{\frac{k(k-1)}{2}} \left(\frac{b_n}{[n]_{q_n}}\right)^{m-k} \frac{1}{[k]_{q_n}!} v^k_{n,q_n}(x)$$

$$= \sum_{k=0}^{\infty}\left(\frac{b_n}{[n]_{q_n}}\right)^{m-k} S_{q_n}(m,k)q_n^{\frac{k(k-1)}{2}} v^k_{n,q_n}(x)$$

$$= S_{q_n}(m,m)q_n^{\frac{m(m-1)}{2}} v^m_{n,q_n}(x)$$

$$+ \sum_{k=0}^{m-1}\left(\frac{b_n}{[n]_{q_n}}\right)^{m-k} S_{q_n}(m,k)q_n^{\frac{k(k-1)}{2}} v^k_{n,q_n}(x)$$

Knowing that $S_{q_n}(m,m) = 1$ and $S_{q_n}(m,0) = 0$ $(m \geq 1)$, one obtains (6.48). ∎

We mention that A. Aral proved a similar result [29, Lemma 1] for the operators given at (6.36). His proof is based on the forward q-differences up to order m.

Set $A_m(n;q_n,b_n) := \sum_{k=1}^{m-1} S_{q_n}(m,k)\left(\frac{b_n}{[n]_{q_n}}\right)^{m-k}$.

Under the hypothesis $\lim_n q_n = 1$ and $\lim_n \frac{b_n}{[n]_{q_n}}$, we get $\lim_n S_{q_n}(m,k) = S_1(m,k)$, $1 \leq k \leq m-1$, and a real constant B_m depending only on m exists such that

$$\sup_{n\in\mathbb{N}} A_m(n;q_n,b_n) = B_m \tag{6.49}$$

Lemma 6.6. *Let the sequence* $(q_n)_{n\geq1}$, $0 < q_n < 1$, *be given and let* S^*_{n,q_n}, $n \in \mathbb{N}$, *be operators defined as in (6.38). If the condition (6.41) are fulfilled, then one has*

(*i*)

$$(S^*_{n,q_n} e_m)(x) \leq (1 + B_m)(1 + x^m), \; x \in J_n(q_n) \tag{6.50}$$

where B_m is given at (6.49.).

(*ii*)   *For each $n \in \mathbb{N}$, the operator S^*_{n,q_n} maps the space $B_{\rho_{\lceil \lambda \rceil}}$ into $B_{\rho_{\lceil \lambda \rceil}}$, $\lambda > 0$. Here $\lceil \lambda \rceil$ represents the ceiling of number λ.*

Setting $\mu_x(t) = 1 + (x + |t - x|)^{2 + \lceil \lambda \rceil}$, $t \geq 0$, the following inequalities hold:

(*iii*)

$$S^*_{n,q_n} \mu_x \leq c_\lambda (1 + e_{2 + \lceil \lambda \rceil}) \tag{6.51}$$

(*iv*)

$$\sqrt{S^*_{n,q_n} \mu_x^2} \leq \tilde{c}_\lambda (1 + e_{2 + \lceil \lambda \rceil}) \tag{6.52}$$

where $c_\lambda, \tilde{c}_\lambda$ are constants independent on x and n.

Proof.   (*i*)   Based on Lemma 6.5, relation (6.45), and knowing that $q_n \in (0,1)$, for each $x \in J_n(q_n)$, we can write

$$(S^*_{n,q_n} e_m)(x) \leq x^m + \sum_{k=1}^{m-1} S_{q_n}(m,k) \left(\frac{b_n}{[n]_{q_n}} \right)^{m-k} x^k$$

$$\leq x^m + A_m(n; q_n, b_n) \max\{1, x^m\}$$

$$\leq x^m + B_m \max\{1, x^m\}$$

and (6.50) follows:

(*ii*)   If $f \in B_{\rho_{\lceil \lambda \rceil}}$, then $|f| \leq M_f (1 + x^{2 + \lceil \lambda \rceil})$. S^*_{n,q_n} being linear and positive is monotone. Relation (6.50) implies our statement.

(*iii*)   For each $t \geq 0$ and $x \in J_n(q_n)$ we get

$$\mu_x(t) \leq 1 + (2x + t)^{2 + \lceil \lambda \rceil} \leq 1 + 2^{1 + \lceil \lambda \rceil} \left((2x)^{2 + \lceil \lambda \rceil} + t^{2 + \lceil \lambda \rceil} \right)$$

By using (6.50) and (6.39) we obtain (6.51).

(*iv*)   Since $\mu_x^2(t) \leq 2 \left(1 + (2x + t)^{4 + 2\lceil \lambda \rceil} \right)$, the same relations (6.50) and (6.39) imply (6.52).   ∎

We proceed with estimation of the errors $\left| S^*_{n,q_n} f - f \right|$, $n \in \mathbb{N}$, involving unbounded functions, by using a weighted modulus of smoothness associated to the space B_{ρ_λ}. In this respect, we consider

$$\Omega_{\rho_\lambda}(f; \delta) = \sup_{\substack{x \geq 0 \\ 0 < h \leq \delta}} \frac{|f(x+h) - f(x)|}{1 + (x+h)^{2 + \lambda}}, \quad \delta > 0 \tag{6.53}$$

Clearly, $\Omega_{\rho_\lambda}(f;\cdot) < 2\|f\|_{\rho_\lambda}$ for each $f \in B_{\rho_\lambda}$. Among some basic properties of this modulus, we recall

$$\Omega_{\rho_\lambda}(f;\alpha\delta) = (\alpha+1)\Omega_{\rho_\lambda}(f;\delta), \quad \delta > 0, \quad \alpha > 0 \qquad (6.54)$$

Theorem 6.6. *Let the sequence $(q_n)_{n\geq1}$, $0 < q_n < 1$, be given and let S^*_{n,q_n}, $n \in \mathbb{N}$, be operators defined as in (6.38) such that the conditions (6.41) are fulfilled. For each $f \in B_{\rho_{\lceil\lambda\rceil}}$, the following inequality*

$$\left|\left(S^*_{n,q_n}f\right)(x) - f(x)\right| \leq k_\lambda\left(1+x^{2+\lceil\lambda\rceil}\right)\Omega_{\rho_{\lceil\lambda\rceil}}\left(f;\sqrt{\frac{b_n}{[n]_{q_n}}}\right), \quad x \in J_n(q_n)$$

holds, where k_λ is a constant independent of f and n.

Proof. Let $n \in \mathbb{N}$ and $f \in B_{\rho_{\lceil\lambda\rceil}}$ be fixed. For each $t \geq 0$ and $\delta > 0$, based both on definition (6.53) and on property (6.54) with $\alpha := |t-x|\delta^{-1}$, we get

$$|f(t) - f(x)| \leq \left(1 + (x+|t-x|)^{2+\lceil\lambda\rceil}\right)\left(\frac{|t-x|}{\delta}+1\right)\Omega_{\rho_{\lceil\lambda\rceil}}(f;\delta)$$

$$= \left(\mu_x(t) + \frac{1}{\delta}\mu_x(t)|t-x|\right)\Omega_{\rho_{\lceil\lambda\rceil}}(f;\delta)$$

where μ_x was introduced at Lemma 6.6.

Taking into account that S^*_{n,q_n} is a linear positive operator preserving the constants, we can write

$$\left|\left(S^*_{n,q_n}f\right)(x) - f(x)\right| = \left|S^*_{n,q_n}\left(f-f(x);x\right)\right|$$

$$\leq S^*_{n,q_n}\left(|f-f(x)|;x\right)$$

$$\leq S^*_{n,q_n}\left(\mu_x + \delta^{-1}\mu_x|\psi_x|;x\right)\Omega_{\rho_{\lceil\lambda\rceil}}(f;\delta)$$

$$= \left\{\left(S^*_{n,q_n}\mu_x\right)(x) + \frac{1}{\delta}\left(S^*_{n,q_n}\mu_x|\psi_x|\right)(x)\right\}\Omega_{\rho_{\lceil\lambda\rceil}}(f;\delta)$$

$$\leq \left\{\left(S^*_{n,q_n}\mu_x\right)(x) + \frac{1}{\delta}\sqrt{\left(S^*_{n,q_n}\mu_x^2\right)(x)}\sqrt{\left(S^*_{n,q_n}\psi_x^2\right)(x)}\right\}\Omega_{\rho_{\lceil\lambda\rceil}}(f;\delta)$$

where ψ_x was introduced at Lemma 6.3.

The last increase is based on Cauchy–Schwarz inequality frequently used for positive operators of discrete type. It was proved by Yuan-Chuan Li and Sen-Yen Shaw [111] that this classical inequality has great and unexpected force.

Relations (6.40) and (6.46) allow us to write $\left(S_{n,q_n}^* \, \psi_x^2\right)(x) \leq \frac{b_n x}{[n]_{q_n}}$. Further on, by using inequalities (6.51) and, (6.52), we get

$$\left|\left(S_{n,q_n}^* f\right)(x) - f(x)\right| \leq \left\{ c_\lambda \left(1 + x^{2+\lceil\lambda\rceil}\right) + \frac{1}{\delta}\sqrt{\tilde{c}_\lambda}\sqrt{1 + x^{2+\lceil\lambda\rceil}}\sqrt{x}\sqrt{\frac{b_n}{[n]_{q_n}}} \right\} \Omega_{\rho_{\lceil\lambda\rceil}}(f;\delta)$$

$$\leq \left(c_\lambda + \frac{\sqrt{\tilde{c}_\lambda}}{\delta}\sqrt{\frac{b_n}{[n]_{q_n}}} \right) \left(1 + x^{2+\lceil\lambda\rceil}\right) \Omega_{\rho_{\lceil\lambda\rceil}}(f;\delta)$$

choosing $\delta = \sqrt{\frac{b_n}{[n]_{q_n}}}$ and setting $k_\lambda := c_\lambda + \sqrt{\tilde{c}_\lambda}$, the conclusion follows. ∎

Corollary 6.2. *Under the assumptions of Theorem 6.6 the following global estimate takes place:*

$$\left\|S_{n,q_n}^* f - f\right\|_{\rho_{\lceil\lambda\rceil}} \leq k_\lambda \Omega_{\rho_{\lceil\lambda\rceil}}\left(f; \sqrt{\frac{b_n}{[n]_{q_n}}}\right), \quad f \in B_{\rho_{\lceil\lambda\rceil}}.$$

6.3 *q*-Baskakov–Kantorovich Operators

6.3.1 *Introduction*

Recently the Durrmeyer-type certain integrated operators based on q-integers were studied in [48] and [86].

Lemma 6.7. *Let* $t, s, a \in \mathbb{R}$, *then we have*

$$D_q(1 + ax)_q^t = [t]_q \, a\,(1 + aqx)_q^{t-1}, \tag{6.55}$$

$$(1 + x)_q^{s+t} = (1 + x)_q^s(1 + q^s x)_q^t,$$

$$(1 + x)_q^{-t} = \frac{1}{(1 + q^{-t}x)_q^t}. \tag{6.56}$$

The proof follows immediately from (6.1)–(6.3), (see [104, pp. 106–107]).

Aral and Gupta [30] introduced a *q*-analogue of Baskakov operators, which for $q \in (0,1)$, $f \in C([0,\infty))$, $x \in \mathbb{R}_+ := [0,\infty)$, $n \in \mathbb{N}$ is defined as

$$V_{n,q}(f;x) = \sum_{k=0}^{\infty} b_{n,k}(q;x) f\left(\frac{[k]_q}{q^{k-1}[n]_q}\right), \tag{6.57}$$

where $b_{n,k}(q;x) := \begin{bmatrix} n+k-1 \\ k \end{bmatrix}_q q^{\frac{k(k-1)}{2}} \frac{x^k}{(1+x)_q^{n+k}}.$

Remark 6.6. It is obvious that $V_{n,q}$, $n \in \mathbb{N}$, are positive and linear operators. Furthermore, when $q \to 1^-$, the operators given by (6.57) reduce to the classical Baskakov operators (see, e.g., [37]).

We set e_i, $e_i(x) = x^i$, $i \geq 0$.

Lemma 6.8. *For all $n \in \mathbb{N}$, $x \in \mathbb{R}_+$, and $0 < q < 1$, we have*

$$V_{n,q}(e_0;x) = 1, \tag{6.58}$$

$$V_{n,q}(e_1;x) = x, \tag{6.59}$$

$$V_{n,q}(e_2;x) = \frac{[n+1]_q}{q[n]_q}x^2 + \frac{1}{[n]_q}x. \tag{6.60}$$

Proof. For $n \in \mathbb{N}$, we consider the function $g(x) := (1+q^n x)_q^{-n}$, $x \in \mathbb{R}_+$.
By using (6.56), (6.55), and (1.4), we get

$$D_q^k g(x) = (-1)^k [n]_q [n+1]_q \ldots [n+k-1]_q (1+q^{n+k}x)_q^{-n-k}, \quad k \in \mathbb{N}_0 := \{0\} \cup \mathbb{N}.$$

For a fixed $x \in \mathbb{R}_+$, by Theorem 6.1, we obtain

$$(1+q^n t)_q^{-n} = \sum_{k=0}^{\infty} (-1)^k \frac{(t-x)_q^k}{[k]_q!} \frac{[n+k-1]_q!}{[n-1]_q!} (1+q^{n+k}x)_q^{-n-k}.$$

Choosing $t := 0$ in the above relation and taking into account $(-x)_q^k = (-x)^k q^{\frac{k(k-1)}{2}}$, we get

$$\sum_{k=0}^{\infty} b_{n,k}(q;x) = 1 \tag{6.61}$$

and (6.58) is proved.
By definition (6.57) of the operators $V_{n,q}$ and using (6.61) we have

$$V_{n,q}(e_1;x) = \sum_{k=1}^{\infty} \frac{[n+k-1]_q!}{[k]_q![n-1]_q!} q^{\frac{(k-1)(k-2)}{2}} \frac{x^k}{(1+x)_q^{n+k}} \frac{[k]_q}{[n]_q}$$

$$= x \sum_{k=0}^{\infty} b_{n+1,k}(q;x) = x.$$

Next we estimate $V_{n,q}(e_2;x)$. Taking into account the relation $[k]_q = [k-1]_q + q^{k-1}$, $k \in \mathbb{N}$, we obtain (6.60) as follows:

$$V_{n,q}(e_2;x) = \sum_{k=1}^{\infty} \begin{bmatrix} n+k-1 \\ k \end{bmatrix}_q q^{\frac{(k-2)(k-3)}{2}} \frac{x^k}{(1+x)_q^{n+k}} \frac{[k]_q([k-1]_q + q^{k-1})}{q[n]_q^2}$$

$$= \sum_{k=2}^{\infty} \frac{[n+k-1]_q!}{[k]_q![n-1]_q!} q^{\frac{(k-2)(k-3)}{2}} \frac{x^k}{(1+x)_q^{n+k}} \frac{[k]_q[k-1]_q}{q[n]_q^2}$$

$$+ \sum_{k=1}^{\infty} b_{n,k}(q;x) \frac{[k]_q}{q^{k-1}[n]_q^2}$$

$$= \frac{[n+1]_q}{q[n]_q} x^2 \sum_{k=0}^{\infty} b_{n+2,k}(q;x) + \frac{1}{[n]_q} V_{n,q}(e_1;x)$$

$$= \frac{[n+1]_q}{q[n]_q} x^2 + \frac{1}{[n]_q} x.$$

This completes the proof of Lemma 6.8.   ∎

6.3.2   q-Analogue of Baskakov–Kantorovich Operators

Based on the q-integration, Gupta and Radu [91] proposed for $q \in (0,1)$, the Kantorovich variant of the q-Baskakov operators as

$$K_{n,q}(f;x) = [n]_q \sum_{k=0}^{\infty} b_{n,k}(q;x) \int_{q[k]_q/[n]_q}^{[k+1]_q/[n]_q} f(q^{-k+1}t)d_qt, \tag{6.62}$$

$x \in \mathbb{R}_+, n \in \mathbb{N}$.

It can be seen that for $q \to 1^-$, the q-Baskakov–Kantorovich operator becomes the operator studied in [1].

Remark 6.7. By simple computation, it is observed from the definition of q-integration that

$$\int_{q[k]_q/[n]_q}^{[k+1]_q/[n]_q} d_qt = \frac{1}{[n]_q}, \tag{6.63}$$

$$\int_{q[k]_q/[n]_q}^{[k+1]_q/[n]_q} td_qt = \frac{[2]_q[k]_q + q^k}{[2]_q[n]_q^2}, \tag{6.64}$$

$$\int_{q[k]_q/[n]_q}^{[k+1]_q/[n]_q} t^2d_qt = \frac{[3]_q[k]_q^2 + (1+[2]_q)q^k[k]_q + q^{2k}}{[3]_q[n]_q^3}. \tag{6.65}$$

Lemma 6.9. *For all $n \in \mathbb{N}$, $x \in \mathbb{R}_+$, and $0 < q < 1$, we have*

$$K_{n,q}(e_0;x) = 1, \tag{6.66}$$

$$K_{n,q}(e_1;x) = x + \frac{q}{[2]_q[n]_q},\tag{6.67}$$

$$K_{n,q}(e_2;x) = \frac{[n+1]_q}{q[n]_q}x^2 + \frac{q(1+[2]_q)+[3]_q}{[3]_q[n]_q}x + \frac{q^2}{[3]_q[n]_q^2}.\tag{6.68}$$

Proof. Let $q \in (0,1)$. Using the definition (6.62) and the identities (6.63) and (6.58), it is easy to see that (6.66) holds true.

Taking into account (6.64), (6.58) and (6.59), by direct computation, we obtain (6.67) as follows:

$$K_{n,q}(e_1;x) = [n]_q \sum_{k=0}^{\infty} b_{n,k}(q;x)q^{-k+1} \int_{q[k]_q/[n]_q}^{[k+1]_q/[n]_q} t\, d_q t$$

$$= \sum_{k=0}^{\infty} b_{n,k}(q;x)\frac{[k]_q}{q^{k-1}[n]_q} + \frac{q}{[2]_q[n]_q}\sum_{k=0}^{\infty} b_{n,k}(q;x)$$

$$= V_{n,q}(e_1;x) + \frac{q}{[2]_q[n]_q}V_{n,q}(e_0;x) = x + \frac{q}{[2]_q[n]_q}.$$

Based on (6.65) and Lemma 6.8 a similar calculus reveals

$$K_{n,q}(e_2;x) = [n]_q \sum_{k=0}^{\infty} b_{n,k}(q;x)q^{-2k+2} \int_{q[k]_q/[n]_q}^{[k+1]_q/[n]_q} t^2\, d_q t$$

$$= \sum_{k=0}^{\infty} b_{n,k}(q;x)\frac{[k]_q^2}{q^{2k-2}[n]_q^2} + \frac{q(1+[2]_q)}{[3]_q[n]_q}\sum_{k=0}^{\infty} b_{n,k}(q;x)\frac{[k]_q}{q^{k-1}[n]_q}$$

$$+ \frac{q^2}{[3]_q[n]_q^2}\sum_{k=0}^{\infty} b_{n,k}(q;x)$$

$$= V_{n,q}(e_2;x) + \frac{q(1+[2]_q)}{[3]_q[n]_q}V_{n,q}(e_1;x) + \frac{q^2}{[3]_q[n]_q^2}V_{n,q}(e_0;x)$$

$$= \frac{[n+1]_q}{q[n]_q}x^2 + \frac{q(1+[2]_q)+[3]_q}{[3]_q[n]_q}x + \frac{q^2}{[3]_q[n]_q^2}.$$

This completes the proof of Lemma 6.9.   ∎

Remark 6.8. It is observed from the above lemma that for $q \to 1^-$, we get the moments of the Baskakov–Kantorovich operators (see, e.g., [3])

$$K_n(e_0;x) = 1, \quad K_n(e_1;x) = x + \frac{1}{2n},$$

$$K_n(e_2;x) = x^2 + \frac{x(2+x)}{n} + \frac{1}{3n^2}.$$

Examining relations (6.67) and (6.68), it is clear that the sequence of the operators $(K_{n,q})_n$ does not satisfy the conditions of Bohman–Korovkin theorem.

Further on, we consider a sequence $(q_n)_n$, $q_n \in (0,1)$, such that

$$\lim_n q_n = 1. \tag{6.69}$$

Theorem 6.7. *Let $(q_n)_n$ be a sequence satisfying (6.69) and let the operators K_{n,q_n}, $n \in \mathbb{N}$, be defined by (6.62). Then for any compact $J \subset \mathbb{R}_+$ and for each $f \in C(\mathbb{R}_+)$, we have*

$$\lim_{n \to \infty} K_{n,q_n}(f;x) = f(x), \quad \text{uniformly in } x \in J.$$

Proof. Replacing q by a sequence $(q_n)_n$ with the given conditions, the result follows from Lemma 6.9 and the well-known Bohman–Korovkin theorem (see [113, pp. 8–9]). ∎

6.3.3 Weighted Statistical Approximation Properties

In this section, by using a Bohman–Korovkin-type theorem proved in [57], we present the statistical approximation properties of the operator $K_{n,q}$ given by (6.62).

A real function ρ is called a weight function if it is continuous on $\mathbb{R}$ and $\lim_{|x| \to \infty} \rho(x) = \infty$, $\rho(x) \geq 1$ for all $x \in \mathbb{R}$.

Let us denote by $B_\rho(\mathbb{R})$ the weighted space of real-valued functions f defined on $\mathbb{R}$ with the property $|f(x)| \leq M_f \rho(x)$ for all $x \in \mathbb{R}$, where M_f is a constant depending on the function f. We also consider the weighted subspace $C_\rho(\mathbb{R})$ of $B_\rho(\mathbb{R})$ given by

$$C_\rho(\mathbb{R}) := \{ f \in B_\rho(\mathbb{R}) : f \text{ continuous on } \mathbb{R} \}.$$

Endowed with the norm $\|\cdot\|_\rho$, where $\|f\|_\rho := \sup_{x \in \mathbb{R}} \frac{|f(x)|}{\rho(x)}$, $B_\rho(\mathbb{R})$ and $C_\rho(\mathbb{R})$ are Banach spaces.

Examining this result, clearly, replacing $\mathbb{R}$ by $\mathbb{R}_+$, the theorem holds true.

Let us consider the weight functions $\rho_0(x) = 1 + x^2$, $\rho_\alpha(x) = 1 + x^{2+\alpha}$, $x \in \mathbb{R}_+$, $\alpha > 0$.

Further on, we consider a sequence $(q_n)_n$, $q_n \in (0,1)$, such that

$$st - \lim_n q_n = 1. \tag{6.70}$$

Theorem 6.8. *Let $(q_n)_n$ be a sequence satisfying (6.70). Then for all $f \in C_{\rho_0}(\mathbb{R}_+)$, we have*

$$st - \lim_n \|K_{n,q}(f;\cdot) - f\|_{\rho_\alpha} = 0, \quad \alpha > 0.$$

Proof. It is clear that

$$st - \lim_n \|K_{n,q_n}(e_0;\cdot) - e_0\|_{\rho_0} = 0. \tag{6.71}$$

Based on (6.67) we have

$$\frac{\left|K_{n,q_n}(e_1;x) - e_1(x)\right|}{1+x^2} \leq \|e_0\|_{\rho_0} \frac{q_n}{[2]_{q_n}[n]_{q_n}} \leq \frac{1}{[n]_{q_n}}.$$

Since $st - \lim_n q_n = 1$ we get $st - \lim_n \frac{1}{[n]_{q_n}} = 0$, and thus,

$$st - \lim_n \left\|K_{n,q_n}(e_1;\cdot) - e_1\right\|_{\rho_0} = 0. \tag{6.72}$$

By using (6.68) we obtain

$$\frac{\left|K_{n,q_n}(e_2;x) - e_2(x)\right|}{1+x^2} \leq \|e_2\|_{\rho_0} \left|\frac{[n+1]_{q_n}}{q_n[n]_{q_n}} - 1\right| + \|e_0\|_{\rho_0} \frac{q_n(1+[2]_{q_n}) + [3]_{q_n}}{[3]_{q_n}[n]_{q_n}}$$

$$+ \|e_0\|_{\rho_0} \frac{q_n^2}{[3]_{q_n}[n]_{q_n}^2}$$

$$\leq \frac{1}{q_n[n]_{q_n}} + \frac{3}{[n]_{q_n}} + \frac{1}{[n]_{q_n}^2} \leq \frac{4}{q_n[n]_{q_n}} + \frac{1}{[n]_{q_n}^2}.$$

Consequently,

$$st - \lim_n \left\|K_{n,q_n}(e_2;\cdot) - e_2\right\|_{\rho_0} = 0. \tag{6.73}$$

Finally, using (6.71)–(6.73), the proof follows from Theorem 6.2 by choosing $A = C_1$, the Cesàro matrix of order one and $\rho_1(x) = 1+x^2$, $\rho_2(x) = 1+x^{2+\alpha}$, $x \in \mathbb{R}_+$, $\alpha > 0$. ∎

6.3.4   Rate of Convergence

We can give estimates of the errors $\left|K_{n,q}(f;\cdot) - f\right|$, $n \in \mathbb{N}$, for unbounded functions by using a weighted modulus of smoothness associated to the space $B_{\rho_\alpha}(\mathbb{R}_+)$.

We consider

$$\Omega_{\rho_\alpha}(f;\delta) := \sup_{\substack{x \geq 0 \\ 0 < h \leq \delta}} \frac{|f(x+h) - f(x)|}{1+(x+h)^{2+\alpha}}, \delta > 0,\ \alpha \geq 0. \tag{6.74}$$

It is evident that for each $f \in B_{\rho_\alpha}(\mathbb{R}_+)$, $\Omega_{\rho_\alpha}(f;\cdot)$ is well defined, and

$$\Omega_{\rho_\alpha}(f;\delta) \leq 2\|f\|_{\rho_\alpha}, \delta > 0,\ f \in B_{\rho_\alpha}(\mathbb{R}_+),\ \alpha \geq 0.$$

The weighted modulus of smoothness $\Omega_{\rho\alpha}(f;\cdot)$ possesses the following properties (see [112]):

$$\Omega_{\rho\alpha}(f;\lambda\delta) \leq (\lambda+1)\Omega_{\rho\alpha}(f;\delta), \quad \delta > 0, \lambda > 0, \tag{6.75}$$

$$\Omega_{\rho\alpha}(f;n\delta) \leq n\Omega_{\rho\alpha}(f;\delta), \quad \delta > 0, n \in \mathbb{N},$$

$$\lim_{\delta\to 0^+} \Omega_{\rho\alpha}(f;\delta) = 0.$$

Theorem 6.9. *Let* $q \in (0,1)$ *and* $\alpha \geq 0$. *For all* $f \in B_{\rho\alpha}(\mathbb{R}_+)$, *we have*

$$\left|K_{n,q}(f;x) - f(x)\right| \leq \sqrt{K_{n,q}(\mu_{x,\alpha}^2;x)} \left(1 + \frac{1}{\delta}\sqrt{K_{n,q}(\psi_x^2;x)}\right)\Omega_{\rho\alpha}(f;\delta), \quad x \geq 0, \delta > 0, n \in \mathbb{N},$$

where $\mu_{x,\alpha}(t) := 1 + (x+|t-x|)^{2+\alpha}$, $\psi_x(t) := |t-x|$, $t \geq 0$.

Proof. Let $n \in \mathbb{N}$ and $f \in B_{\rho\alpha}(\mathbb{R}_+)$. Based on (6.74) and (6.75), we can write

$$|f(t) - f(x)| \leq \left(1 + (x+|t-x|)^{2+\alpha}\right)\left(\frac{1}{\delta}|t-x| + 1\right)\Omega_{\rho\alpha}(f;\delta)$$

$$= \mu_{x,\alpha}(t)\left(1 + \frac{1}{\delta}\psi_x(t)\right)\Omega_{\rho\alpha}(f;\delta).$$

Taking into account the definition of integration (1.12) and the relation $\int_a^b f(x)d_qx = \int_0^b f(x)d_qx - \int_0^a f(x)d_qx$, we get

$$\int_{q[k]_q/[n]_q}^{[k+1]_q/[n]_q} f(q^{-k+1}t)d_qt = q^{k-1}\int_{[k]_q/q^{k-2}[n]_q}^{[k+1]_q/q^{k-1}[n]_q} f(t)d_qt.$$

Consequently, the operators $K_{n,q}$ can be expressed as follows:

$$K_{n,q}(f;x) = [n]_q\sum_{k=0}^{\infty} b_{n,k}(q;x)q^{k-1}\int_{[k]_q/q^{k-2}[n]_q}^{[k+1]_q/q^{k-1}[n]_q} f(t)d_qt.$$

By using the Cauchy inequality for linear positive operators, we obtain

$$\left|K_{n,q}(f;x) - f(x)\right| \leq [n]_q\sum_{k=0}^{\infty} b_{n,k}(q;x)q^{k-1}\int_{[k]_q/q^{k-2}[n]_q}^{[k+1]_q/q^{k-1}[n]_q} |f(t) - f(x)|d_qt$$

$$\leq \left(K_{n,q}(\mu_{x,\alpha};x) + \frac{1}{\delta}K_{n,q}(\mu_{x,\alpha}\psi_x;x)\right)\Omega_{\rho\alpha}(f;\delta)$$

$$\leq \sqrt{K_{n,q}(\mu_{x,\alpha}^2;x)}\left(1 + \frac{1}{\delta}\sqrt{K_{n,q}(\psi_x^2;x)}\right)\Omega_{\rho\alpha}(f;\delta). \qquad \blacksquare$$

Lemma 6.10. *For $m \in \mathbb{N}$ we have*

$$V_{n,q}(e_m;x) \leq \frac{x}{[n]_q^{m-1}(1+x)_q^{n+1}} + \frac{2^{m-1}x}{q^{m-1}}V_{n+1,q}(e_{m-1};x), \quad x \in \mathbb{R}_+, n \in \mathbb{N},$$

$$K_{n,q}(e_m;x) \leq \frac{q^m}{[m+1]_q[n]_q^m}\frac{1}{(1+x)_q^n} + \frac{2^m(m+1)}{[m+1]_q}V_{n,q}(e_m;x), \quad x \in \mathbb{R}_+, n \in \mathbb{N},$$

Proof. For $k \in \mathbb{N}$ and $0 < q < 1$ the following inequality holds true:

$$1 \leq [k+1]_q \leq 2[k]_q \tag{6.76}$$

Let $m \in \mathbb{N}$. By (6.62) and (6.76), we get

$$\begin{aligned}
V_{n,q}(e_m;x) &= \sum_{k=0}^{\infty} b_{n,k}(q;x)\frac{[k]_q^m}{q^{mk-m}[n]_q^m} \\
&= \sum_{k=1}^{\infty} x b_{n+1,k-1}(q;x)\frac{[k]_q^{m-1}}{q^{(m-1)(k-1)}[n]_q^{m-1}} \\
&= \sum_{k=0}^{\infty} x b_{n+1,k}(q;x)\frac{[k+1]_q^{m-1}}{q^{(m-1)k}[n]_q^{m-1}} \\
&= \frac{x}{[n]_q^{m-1}(1+x)_q^{n+1}} + x\sum_{k=1}^{\infty} b_{n+1,k}(q,x)\frac{[k+1]_q^{m-1}}{q^{(m-1)k}[n]_q^{m-1}} \\
&\leq \frac{x}{[n]_q^{m-1}(1+x)_q^{n+1}} + \frac{2^{m-1}x}{q^{m-1}}V_{n+1,q}(e_{m-1};x).
\end{aligned}$$

$$\begin{aligned}
K_{n,q}(e_m;x) &= [n]_q\sum_{k=0}^{\infty} b_{n,k}(q;x)\int_{q[k]_q/[n]_q}^{[k+1]_q/[n]_q} e_m(q^{-k+1}t)d_qt \\
&= [n]_q\sum_{k=0}^{\infty} b_{n,k}(q;x)\frac{q^{-mk+m}}{[n]_q^{m+1}[m+1]_q}\left([k+1]_q^{m+1} - q^{m+1}[k]_q^{m+1}\right)
\end{aligned}$$

Since

$$\begin{aligned}
[k+1]_q^{m+1} - q^{m+1}[k]_q^{m+1} &= \left([k+1]_q - q[k]_q\right)\left([k+1]_q^m + q[k]_q[k+1]_q^{m-1} + \ldots + q^m[k]_q^m\right) \\
&\leq (m+1)[k+1]_q^m \leq 2^m(m+1)[k]_q^m, \quad k \in \mathbb{N}
\end{aligned}$$

we can write

$$K_{n,q}(e_m;x) \le \frac{q^m}{[m+1]_q [n]_q^m} \frac{1}{(1+x)_q^n} + \frac{2^m(m+1)}{[m+1]_q} V_{n,q}(e_m;x)$$

which completes the proof.   ∎

Remark 6.9. Since any linear and positive operator is monotone, Lemma 6.10 guarantees that $K_{n,q}(f,\cdot) \in B_{\rho_\alpha}(\mathbb{R}_+)$ for each $f \in B_{\rho_\alpha}(\mathbb{R}_+)$, $\alpha \in \mathbb{N}_0$.

Theorem 6.10. *Let $(q_n)_n$ be a sequence satisfying (6.70) and $\alpha \in \mathbb{N}_0$. For all $f \in B_{\rho_\alpha}(\mathbb{R}_+)$, we have*

$$\left\| K_{n,q_n}(f;\cdot) - f \right\|_{\rho_\alpha} \le 3C_\alpha \Omega_{\rho_\alpha}(f;\delta_n),$$

where $\delta_n := \sqrt{\frac{1}{q_n[n]_{q_n}}}$ and C_α is a positive constant.

Proof. The identities (6.66)–(6.68) imply

$$
\begin{aligned}
K_{n,q_n}(\psi_x^2;x) &= \left(\frac{[n+1]_{q_n}}{q_n[n]_{q_n}} - 1 \right) x^2 + \frac{1}{[n]_{q_n}} \left(\frac{q_n(1+[2]_{q_n})+[3]_{q_n}}{[3]_{q_n}} - \frac{2q_n}{[2]_{q_n}} \right) x + \frac{q_n^2}{[3]_{q_n}[n]_{q_n}^2} \\
&\le \frac{1}{q_n[n]_{q_n}} x^2 + \frac{1}{[n]_{q_n}} \left(\frac{[2]_{q_n}q_n+[3]_{q_n}-q_n}{[3]_{q_n}} \right) x + \frac{q_n^2}{[3]_{q_n}[n]_{q_n}^2} \\
&\le \frac{1}{q_n[n]_{q_n}} x^2 + \frac{2}{[n]_{q_n}} x + \frac{1}{[n]_{q_n}^2}.
\end{aligned}
$$

Let $\alpha \ge 0$ be fixed and $f \in B_{\rho_\alpha}(\mathbb{R}_+)$. Based on Theorem 6.9 and the above inequality, we can write

$$
\begin{aligned}
\frac{|K_{n,q_n}(f;x) - f(x)|}{1+x^{2+\alpha}} &\le \sqrt{\frac{K_{n,q_n}(\mu_{x,\alpha}^2;x)}{1+x^{2+\alpha}}} \left(1 + \frac{1}{\delta} \sqrt{\frac{K_{n,q_n}(\psi_x^2;x)}{1+x^{2+\alpha}}} \right) \Omega_{\rho_\alpha}(f;\delta) \\
&\le \sqrt{\frac{K_{n,q_n}(\mu_{x,\alpha}^2;x)}{1+x^{2+\alpha}}} \left(1 + \frac{1}{\delta} \sqrt{\frac{1}{q_n}[n]_{q_n} \|e_2\|_{\rho_\alpha} + \frac{2}{[n]_{q_n}} \|e_2\|_{\rho_\alpha} + \frac{1}{[n]_{q_n}^2}} \right) \\
&\quad \Omega_{\rho_\alpha}(f;\delta).
\end{aligned}
$$

Since $\mu_{x,\alpha}^2 \in B_{\rho_\alpha}(\mathbb{R}_+)$, Lemma 6.10 and Remark 6.9 assure that $K_{n,q_n}(\mu_{x,\alpha}^2;\cdot) \in B_{\rho_\alpha}(\mathbb{R}_+)$. Hence we get

$$\frac{|K_{n,q_n}(f;x) - f(x)|}{1+x^{2+\alpha}} \le C_\alpha \left(1 + \frac{2}{\delta} \sqrt{\frac{1}{q_n[n]_{q_n}}} \right) \Omega_{\rho_\alpha}(f;\delta),$$

where $C_\alpha := \left\| K_{n,q_n}(\mu_{x,\alpha}^2;x) \right\|_{\delta_\alpha}$. Choosing $\delta := \sqrt{\frac{1}{q_n[n]_{q_n}}}$, the proof is finished. ∎

Since $(q_n)_n$ satisfies (6.70), the sequence $(\delta_n)_n$ is statistically null, which yields that $st - \lim_n \Omega_{\rho_\alpha}(f; \delta_n) = 0$. Therefore, Theorem 6.10 gives the rate of statistical convergence of $K_{n,q_n}(f; \cdot)$ to f.

Chapter 7
q-Complex Operators

In the recent years applications of q-calculus in the area of approximation theory and number theory are an active area of research. Several researchers have proposed the q-analogue of exponential, Kantorovich- and Durrmeyer-type operators. Also Kim [106] and [105] used q-calculus in the area of number theory. Recently, Gupta and Wang [94] proposed certain q-Durrmeyer operators in the case of real variables. The aim of this present chapter is to present the recent results [5] on q-Durrmeyer operators to the complex case. The main contributions for the complex operators are due to Sorin G. Gal; in fact, several important results have been complied in his recent monograph [76]. Also very recently, Gal and Gupta [78, 79], and [80] have studied some other complex Durrmeyer-type operators, which are different from the operators considered in the present article.

7.1 Summation-Integral-Type Operators in Compact Disks

In this section we shall study approximation results for the complex q-Durrmeyer operators (introduced and studied in the case of real variable by Gupta–Wang [94]), defined by

$$M_{n,q}(f;z) = [n+1]_q \sum_{k=1}^{n} q^{1-k} p_{n,k}(q;z) \int_0^1 f(t) p_{n,k-1}(q;qt) d_q t + f(0) p_{n,0}(q;z),$$

(7.1)

where $z \in \mathbb{C}, n = 1, 2, \dots; q \in (0,1)$ and $(a-b)_q^m = \Pi_{j=0}^{m-1}(a-q^j b)$, q-Bernstein basis functions are defined as

$$p_{n,k}(q;z) := \begin{bmatrix} n \\ k \end{bmatrix}_q z^k (1-z)_q^{n-k}$$

A. Aral et al., *Applications of q-Calculus in Operator Theory*,
DOI 10.1007/978-1-4614-6946-9_7, © Springer Science+Business Media New York 2013

and also in the above q-beta functions [104] are given as

$$B_q(m,n) = \int_0^1 t^{m-1}(1-qt)_q^{n-1}d_qt, \ m,n > 0.$$

This section is based on [94]. Throughout the present section we use the notation $D_R = \{z \in \mathbb{C} : |z| < R\}$, and by $H(D_R)$, we mean the set of all analytic functions on $f : D_R \to \mathbb{C}$ with $f(z) = \sum_{k=0}^{\infty} a_k z^k$ for all $z \in D_R$. The norm $\|f\|_r = \max\{|f(z)| : |z| \leq r\}$. We denote $\pi_{p,n}(q;z) = M_{n,q}(e_p;z)$ for all $e_p = t^p, p \in \mathbb{N} \cup \{0\}$.

7.1.1 Basic Results

To prove the results of next subsections, we need the following basic results.

Lemma 7.1. *Let* $q \in (0,1)$. *Then,* $\pi_{m,n}(q;z)$ *is a polynomial of degree* $\leq \min(m,n)$, *and*

$$\pi_{m,n}(q;z) = \frac{[n+1]_q!}{[n+m+1]_q!} \sum_{s=1}^{m} c_s(m) \, [n]_q^s B_{n,q}(e_s;z),$$

where $c_s(m) \geq 0$ *are constants depending on* m *and* q, *and* $B_{n,q}(f;z)$ *is the* q *Bernstein polynomials given by* $B_{n,q}(f;z) = \sum_{k=0}^{n} p_{n,k}(q;z) f\left([k]_q/[n]_q\right)$.

Proof. By definition of q-beta function, with $B_q(m,n) = \frac{[m-1]_q![n-1]_q!}{[m+n-1]_q!}$, we have

$$\pi_{m,n}(q;z) = [n+1]_q \sum_{k=1}^{n} q^{1-k} p_{n,k}(q;z) \int_0^1 p_{n,k-1}(q;qt)t^m d_qt$$

$$= [n+1]_q \sum_{k=1}^{n} q^{1-k} p_{n,k}(q;z) \int_0^1 \begin{bmatrix} n \\ k-1 \end{bmatrix}_q (qt)^{k-1}(1-qt)_q^{n-k+1}t^m d_qt$$

$$= [n+1]_q \sum_{k=1}^{n} p_{n,k}(q;z) \frac{[n]_q!}{[k-1]_q![n-k+1]_q!} B_q(k+m, n-k+2)$$

$$= \frac{[n+1]_q!}{[n+m+1]_q!} \sum_{k=1}^{n} p_{n,k}(q;z) \frac{[k+m-1]_q!}{[k-1]_q!}.$$

For $m = 1$, we find

$$\pi_{1,n}(q;z) = \frac{[n+1]_q!}{[n+2]_q!} \sum_{k=1}^{n} p_{n,k}(q;z)[k]_q = \frac{1}{[n+2]_q} \sum_{k=0}^{n} p_{n,k}(q;z)[n]_q \frac{[k]_q}{[n]_q}$$

$$= \frac{1}{[n+2]_q} \sum_{s=1}^{1} [n]_q^s B_{n,q}(e_s;z);$$

thus, the result is true for $m = 1$ with $c_1(1) = 1 > 0$.

Next for $m = 2$, with $[k+1]_q = 1 + q[k]_q$, we get

$$\pi_{2,n}(q;z) = \frac{[n+1]_q!}{[n+3]_q!} \sum_{k=0}^{n} p_{n,k}(q;z)(1 + q[k]_q)[k]_q$$

$$= \frac{[n+1]_q!}{[n+3]_q!} \left[[n]_q B_{n,q}(e_1;z) + q[n]_q^2 B_{n,q}(e_2;z) \right]$$

$$= \frac{[n+1]_q!}{[n+3]_q!} \sum_{s=1}^{2} c_s(2)[n]_q^s B_{n,q}(e_s;z);$$

thus the result is true for $m = 2$ with $c_1(2) = 1 > 0$, $c_2(2) = q > 0$.

Similarly for $m = 3$, using $[k+2]_q = [2]_q + q^2[k]_q$ and $[k+1]_q = 1 + q[k]_q$, we have

$$\pi_{3,n}(q;z) = \frac{[n+1]_q!}{[n+4]_q!} \sum_{s=1}^{3} c_s(3)[n]_q^s B_{n,q}(e_s;z),$$

where $c_1(3) = [2]_q > 0$, $c_2(3) = 2q^2 + q > 0$, and $c_3(3) = q^3 > 0$.

Continuing in this way the result follows immediately for all $m \in N$. ∎

Lemma 7.2. *Let* $q \in (0,1)$. *Then, for all* $m, n \in \mathbb{N}$, *we have the inequality*

$$\frac{[n+1]_q!}{[n+m+1]_q!} \sum_{s=1}^{m} c_s(m) [n]_q^s \leq 1.$$

Proof. By Lemma 7.1, with $e_m = t^m$, we have

$$\pi_{m,n}(q;1) = \frac{[n+1]_q!}{[n+m+1]_q!} \sum_{s=1}^{m} c_s(m) [n]_q^s B_{n,q}(e_s;1) = \frac{[n+1]_q!}{[n+m+1]_q!} \sum_{s=1}^{m} c_s(m) [n]_q^s.$$

Also

$$p_{n,k}(q;z) = \begin{bmatrix} n \\ k \end{bmatrix}_q z^k (1-z)(1-qz)(1-q^2z)\dots(1-q^{n-k-1}z).$$

It immediately follows that $p_{n,k}(q;1) = 0,\quad k = 0,1,2,\ldots,n-1$, and $p_{n,n}(q;1) = 1$. Thus, we obtain

$$\pi_{m,n}(q;1) = [n+1]_q \, p_{n,n}(q;1) \, q^{1-n} \int_0^1 p_{n,n-1}(q;qt) \, t^m d_q t$$

$$= [n+1]_q \int_0^1 [n]_q t^{n+m-1}(1-qt) d_q t$$

$$= [n+1]_q [n]_q \left[\frac{t^{n+m}}{[n+m]_q} - q\frac{t^{n+m+1}}{[n+m+1]_q} \right]_0^1$$

$$= \frac{[n+1]_q [n]_q}{[n+m]_q [n+m+1]_q} \le 1.$$

∎

Corollary 7.1. *Let $r \ge 1$ and $q \in (0,1)$. Then, for all $m,n \in \mathbb{N} \cup \{0\}$ and $|z| \le r$, we have $|\pi_{m,n}(q;z)| \le r^m$.*

Proof. By using the methods [76], p. 61, proof of Theorem 1.5.6, we have $|B_{n,q}(e_s;z)| \le r^s$. By Lemma 7.2 and for all $m \in \mathbb{N}$ and $|z| \le r$,

$$|\pi_{m,n}(q;z)| \le \frac{[n+1]_q!}{[n+m+1]_q!} \sum_{s=1}^m c_s(m) \, [n]_q^s \, |B_{n,q}(e_s;z)|$$

$$\le \frac{[n+1]_q!}{[n+m+1]_q!} \sum_{s=1}^m c_s(m) \, [n]_q^s \, r^s \le r^m.$$

∎

Lemma 7.3. *Let $q \in (0,1)$; then for $z \in \mathbb{C}$, we have the following recurrence relation:*

$$\pi_{p+1,n}(q;z) = \frac{q^p z(1-z)}{[n+p+2]_q} D_q \pi_{p,n}(q;z) + \frac{q^p [n]_q z + [p]_q}{[n+p+2]_q} \pi_{p,n}(q;z).$$

Proof. By simple computation, we have

$$z(1-z) D_q \left(p_{n,k}(q;z) \right) = \left([k]_q - [n]_q z \right) p_{n,k}(q;z)$$

and

$$t(1-qt) D_q \left(p_{n,k-1}(q;qt) \right) = \left([k-1]_q - [n]_q qt \right) p_{n,k-1}(q;qt).$$

Using these identities, it follows that

$$z(1-z)D_q\left(\pi_{p,n}(q;z)\right) = [n+1]_q \sum_{k=1}^{n} q^{1-k}\left([k]_q - [n]_q z\right) p_{n,k}(q;z) \int_0^1 p_{n,k-1}(q;qt)t^p d_q t$$

$$= [n+1]_q \sum_{k=1}^{n} q^{1-k} p_{n,k}(q;z) \int_0^1 \left(1+q[k-1]_q - [n]_q q^2 t + [n]_q q^2 t\right) p_{n,k-1}(q;qt)t^p d_q t$$

$$-z[n]_q[n+1]_q \sum_{k=1}^{n} q^{1-k} p_{n,k}(q;z) \int_0^1 p_{n,k-1}(q;qt)t^p d_q t$$

$$= q[n+1]_q \sum_{k=1}^{n} q^{1-k} p_{n,k}(q;z) \int_0^1 \left(D_q p_{n,k-1}(q;qt)\right) t(1-qt)t^p d_q t$$

$$+ \pi_{p,n}(q;z) + [n]_q q^2 \pi_{p+1,n}(q;z) - z[n]_q \pi_{p,n}(q;z).$$

Let us denote $\delta(t) = \frac{t}{q}(1-t)\left(\frac{t}{q}\right)^p = \frac{1}{q^{p+1}}\left(t^{p+1} - t^{p+2}\right)$. Then, the last q-integral becomes

$$\int_0^1 D_q\left(p_{n,k-1}(q;qt)\right) t(1-qt)t^p d_q t = \int_0^1 D_q\left(p_{n,k-1}(q;qt)\right) \delta(qt) d_q t$$

$$= \delta(t) p_{n,k-1}(q;qt) \mid_0^1 - \int_0^1 p_{n,k-1}(q;qt) D_q \delta(t) d_q t$$

$$= -q^{-p-1} \int_0^1 p_{n,k-1}(q;qt) D_q\left(t^{p+1} - t^{p+2}\right) d_q t$$

$$= -q^{-p-1}[p+1]_q \int_0^1 p_{n,k-1}(q;qt)t^p d_q t$$

$$+ q^{-p-1}[p+2]_q \int_0^1 p_{n,k-1}(q;qt)t^{p+1} d_q t,$$

and hence,

$$z(1-z)D_q\pi_{p,n}(q;z) = -q^{-p}[p+1]_q \pi_{p,n}(q;z) + q^{-p}[p+2]_q \pi_{p+1,n}(q;z)$$

$$+ \pi_{p,n}(q;z) + [n]_q q^2 \pi_{p+1,n}(q;z) - z[n]_q \pi_{p,n}(q;z).$$

Therefore,

$$\pi_{p+1,n}(q;z) = \frac{z(1-z)}{q^{-p}[p+2]_q + [n]_q q^2} D_q\pi_{p,n}(q;z) + \frac{[n]_q z + q^{-p}[p+1]_q - 1}{q^{-p}[p+2]_q + [n]_q q^2} \pi_{p,n}(q;z)$$

$$= \frac{q^p z(1-z)}{[p+2]_q + [n]_q q^{p+2}} D_q\pi_{p,n}(q;z) + \frac{q^p[n]_q z + [p]_q}{[p+2]_q + [n]_q q^{p+2}} \pi_{p,n}(q;z).$$

Finally, using the identity $[p+2]_q + [n]_q q^{p+2} = [n+p+2]_q$, we get the required recurrence relation.   ∎

7.1.2   Upper Bound

If $P_m(z)$ is a polynomial of degree m, then by the Bernstein inequality and the complex mean value theorem, we have

$$\left|D_q P_m(z)\right| \leq \left\|P'_m\right\|_r \leq \frac{m}{r}\left\|P_m\right\|_r \quad \text{for all } |z| \leq r.$$

The following theorem gives the upper bound for the operators (7.1):

Theorem 7.1. *Let* $f(z) = \sum_{p=0}^{\infty} a_p z^p$ *for all* $|z| < R$ *and let* $1 \leq r \leq R$*; then for all* $|z| \leq r$, $q \in (0,1)$ *and* $n \in \mathbb{N}$,

$$\left|M_{n,q}(f;z) - f(z)\right| \leq \frac{K_r(f)}{[n+2]_q},$$

where $K_r(f) = (1+r)\sum_{p=1}^{\infty} |a_p| p(p+1) r^{p-1} < \infty$.

Proof. First we shall show that $M_{n,q}(f;z) = \sum_{p=0}^{\infty} a_p \pi_{p,n}(q;z)$. If we denote $f_m(z) = \sum_{j=0}^{m} a_j z^j$, $|z| \leq r$ with $m \in \mathbb{N}$, then by the linearity of $M_{n,q}$, we have

$$M_{n,q}(f_m;z) = \sum_{p=0}^{m} a_p \pi_{p,n}(q;z).$$

Thus, it suffice to show that for any fixed $n \in \mathbb{N}$ and $|z| \leq r$ with $r \geq 1$, $\lim_{m\to\infty} M_{n,q}(f_m,z) = M_{n,q}(f;z)$. But this is immediate from $\lim_{m\to\infty}\|f_m - f\|_r = 0$ and by the inequality

$$\left|M_{n,q}(f_m;z) - M_{n,q}(f;z)\right|$$

$$\leq |f_m(0) - f(0)| \cdot |(1-z)^n| + [n+1]_q \sum_{k=1}^{n} |p_{n,k}(q;z)| q^{1-k} \int_0^1 p_{n,k-1}(q,qt) |f_m(t) - f(t)| d_q t$$

$$\leq C_{r,n}\|f_m - f\|_r,$$

where

$$C_{r,n} = (1+r)^n + [n+1]_q \sum_{k=1}^{n} \begin{bmatrix} n \\ k \end{bmatrix}_q (1+r)^{n-k} r^k \int_0^1 p_{n,k-1}(q;qt) d_q t.$$

Since, $\pi_{0,n}(q;z) = 1$, we have

$$|M_{n,q}(f;z) - f(z)| \leq \sum_{p=1}^{\infty} |a_p| \cdot |\pi_{p,n}(q;z) - e_p(z)|.$$

Now using Lemma 7.3, for all $p \geq 1$, we find

$$\pi_{p,n}(q;z) - e_p(z) = \frac{q^{p-1}z(1-z)}{[n+p+1]_q} D_q(\pi_{p-1,n}(q;z))$$

$$+ \frac{q^{p-1}[n]_q z + [p-1]_q}{[n+p+1]_q}(\pi_{p-1,n}(q;z) - e_{p-1}(z))$$

$$+ \frac{q^{p-1}[n]_q z + [p-1]_q}{[n+p+1]_q} z^{p-1} - z^p$$

$$= \frac{q^{p-1}z(1-z)}{[n+p+1]_q} D_q(\pi_{p-1,n}(q;z))$$

$$+ \frac{q^{p-1}[n]_q z + [p-1]_q}{[n+p+1]_q}(\pi_{p-1,n}(q;z) - e_{p-1}(z))$$

$$+ \frac{[p-1]_q}{[n+p+1]_q} z^{p-1} + \frac{q^{p-1}[n]_q - [n+p+1]_q}{[n+p+1]_q} z^p.$$

However,

$$\left| \frac{q^{p-1}[n]_q - [n+p+1]_q}{[n+p+1]_q} z^p \right| = \left| \frac{q^{p-1}[n]_q - [p-1]_q - q^{p-1}[n]_q - q^{n+p-1} - q^{n+p}}{[n+p+1]_q} z^p \right|$$

$$\leq \frac{[p+1]_q}{[n+p+1]_q} r^p.$$

Combining the above relations and inequalities, we find

$$|\pi_{p,n}(q;z) - e_p(z)| \leq \frac{r(1+r)}{[n+2]_q} \cdot \frac{p-1}{r} \|\pi_{p-1,n}(q;z)\|_r$$

$$+ r |\pi_{p-1,n}(q;z) - e_{p-1}(z)| + \frac{[p+1]_q}{[n+2]_q} r^{p-1}(1+r)$$

$$\leq \frac{(1+r)(p-1)}{[n+2]_q} r^{p-1} + r |\pi_{p-1,n}(q;z) - e_{p-1}(z)|$$

$$+ \frac{[p+1]_q}{[n+2]_q} r^{p-1} (1+r)$$

$$\leq 2p \frac{(1+r)}{[n+2]_q} r^{p-1} + r \left| \pi_{p-1,n}(q;z) - e_{p-1}(z) \right|.$$

From the last inequality, inductively it follows that

$$\left| \pi_{p,n}(q;z) - e_p(z) \right| \leq r \left(r \left| \pi_{p-2,n}(q;z) - e_{p-2}(z) \right| + \frac{2(p-1)}{[n+2]_q}(1+r) r^{p-2} \right)$$

$$+ 2p \frac{(1+r)}{[n+2]_q} r^{p-1}$$

$$= r^2 \left| \pi_{p-2,n}(q;z) - e_{p-2}(z) \right| + 2 \frac{(1+r)}{[n+2]_q} r^{p-1} (p-1+p)$$

$$\leq \dots \leq \frac{(1+r)}{[n+2]_q} p(p+1) r^{p-1}.$$

Thus, we obtain

$$\left| M_{n,q}(f;z) - f(z) \right| \leq \sum_{p=1}^{\infty} |a_p| \cdot \left| \pi_{p,n}(q;z) - e_p(z) \right| \leq \frac{1+r}{[n+2]_q} \sum_{p=1}^{\infty} |a_p| p(p+1) r^{p-1},$$

which proves the theorem.   ∎

Remark 7.1. Let $q \in (0,1)$ be fixed. As, $\lim_{n\to\infty} \frac{1}{[n+2]_q} = 1-q$, Theorem 7.1 is not a convergence result. To obtain the convergence, one can choose $0 < q_n < 1$ with $q_n \nearrow 1$ as $n \to \infty$. In that case, $\frac{1}{[n+2]_{q_n}} \to 0$ as $n \to \infty$ (see Videnskii [152], formula (2.7)); from Theorem 7.1 we get $M_{n,q_n}(f;z) \to f(z)$, uniformly for $|z| \leq r$ and for any $1 \leq r < R$.

7.1.3 Asymptotic Formula and Exact Order

The following result is the quantitative Voronovskaja-type asymptotic result:

Theorem 7.2. *Suppose that $f \in H(D_R), R > 1$. Then, for any fixed $r \in [1,R]$ and for all $n \in \mathbb{N}, |z| \leq r$ and $q \in (0,1)$, we have*

$$\left| M_{n,q}(f;z) - f(z) - \frac{z(1-z)f''(z) - 2zf'(z)}{[n]_q} \right| \leq \frac{M_r(f)}{[n]_q^2} + 2(1-q) \sum_{k=1}^{\infty} |a_k| k r^k,$$

where $M_r(f) = \sum_{k=1}^{\infty} |a_k| k B_{k,r} r^k < \infty$, and

$$B_{k,r} = (k-1)(k-2)(2k-3) + 8k(k-1)^2 + 6(k-1)k^2 + +4k(k-1)^2(1+r).$$

Proof. In view of the proof of Theorem 7.1, we can write $M_{n,q}(f;z) = \sum_{k=0}^{\infty} a_k$ $\pi_{k,n}(q;z)$. Thus,

$$\left| M_{n,q}(f;z) - f(z) - \frac{z(1-z)f''(z) - 2zf'(z)}{[n]_q} \right|$$

$$\leq \sum_{k=1}^{\infty} |a_k| \left| \pi_{k,n}(q;z) - e_k(z) - \frac{(k(k-1) - k(k+1)z)z^{k-1}}{[n]_q} \right|,$$

for all $z \in D_R, n \in \mathbb{N}$. If we denote

$$E_{k,n}(q;z) = \pi_{k,n}(q;z) - e_k(z) - \frac{(k(k-1) - k(k+1)z)z^{k-1}}{[n]_q},$$

then $E_{k,n}(q;z)$ is a polynomial of degree $\leq k$, and by simple calculation and using Lemma 7.3, we have

$$E_{k,n}(q;z) = \frac{q^{k-1}z(1-z)}{[n+k+1]_q} D_q E_{k-1,n}(q;z) + \frac{q^{k-1}[n]_q z + [k-1]_q}{[n+k+1]_q} E_{k-1,n}(q;z) + X_{k,n}(q;z),$$

where

$$X_{k,n}(q;z) = \frac{z^{k-2}}{[n]_q[n+k+1]_q} \left[q^{k-1}(k-1)(k-2)[k-2]_q + [k-1]_q(k-1)(k-2) \right.$$

$$+ z \left(q^{k-1}[n]_q[k-1]_q - q^{k-1}(k-1)(k-2)[k-2]_q - q^{k-1}k(k-1)[k-1]_q \right.$$

$$\left. + q^{k-1}[n]_q(k-1)(k-2) + [k-1]_q[n]_q - [k-1]_q k(k-1) - k(k-1)[n+k+1]_q \right)$$

$$z^2 \left(k(k+1)[n+k+1]_q - [n]_q[n+k+1]_q - q^{k-1}[n]_q k(k-1) \right.$$

$$\left. \left. + q^{k-1}[n]_q^2 + q^{k-1}k(k-1)[k-1]_q - q^{k-1}[n]_q[k-1]_q \right) \right]$$

$$=: \frac{z^{k-2}}{[n]_q[n+k+1]_q} \left(X_{1,q,n}(k) + zX_{2,q,n}(k) + z^2 X_{3,q,n}(k) \right).$$

Obviously as $0 < q < 1$, it follows that

$$\left| X_{1,q,n}(k) \right| \leq (k-1)(k-2)(2k-3).$$

Next with $[n+k+1]_q = [k-1]_q + q^{k-1}[n]_q + q^{n+k-1} + q^{n+k}$, we have

$$X_{2,q,n}(k) = [n]_q \left(q^{k-1}[k-1]_q + [k-1]_q - 2q^{k-1}(k-1) \right)$$
$$- q^{k-1}(k-1)(k-2)[k-2]_q - q^{k-1}k(k-1)[k-1]_q$$
$$- [k-1]_q k(k-1) - k(k-1)[k-1]_q - k(k-1)q^{n+k-1} - k(k-1)q^{n+k}$$

and

$$[n]_q \left(q^{k-1}[k-1]_q + [k-1]_q - 2q^{k-1}(k-1) \right)$$
$$= [n]_q q^{k-1} \left([k-1]_q - (k-1) \right) + [n]_q \left([k-1]_q - q^{k-1}(k-1) \right)$$
$$= [n]_q q^{k-1} (q-1) \sum_{j=0}^{k-2} [j]_q + [n]_q (1-q) \sum_{j=1}^{k-1} [j]_q q^{k-1-j}$$
$$= q^{k-1} (q^n - 1) \sum_{j=0}^{k-2} [j]_q + (1-q^n) \sum_{j=1}^{k-1} [j]_q q^{k-1-j}.$$

Thus,

$$|X_{2,q,n}(k)| \le (k-1)[k-2]_q + (k-1)[k-1]_q$$
$$+ (k-1)(k-2)[k-2]_q + k(k-1)[k-1]_q + [k-1]_q k(k-1)$$
$$+ k(k-1)[k-1]_q + k(k-1) + k(k-1)$$
$$\le 8k(k-1)^2.$$

Now we will estimate $X_{3,q,n}(k)$:

$$X_{3,q,n}(k) = k(k+1)[n+k+1]_q - [n]_q[n+k+1]_q - q^{k-1}[n]_q k(k-1)$$
$$+ q^{k-1}[n]_q^2 + q^{k-1}k(k-1)[k-1]_q - q^{k-1}[n]_q[k-1]_q$$
$$= k(k+1) \left([k-1]_q + q^{k-1}[n]_q + q^{n+k-1} + q^{n+k} \right)$$
$$- [n]_q \left([k-1]_q + q^{k-1}[n]_q + q^{n+k-1} + q^{n+k} \right) - q^{k-1}[n]_q k(k-1)$$
$$+ q^{k-1}[n]_q^2 + q^{k-1}k(k-1)[k-1]_q - q^{k-1}[n]_q[k-1]_q$$
$$= k(k+1)[k-1]_q + k(k+1) \left(q^{n+k-1} + q^{n+k} \right) - [n]_q[k-1]_q$$
$$- [n]_q \left(q^{n+k-1} + q^{n+k} \right) + 2kq^{k-1}[n]_q$$
$$+ q^{k-1}k(k-1)[k-1]_q - q^{k-1}[n]_q[k-1]_q$$

$$= [n]_q \left(-q^{k-1}[k-1]_q - [k-1]_q + q^{k-1}(2k) - q^{n+k-1} - q^{n+k} \right)$$

$$+ k(k+1)[k-1]_q + k(k+1) \left(q^{n+k-1} + q^{n+k} \right) + q^{k-1}k(k-1)[k-1]_q$$

$$= -[n]_q q^{k-1} \left([k-1]_q - (k-1) \right) + [n]_q q^{k-1}(1-q^n) + [n]_q \left(kq^{k-1} - [k-1]_q - q^{n+k} \right)$$

$$+ k(k+1)[k-1]_q + k(k+1) \left(q^{n+k-1} + q^{n+k} \right) + q^{k-1}k(k-1)[k-1]_q$$

$$= -[n]_q q^{k-1} \left([k-1]_q - (k-1) \right) + [n]_q q^{k-1}(1-q^n) - [n]_q \left([k-1]_q - (k-1)q^{k-1} \right)$$

$$- [n]_q \left(q^{n+k} - q^{k-1} \right) + k(k+1)[k-1]_q + k(k+1) \left(q^{n+k-1} + q^{n+k} \right) + q^{k-1}k(k-1)[k-1]_q$$

$$= -q^{k-1}(q^n - 1) \sum_{j=0}^{k-2} [j]_q - (1-q^n) \sum_{j=1}^{k-1} [j]_q q^{k-1-j} + q^{k-1}(1-q^n)[n]_q$$

$$- [n]_q \left(q^{n+k} - q^{k-1} \right) + k(k+1)[k-1]_q + k(k+1) \left(q^{n+k-1} + q^{n+k} \right) + q^{k-1}k(k-1)[k-1]_q.$$

Hence, it follows that

$$\left| X_{3,q,n}(k) \right| \leq (k-1)[k-2]_q + (k-1)[k-1]_q + (1-q^n)[n]_q$$

$$+ (1-q^{n+1})[n]_q + k(k+1)[k-1]_q + 2k(k+1) + k(k-1)[k-1]_q$$

$$\leq 6(k-1)k^2 + (1-q^n)[n]_q + (1-q^{n+1})[n]_q.$$

Thus,

$$|X_{k,n}(q;z)| \leq \frac{r^{k-2}}{[n]_q^2} \left((k-1)(k-2)(2k-3) + r8k(k-1)^2 + r^26(k-1)k^2 \right)$$

$$+ \frac{r^k}{[n]_q} (1-q^n) + \frac{r^k}{[n+1]_q} (1-q^{n+1})$$

$$= \frac{r^{k-2}}{[n]_q^2} \left((k-1)(k-2)(2k-3) + r8k(k-1)^2 + r^26(k-1)k^2 \right) + 2r^k(1-q)$$

for all $k \geq 1, n \in \mathbb{N}$ and $|z| \leq r$.

Next, using the estimate in the proof of Theorem 7.1, we have

$$\left| \pi_{k,n}(q;z) - e_k(z) \right| \leq \frac{(1+r)k(k+1)r^{k-1}}{[n+2]_q},$$

for all $k, n \in \mathbb{N}, |z| \leq r$, with $1 \leq r$.

Hence, for all $k, n \in \mathbb{N}, k \geq 1$ and $|z| \leq r$, we have

$$|E_{k,n}(q;z)| \leq \frac{q^{k-1}r(1+r)}{[n+k+1]_q}|E'_{k-1,n}(q;z)| + \frac{q^{k-1}[n]_q r + [k-1]_q}{[n+k+1]_q}|E_{k-1,n}(q;z)| + |X_{k,n}(q;z)|.$$

However, since $\frac{q^{k-1}r(1+r)}{[n+k+1]_q} \leq \frac{r(1+r)}{[n+k+1]_q}$ and $\frac{q^{k-1}[n]_q r + [k-1]_q}{[n+k+1]_q} \leq r$, it follows that

$$|E_{k,n}(q;z)| \leq \frac{r(1+r)}{[n+k+1]_q}|E'_{k-1,n}(q;z)| + r|E_{k-1,n}(q;z)| + |X_{k,n}(q;z)|.$$

Now we shall compute an estimate for $|E'_{k-1,n}(q;z)|$, $k \geq 1$. For this, taking into account the fact that $E_{k-1,n}(q;z)$ is a polynomial of degree $\leq k-1$, we have

$$|E'_{k-1,n}(q;z)| \leq \frac{k-1}{r}||E_{k-1,n}||_r$$

$$\leq \frac{k-1}{r}\left[||\pi_{k-1,n} - e_{k-1}||_r + \left|\left|\frac{\{(k-1)(k-2) - k(k-1)e_1\}e_{k-2}]}{[n]_q}\right|\right|_r\right]$$

$$\leq \frac{k(k-1)}{r}\left[\frac{(1+r)(k-1)kr^{k-2}}{[n+2]_q} + \frac{r^{k-2}k(k-1)(1+r)}{[n]_q}\right]$$

$$\leq \frac{k(k-1)^2}{[n]_q}\left[2r^{k-2} + 2r^{k-2}\right] = \frac{4k(k-1)^2 r^{k-2}}{[n]_q}.$$

Thus,

$$\frac{r(1+r)}{[n+k+1]_q}|E'_{k-1,n}(q;z)| \leq \frac{4k(k-1)^2(1+r)r^{k-1}}{[n]_q^2}$$

and

$$|E_{k,n}(q;z)| \leq \frac{4k(k-1)^2(1+r)r^k}{[n]_q^2} + r|E_{k-1,n}(q;z)| + |X_{k,n}(q;z)|,$$

where

$$|X_{k,n}(q;z)| \leq \frac{r^k}{[n]_q^2}A_k + 2r^k(1-q),$$

for all $|z| \leq r, k \geq 1, n \in \mathbb{N}$, where

$$A_k = (k-1)(k-2)(2k-3) + 8k(k-1)^2 + 6(k-1)k^2.$$

Hence, for all $|z| \leq r, k \geq 1, n \in \mathbb{N}$,

$$|E_{k,n}(q;z)| \leq r|E_{k-1,n}(q;z)| + \frac{r^k}{[n]_q^2}B_{k,r} + 2r^k(1-q),$$

where $B_{k,r}$ is a polynomial of degree 3 in k defined as

$$B_{k,r} = A_k + 4k(k-1)^2(1+r).$$

But $E_{0,n}(q;z) = 0$, for any $z \in C$, and therefore by writing the last inequality for $k = 1, 2, \ldots$, we easily obtain step by step the following:

$$|E_{k,n}(q;z)| \le \frac{r^k}{[n]_q^2} \sum_{j=1}^{k} B_{j,r} + 2r^k(1-q) \le \frac{kr^k}{[n]_q^2} B_{k,r} + 2r^k k(1-q).$$

Therefore, we can conclude that

$$\left| M_{n,q}(f;z) - f(z) - \frac{z(1-z)f''(z) - 2zf'(z)}{[n]_q} \right| \le \sum_{k=1}^{\infty} |a_k| |E_{k,n}(q;z)|$$

$$\le \frac{1}{[n]_q^2} \sum_{k=1}^{\infty} |a_k| k B_{k,r} r^k + 2(1-q) \sum_{k=1}^{\infty} |a_k| k r^k.$$

As $f^{(4)}(z) = \sum_{k=4}^{\infty} a_k k(k-1)(k-2)(k-3)z^{k-4}$ and the series is absolutely convergent in $|z| \le r$, it easily follows that $\sum_{k=4}^{\infty} |a_k| k(k-1)(k-2)(k-3)r^{k-4} < \infty$, which implies that $\sum_{k=1}^{\infty} |a_k| k B_{k,r} r^k < \infty$. This completes the proof of theorem. ∎

Remark 7.2. For $q \in (0,1)$ fixed, we have $\frac{1}{[n]_q} \to 1-q$ as $n \to \infty$; thus Theorem 7.2 does not provide convergence. But this can be improved by choosing $1 - \frac{1}{n^2} \le q_n < 1$ with $q_n \nearrow 1$ as $n \to \infty$. Indeed, since in this case $\frac{1}{[n]_{q_n}} \to 0$ as $n \to \infty$ and $1 - q_n \le \frac{1}{n^2} \le \frac{1}{[n]_{q_n}^2}$ from Theorem 7.2, we get

$$\left| M_{n,q_n}(f;z) - f(z) - \frac{z(1-z)f''(z) - 2zf'(z)}{[n]_{q_n}} \right| \le \frac{M_r(f)}{[n]_{q_n}^2} + \frac{2}{[n]_{q_n}^2} \sum_{k=1}^{\infty} |a_k| k r^k.$$

Our next main result is the exact order of approximation for the operator (7.1).

Theorem 7.3. *Let* $1 - \frac{1}{n^2} \le q_n < 1$, $n \in \mathbb{N}$, $R > 1$, *and let* $f \in H(D_R), R > 1$. *If* f *is not a polynomial of degree 0, then for any* $r \in [1, R)$, *we have*

$$\|M_{n,q_n}(f; \cdot) - f\|_r \ge \frac{C_r(f)}{[n]_{q_n}}, \quad n \in \mathbb{N},$$

where the constant $C_r(f) > 0$ *depends on* f, r *and on the sequence* $(q_n)_{n \in \mathbb{N}}$, *but it is independent of* n.

Proof. For all $z \in \mathbb{D}_R$ and $n \in \mathbb{N}$, we have

$$M_{n,q_n}(f;z) - f(z) = \frac{1}{[n]_{q_n}} \left[z(1-z)f''(z) - 2zf'(z) \right.$$

$$\left. + \frac{1}{[n]_{q_n}} \left\{ [n]_{q_n}^2 \left(M_{n,q_n}(f;z) - f(z) - \frac{z(1-z)f''(z) - 2zf'(z)}{[n]_{q_n}} \right) \right\} \right].$$

We use the following property:

$$\|F+G\|_r \geq \left| \|F\|_r - \|G\|_r \right| \geq \|F\|_r - \|G\|_r$$

to obtain

$$\|M_{n,q_n}(f;\cdot) - f\|_r$$

$$\geq \frac{1}{[n]_{q_n}} \left[\|e_1(1-e_1)f'' - 2e_1 f'\|_r \right.$$

$$\left. - \frac{1}{[n]_{q_n}} \left\{ [n]_{q_n}^2 \left\| M_{n,q_n}(f;\cdot) - f - \frac{e_1(1-e_1)f'' - 2e_1 f'}{[n]_{q_n}} \right\|_r \right\} \right].$$

By the hypothesis, f is not a polynomial of degree 0 in D_R; we get $\|e_1(1-e_1)f'' - 2e_1 f'\|_r > 0$. Supposing the contrary, it follows that $z(1-z)f''(z) - 2zf'(z) = 0$ for all $|z| \leq r$, that is, $(1-z)f''(z) - 2f'(z) = 0$ for all $|z| \leq r$ with $z \neq 0$. The last equality is equivalent to $[(1-z)f'(z)]' - f'(z) = 0$, for all $|z| \leq r$ with $z \neq 0$. Therefore, $(1-z)f'(z) - f(z) = C$, where C is a constant, that is, $f(z) = \frac{Cz}{1-z}$, for all $|z| \leq r$ with $z \neq 0$. But since f is analytic in $\overline{D_r}$ and $r \geq 1$, we necessarily have $C = 0$, a contradiction to the hypothesis.

But by Remark 7.2, we have

$$[n]_{q_n}^2 \left\| M_{n,q_n}(f;\cdot) - f - \frac{e_1(1-e_1)f'' - 2e_1 f'}{[n]_{q_n}} \right\|_r \leq M_r(f) + 2\sum_{k=1}^{\infty} |a_k| k r^k,$$

with $\frac{1}{[n]_{q_n}} \to 0$ as $n \to \infty$. Therefore, it follows that there exists an index n_0 depending only on f, r and on the sequence $(q_n)_n$, such that for all $n \geq n_0$, we have

$$\|e_1(1-e_1)f'' - 2e_1 f'\|_r$$

$$- \frac{1}{[n]_{q_n}} \left\{ [n]_{q_n}^2 \left\| M_{n,q_n}(f;\cdot) - f - \frac{e_1(1-e_1)f'' - 2e_1 f'}{[n]_{q_n}} \right\|_r \right\}$$

$$\geq \frac{1}{2} \|e_1(1-e_1)f'' - 2e_1 f'\|_r,$$

which implies that

$$\|M_{n,q_n}(f;\cdot) - f\|_r \geq \frac{1}{2[n]_{q_n}}\|e_1(1 - e_1)f'' - 2e_1f'\|_r, \forall n \geq n_0.$$

For $1 \leq n \leq n_0 - 1$, we clearly have

$$\|M_{n,q_n}(f;\cdot) - f\|_r \geq \frac{c_{r,n}(f)}{[n]_{q_n}},$$

where $c_{r,n}(f) = [n]_{q_n} \cdot \|M_{n,q_n}(f;\cdot) - f\|_r > 0$, which finally implies

$$\|M_{n,q_n}(f;\cdot) - f\|_r \geq \frac{C_r(f)}{[n]_{q_n}}, \text{ for all } n \in \mathbb{N},$$

where

$$C_r(f) = \min\{c_{r,1}(f), c_{r,2}(f)\ldots, c_{r,n_0-1}(f), \frac{1}{2}\|e_1(1 - e_1)f'' - 2e_1f'\|_r\}. \quad \blacksquare$$

Combining Theorem 7.3 with Theorem 7.1, we get the following.

Corollary 7.2. *Let* $1 - \frac{1}{n^2} < q_n < 1$ *for all* $n \in \mathbb{N}$, $R > 1$ *and suppose that* $f \in H(D_R)$. *If* f *is not a polynomial of degree 0, then for any* $r \in [1,R)$, *we have*

$$\|M_{n,q_n}(f;\cdot) - f\|_r \sim \frac{1}{[n]_{q_n}}, \quad n \in \mathbb{N},$$

where the constants in the above equivalence depend on $f, r, (q_n)_n$, *but are independent of* n.

The proof follows along the lines of [80].

Remark 7.3. For $0 \leq \alpha \leq \beta$, we can define the Stancu-type generalization of the operators (7.1) as

$$M_{n,q}^{\alpha,\beta}(f;z) = [n+1]_q \sum_{k=1}^{n} q^{1-k} p_{n,k}(q;z) \int_0^1 f\left(\frac{[n]_q t + \alpha}{[n]_q + \beta}\right) p_{n,k-1}(q;qt)d_q t$$

$$+ f\left(\frac{\alpha}{[n]_q + \beta}\right) p_{n,0}(q;z).$$

The analogous results can be obtained for such operators. As analysis is different, it may be considered elsewhere.

7.2 q-Gauss–Weierstrass Operator

In this section we study a complex q-Gauss–Weierstrass integral operators taking into consideration the operators introduced by Anastassiou and Aral in [17]. We show that these operators are an approximation process in some subclasses of analytic functions giving Jackson-type estimates in approximation. Furthermore, we give q-calculus analogues of some shape-preserving properties for these operators satisfied by classical complex Gauss–Weierstrass integral operators. The results of this section were discussed in [36].

7.2.1 Introduction

In a recent study, Anastassiou and Aral [17] introduced a new q-analogue of Gauss–Weierstrass operators, which for $n \in \mathbb{N}$, $q \in (0, 1)$, $x \in \mathbb{R}$, and $f : \mathbb{R} \to \mathbb{R}$ be a function, defined as

$$W_n(f;q,x) := \frac{\sqrt{[n]_q}(q+1)}{2\Gamma_{q^2}\left(\frac{1}{2}\right)} \int_0^{\frac{2}{\sqrt{[n]_q}\sqrt{1-q^2}}} f(x+t) E_{q^2}\left(-q^2[n]_q\frac{t^2}{4}\right) d_qt. \quad (7.2)$$

The goal of the present section is to introduce complex q-Gauss–Weierstrass operators and to obtain Jackson-type estimates in approximation by these operators. Also, we prove shape-preserving properties and some geometric properties of the operators using q-derivative.

Note that geometric and approximation properties of some complex convolution polynomials, complex singular integrals, and complex variant of well known operators were studied in detail in [76]. Also shape-preserving approximation by real or complex polynomials in one or several variables was given in [75].

Definition 7.1. Let $\mathbb{D} = \{z \in \mathbb{C}; |z| < 1\}$ be the open unit disk and $A\left(\overline{\mathbb{D}}\right) = \{f : \overline{\mathbb{D}} \to \mathbb{C}; \ f \text{ is analytic on } \mathbb{D}, \text{ continuous on } \overline{\mathbb{D}}, f(0) = 0, D_qf(0) = 1\}$. For $\xi > 0$, $q \in (0, 1)$, the complex q-Gauss–Weierstrass integral of $f \in A\left(\overline{\mathbb{D}}\right)$ is defined as

$$W_\xi(f;q,z) := \frac{(q+1)}{\sqrt{\xi}\Gamma_{q^2}\left(\frac{1}{2}\right)} \int_0^{\frac{\sqrt{\xi}}{\sqrt{1-q^2}}} f\left(ze^{-it}\right) E_{q^2}\left(-q^2\frac{t^2}{\xi}\right) d_qt \quad (7.3)$$

for $z \in \overline{\mathbb{D}}$.

Remark 7.4. Noting that the complex q-Gauss–Weierstrass operators $W_\xi(f)(z)$ given by (7.3) can be rewritten via an improper integral, we can easily see that

$E_q\left(-\frac{q^n}{1-q}\right) = 0$ for $n \leq 0$. Thus, we may write

$$\mathcal{W}_\xi\left(f;q,z\right) = \frac{(q+1)}{\sqrt{\xi}\Gamma_{q^2}\left(\frac{1}{2}\right)} \int_0^{\sqrt{(1-q^2)/\xi}} \overset{\infty}{} f\left(ze^{-it}\right) E_{q^2}\left(-q^2\frac{t^2}{\xi}\right) d_q t$$

7.2.2 Approximation Properties

In this section, we obtain Jackson-type rate in approximation by complex operators given (7.3) and global smoothness preservation properties of them.

Lemma 7.4. *We have*

$$\mathcal{W}_\xi\left(1;q,z\right) = 1.$$

Proof. We can write the q-derivative of the equality $t = \sqrt{\xi}\sqrt{u}$ as

$$D_{q^2}(t) = \sqrt{\xi}\frac{\sqrt{u}-\sqrt{q^2 u}}{(1-q^2)u}$$

$$= \sqrt{\xi}\frac{1}{(q+1)\sqrt{u}}. \tag{7.4}$$

Also, using the change of variable formula for q-integral with $\beta = \frac{1}{2}$, we have

$$\int_0^{\frac{\sqrt{\xi}}{\sqrt{1-q^2}}} E_{q^2}\left(-q^2\frac{t^2}{\xi}\right) d_q t = \frac{\sqrt{\xi}}{(q+1)} \int_0^{\frac{1}{1-q^2}} u^{-\frac{1}{2}} E_{q^2}\left(-q^2 u\right) d_{q^2} u$$

$$= \frac{\sqrt{\xi}}{(q+1)}\Gamma_{q^2}\left(\frac{1}{2}\right),$$

which proves $\mathcal{W}_\xi\left(1;q,z\right) = 1$.   ∎

Theorem 7.4. *Let $f \in A\left(\overline{\mathbb{D}}\right)$.*

(i) *For $z \in \overline{\mathbb{D}}$, $\xi \in (0,1]$, we have*

$$\left|\mathcal{W}_\xi\left(f;q,z\right) - f(z)\right| \leq \omega_1\left(f;\sqrt{\xi}\right)_{\partial\mathbb{D}}\left(1 + \frac{1}{\Gamma_{q^2}\left(\frac{1}{2}\right)}\right),$$

where

$$\omega_1\left(f;\xi\right)_{\partial\mathbb{D}} = \sup\left\{\left|f\left(e^{i(x-t)}\right) - f\left(e^{-it}\right)\right|; x \in \mathbb{R}, 0 \leq t \leq \xi\right\}.$$

(*ii*) *We have*

$$\omega_1 \left(\mathcal{W}_\xi \left(f; q, z \right); \delta \right)_{\overline{\mathbb{D}}} \leq C \omega_1 \left(f; \delta \right)_{\overline{\mathbb{D}}}, \forall \delta > 0, \ \xi > 0,$$

where

$$\omega_1 \left(f; \delta \right)_{\overline{\mathbb{D}}} = \sup \left\{ |f(z_1) - f(z_2)|; \ z_1, z_2 \in \overline{\mathbb{D}}, \ |z_1 - z_2| \leq \delta \right\}.$$

Proof.

(*i*) Since $\mathcal{W}_\xi \left(1; q, z \right) = 1$, for $z \in \overline{\mathbb{D}}$, we get

$$\left| \mathcal{W}_\xi \left(f; q, z \right) - f(z) \right| \leq \frac{(q+1)}{\sqrt{\xi} \Gamma_{q^2} \left(\frac{1}{2} \right)} \int_0^{\frac{\sqrt{\xi}}{\sqrt{1-q^2}}} \left| f\left(z e^{-it} \right) - f(z) \right| E_{q^2} \left(-q^2 \frac{t^2}{\xi} \right) d_q t.$$

By the maximum modulus principle we can restrict our considerations to $|z| = 1$, and we can write

$$\left| \mathcal{W}_\xi \left(f; q, z \right) - f(z) \right|$$

$$\leq \frac{(q+1)}{\sqrt{\xi} \Gamma_{q^2} \left(\frac{1}{2} \right)} \int_0^{\frac{\sqrt{\xi}}{\sqrt{1-q^2}}} \omega_1 \left(f; |z| \left| 1 - e^{-it} \right| \right)_{\partial \mathbb{D}} E_{q^2} \left(-q^2 \frac{t^2}{\xi} \right) d_q t$$

Combined this with the inequality

$$|z| \left| 1 - e^{-it} \right| \leq 2 \left| \sin \frac{t}{2} \right| \leq t, \ \forall t > 0,$$

it follows that

$$\left| \mathcal{W}_\xi \left(f; q, z \right) - f(z) \right|$$

$$= \frac{(q+1)}{\sqrt{\xi} \Gamma_{q^2} \left(\frac{1}{2} \right)} \int_0^{\frac{\sqrt{\xi}}{\sqrt{1-q^2}}} \omega_1 \left(f; 2 \left| \sin \frac{t}{2} \right| \right)_{\partial \mathbb{D}} E_{q^2} \left(-q^2 \frac{t^2}{\xi} \right) d_q t$$

$$\leq \frac{(q+1)}{\sqrt{\xi} \Gamma_{q^2} \left(\frac{1}{2} \right)} \int_0^{\frac{\sqrt{\xi}}{\sqrt{1-q^2}}} \omega_1 \left(f; t \right)_{\partial \mathbb{D}} E_{q^2} \left(-q^2 \frac{t^2}{\xi} \right) d_q t$$

$$\leq \frac{(q+1)}{\sqrt{\xi} \Gamma_{q^2} \left(\frac{1}{2} \right)} \int_0^{\frac{\sqrt{\xi}}{\sqrt{1-q^2}}} \left(1 + \frac{t}{\sqrt{\xi}} \right) \omega_1 \left(f; \sqrt{\xi} \right)_{\partial \mathbb{D}} E_{q^2} \left(-q^2 \frac{t^2}{\xi} \right) d_q t$$

$$= \omega_1 \left(f; \sqrt{\xi} \right)_{\partial \mathbb{D}} \left(1 + \frac{(q+1)}{\xi \Gamma_{q^2} \left(\frac{1}{2} \right)} \int_0^{\frac{\sqrt{\xi}}{\sqrt{1-q^2}}} t E_{q^2} \left(-q^2 \frac{t^2}{\xi} \right) d_q t \right).$$

Also, using the change of variable formula for q-integral with $\beta = \frac{1}{2}$, we have

$$\int_0^{\frac{\sqrt{\xi}}{\sqrt{1-q^2}}} t E_{q^2} \left(-q^2 \frac{t^2}{\xi} \right) d_q t = \frac{\xi}{(q+1)} \int_0^{\frac{1}{1-q^2}} E_{q^2} \left(-q^2 u \right) d_{q^2} u$$

$$= \frac{\xi}{(q+1)} \Gamma_{q^2} (1) = \frac{\xi}{(q+1)}.$$

Thus, we have

$$\left| \mathcal{W}_\xi (f; q, z) - f(z) \right| \leq \omega_1 \left(f; \sqrt{\xi} \right)_{\partial \mathbb{D}} \left(1 + \frac{1}{\Gamma_{q^2} \left(\frac{1}{2} \right)} \right).$$

(*ii*) For $z_1, z_2 \in \overline{\mathbb{D}}$, $|z_1 - z_2| \leq \delta$, we have following:

$$\left| \mathcal{W}_\xi (f; q, z_1) - \mathcal{W}_\xi (f; q, z_2) \right|$$

$$\leq \frac{(q+1)}{\sqrt{\xi} \Gamma_{q^2} \left(\frac{1}{2} \right)} \int_0^{\frac{\sqrt{\xi}}{\sqrt{1-q^2}}} \left| f \left(z_1 e^{-it} \right) - f \left(z_2 e^{-it} \right) \right| E_{q^2} \left(-q^2 \frac{t^2}{\xi} \right) d_q t$$

$$\leq \omega_1 \left(f; |z_1 - z_2| \right)_{\overline{\mathbb{D}}} \mathcal{W}_\xi (1; q, z)$$

$$\leq \omega_1 (f; \delta)_{\overline{\mathbb{D}}}.$$

From which, we derive by passing supremum over $|z_1 - z_2| \leq \delta$

$$\omega_1 \left(\mathcal{W}_\xi (f; q, z); \delta \right)_{\overline{\mathbb{D}}} \leq \omega_1 (f; \delta)_{\overline{\mathbb{D}}} \qquad \blacksquare$$

7.2.3 Shape-Preserving Properties

In this section, we deal with some properties of the complex operators given in Definition 7.1. Firstly we present following function classes:

$$S_2 = \left\{ f \text{ is analytic on } \mathbb{D}, \, f(z) = \sum_{k=1}^\infty a_k z^k, \, z \in \mathbb{D}, \, |a_1| \geq \sum_{k=2}^\infty |a_k| \right\},$$

$$S_3^q = \left\{ f \in A \left(\overline{\mathbb{D}} \right); \, \left| D_q^2 f(z) \right| \leq 1, \text{ for all } z \in \mathbb{D} \right\}$$

and

$$\mathfrak{P} = \{f : \overline{\mathbb{D}} \to \mathbb{C}; \ f \ \text{is analytic on} \ \mathbb{D}, \ f(0) = 1, \ Re[f(z)] > 0, \ \forall z \in \mathbb{D}\}.$$

Theorem 7.5. *If* $f(z) = \sum\limits_{k=0}^{\infty} a_k z^k$ *is analytic in* $\mathbb{D}$, *then for* $\xi > 0$, $\mathcal{W}_\xi(f)(z)$ *is analytic in* $\mathbb{D}$, *and we have*

$$\mathcal{W}_\xi(f; q, z) = \sum_{k=0}^{\infty} a_k d_k(\xi, q) z^k, \qquad \forall z \in \mathbb{D}$$

where

$$d_k(\xi, q) = \frac{(q+1)}{\sqrt{\xi} \Gamma_{q^2}\left(\frac{1}{2}\right)} \int_0^{\frac{\sqrt{\xi}}{\sqrt{1-q^2}}} e^{-ikt} E_{q^2}\left(-q^2 \frac{t^2}{\xi}\right) d_q t. \tag{7.5}$$

Also, if f *is continuous on* $\overline{\mathbb{D}}$, *then* $\mathcal{W}_\xi(f)$ *is continuous on* $\overline{\mathbb{D}}$.

Proof. For the continuity at $z_0 \in \overline{\mathbb{D}}$, let $z_n \in \overline{\mathbb{D}}$ be with $z_n \to z_0$ as $n \to \infty$. From (7.3), we can write

$$\left| \mathcal{W}_\xi(f; q, z_n) - \mathcal{W}_\xi(f; q, z_0) \right|$$

$$\leq \frac{(q+1)}{\sqrt{\xi} \Gamma_{q^2}\left(\frac{1}{2}\right)} \int_0^{\frac{\sqrt{\xi}}{\sqrt{1-q^2}}} \left| f(z_n e^{-it}) - f(z_0 e^{-it}) \right| E_{q^2}\left(-q^2 \frac{t^2}{\xi}\right) d_q t$$

$$\leq \frac{(q+1)}{\sqrt{\xi} \Gamma_{q^2}\left(\frac{1}{2}\right)} \int_0^{\frac{\sqrt{\xi}}{\sqrt{1-q^2}}} \omega_1\left(f; |z_n e^{-it} - z_0 e^{-it}|\right)_{\overline{\mathbb{D}}} E_{q^2}\left(-q^2 \frac{t^2}{\xi}\right) d_q t$$

$$= \frac{(q+1)}{\sqrt{\xi} \Gamma_{q^2}\left(\frac{1}{2}\right)} \int_0^{\frac{\sqrt{\xi}}{\sqrt{1-q^2}}} \omega_1\left(f; |z_n - z_0|\right)_{\overline{\mathbb{D}}} E_{q^2}\left(-q^2 \frac{t^2}{\xi}\right) d_q t$$

$$= \omega_1\left(f; |z_n - z_0|\right)_{\overline{\mathbb{D}}},$$

from which the continuity of f at $z_0 \in \overline{\mathbb{D}}$ immediately implies the continuity of $\mathcal{W}_\xi(f)$ too at z_0.

Since $f(z) = \sum\limits_{k=0}^{\infty} a_k z^k$, $z \in \mathbb{D}$, we get

$$\mathcal{W}_\xi(f; q, z) = \frac{(q+1)}{\sqrt{\xi} \Gamma_{q^2}\left(\frac{1}{2}\right)} \int_0^{\frac{\sqrt{\xi}}{\sqrt{1-q^2}}} \sum_{k=0}^{\infty} a_k z^k e^{-ikt} E_{q^2}\left(-q^2 \frac{t^2}{\xi}\right) d_q t$$

$$= \frac{\sqrt{1-q^2}}{\Gamma_{q^2}\left(\frac{1}{2}\right)} \sum_{n=0}^{\infty} \sum_{k=0}^{\infty} a_k z^k e^{-ik \frac{\sqrt{\xi}}{\sqrt{1-q^2}} q^n} E_{q^2}\left(-q^2 \frac{q^{2n}}{1-q^2}\right) q^n. \tag{7.6}$$

If $g_{n,k}$ is absolutely summable, that is, if $\sum_{n=0}^{\infty} \sum_{k=0}^{\infty} |g_{n,k}| < \infty$, then we know from *Fubini's theorem*:

$$\sum_{n=0}^{\infty} \sum_{k=0}^{\infty} g_{n,k} = \sum_{k=0}^{\infty} \sum_{n=0}^{\infty} g_{n,k}.$$

Since

$$\left| a_k e^{-ik \frac{\sqrt{\xi}}{\sqrt{1-q^2}} q^n} \right| = |a_k|,$$

for all $n \in \mathbb{N}$, the series $\sum_{k=0}^{\infty} a_k z^k$ is convergent, and it follows that the series $\sum_{k=0}^{\infty} a_k z^k e^{-ik \frac{\sqrt{\xi}}{\sqrt{1-q^2}} q^n}$ is uniformly convergent with respect to n. Also, we can write

$$\frac{\sqrt{1-q^2}}{\Gamma_{q^2}\left(\frac{1}{2}\right)} \sum_{n=0}^{\infty} E_{q^2}\left(-q^2 \frac{q^{2n}}{1-q^2}\right) q^n = \frac{1}{\Gamma_{q^2}\left(\frac{1}{2}\right)} \int_0^{\frac{1}{1-q^2}} t^{-\frac{1}{2}} E_{q^2}\left(-q^2 t\right) d_{q^2} t$$

$$= 1.$$

These immediately imply that the series in (7.6) can be interchangeable by Fubini's theorem, that is,

$$\mathcal{W}_\xi\left(f;q,z\right) = \frac{\sqrt{1-q^2}}{\Gamma_{q^2}\left(\frac{1}{2}\right)} \sum_{k=0}^{\infty} a_k z^k \sum_{n=0}^{\infty} e^{-ik \frac{\sqrt{\xi}}{\sqrt{1-q^2}} q^n} E_{q^2}\left(-q^2 \frac{q^{2n}}{1-q^2}\right) q^n$$

$$= \sum_{k=0}^{\infty} a_k d_k\left(\xi,q\right) z^k,$$

where

$$d_k\left(\xi,q\right) = \frac{\sqrt{1-q^2}}{\Gamma_{q^2}\left(\frac{1}{2}\right)} \sum_{n=0}^{\infty} e^{-ik \frac{\sqrt{\xi}}{\sqrt{1-q^2}} q^n} E_{q^2}\left(-q^2 \frac{q^{2n}}{1-q^2}\right) q^n$$

$$= \frac{(q+1)}{\sqrt{\xi} \Gamma_{q^2}\left(\frac{1}{2}\right)} \int_0^{\frac{\sqrt{\xi}}{\sqrt{1-q^2}}} e^{-ikt} E_{q^2}\left(-q^2 \frac{t^2}{\xi}\right) d_q t.$$

■

Theorem 7.6. *For $\xi > 0$, it holds that*

$$\mathcal{W}_\xi\left(S_2\right) \subset S_2 \quad \text{and} \quad \mathcal{W}_\xi\left(\mathfrak{P}\right) \subset \mathfrak{P}.$$

Proof. By Theorem 7.5, we get

$$W_\xi \left(f;q,z\right) = \sum_{k=0}^{\infty} a_k d_k \left(\xi,q\right) z^k,$$

and

$$|d_k \left(\xi,q\right)| \leq \frac{(q+1)}{\sqrt{\xi}\Gamma_{q^2}\left(\frac{1}{2}\right)} \int_0^{\frac{\sqrt{\xi}}{\sqrt{1-q^2}}} \left|e^{-ikt}\right| E_{q^2}\left(-q^2\frac{t^2}{\xi}\right) d_q t$$

$$\leq \frac{(q+1)}{\sqrt{\xi}\Gamma_{q^2}\left(\frac{1}{2}\right)} \int_0^{\frac{\sqrt{\xi}}{\sqrt{1-q^2}}} E_{q^2}\left(-q^2\frac{t^2}{\xi}\right) d_q t$$

$$= 1.$$

Since $f \in S_2$, it follows that

$$\sum_{k=2}^{\infty} |a_k d_k \left(\xi,q\right)| \leq \sum_{k=2}^{\infty} |a_k| \leq a_1.$$

Thus we have,

$$W_\xi \left(f\right) \in S_2.$$

Let $f(z) = \sum_{k=0}^{\infty} a_k z^k \in \mathfrak{P}$, that is, $a_0 = f(0) = 1$ and if $f(z) = U(x,y) + iV(x,y)$, $z = x + iy \in \mathbb{D}$, then $U(x,y) > 0$, for all $z = x + iy \in \mathbb{D}$.

We have

$$W_\xi \left(f\right)(0) = a_0 = 1$$

with the condition $a_0 = f(0) = 1$ and for $\forall z = re^{it}$,

$$W_\xi \left(f;q,z\right)$$

$$= \frac{(q+1)}{\sqrt{\xi}\Gamma_{q^2}\left(\frac{1}{2}\right)} \int_0^{\frac{\sqrt{\xi}}{\sqrt{1-q^2}}} U \left(r\cos\left(t-u\right), r\sin\left(t-u\right)\right) E_{q^2}\left(-q^2\frac{u^2}{\xi}\right) d_q u$$

$$+ i\frac{(q+1)}{\sqrt{\xi}\Gamma_{q^2}\left(\frac{1}{2}\right)} \int_0^{\frac{\sqrt{\xi}}{\sqrt{1-q^2}}} V \left(r\cos\left(t-u\right), r\sin\left(t-u\right)\right) E_{q^2}\left(-q^2\frac{u^2}{\xi}\right) d_q u,$$

which implies that

$$Re \left[W_\xi \left(f;q,z\right)\right]$$

$$= \frac{(q+1)}{\sqrt{\xi}\Gamma_{q^2}\left(\frac{1}{2}\right)} \int_0^{\frac{\sqrt{\xi}}{\sqrt{1-q^2}}} U \left(r\cos\left(t-u\right), r\sin\left(t-u\right)\right) E_{q^2}\left(-q^2\frac{u^2}{\xi}\right) d_q u > 0,$$

that is, $W_\xi \left(f;q,z\right) \in \mathfrak{P}$.   ∎

Remark 7.5. By [11], if $f \in S_2$, then f is starlike (and univalent) on $\mathbb{D}$. According to Theorem 7.6, the operators $\mathcal{W}_\xi$ possess this property.

7.2.4 Applications of q-Derivative to Operators

In this section, we present some properties of the complex operators $\mathcal{W}_\xi f(z), \xi > 0$ via q-derivative.

Lemma 7.5. *The $d_k(\xi, q)$ is defined as (7.5). We have*

$$\lim_{\xi \to 0} d_k(\xi, q) = 1.$$

Proof. We can write

$$\lim_{\xi \to 0} d_k(\xi, q) = \lim_{\xi \to 0} \frac{(q+1)}{\sqrt{\xi}\,\Gamma_{q^2}\left(\frac{1}{2}\right)} \int_0^{\frac{\sqrt{\xi}}{\sqrt{1-q^2}}} e^{-ikt} E_{q^2}\left(-q^2 \frac{t^2}{\xi}\right) d_q t$$

$$= \lim_{\xi \to 0} \frac{\sqrt{1-q^2}}{\Gamma_{q^2}\left(\frac{1}{2}\right)} \sum_{n=0}^{\infty} e^{-ik\frac{\sqrt{\xi}}{\sqrt{1-q^2}}q^n} E_{q^2}\left(-\frac{q^2}{1-q^2}q^{2n}\right) q^n.$$

Since the series of above equality is uniform convergent, it follows that the series can be interchangeable with limit, that is,

$$\lim_{\xi \to 0} d_k(\xi, q) = \frac{\sqrt{1-q^2}}{\Gamma_{q^2}\left(\frac{1}{2}\right)} \sum_{n=0}^{\infty} \lim_{\xi \to 0} e^{-i\frac{\sqrt{\xi}}{\sqrt{1-q^2}}q^n} E_{q^2}\left(-\frac{q^2}{1-q^2}q^{2n}\right) q^n$$

$$= \frac{\sqrt{1-q^2}}{\Gamma_{q^2}\left(\frac{1}{2}\right)} \sum_{n=0}^{\infty} E_{q^2}\left(-\frac{q^2}{1-q^2}q^{2n}\right) q^n$$

$$= \frac{1}{\Gamma_{q^2}\left(\frac{1}{2}\right)} \int_0^{\frac{1}{1-q^2}} u^{-\frac{1}{2}} E_{q^2}\left(-q^2 u\right) d_{q^2} u$$

$$= 1. \qquad \blacksquare$$

Theorem 7.7. *For all $\xi > 0$,*

$$\frac{1}{d_1(\xi, q)} \mathcal{W}_\xi\left(S_{3,d_1(\xi,q)}^q\right) \subset S_3^q, \quad \frac{1}{d_1(\xi, q)} \mathcal{W}_\xi\left(S_M^q\right) \subset S_{\frac{M}{d_1(\xi,q)}}^q,$$

where

$$S_{3,d_1(\xi,q)}^q = \left\{f \in S_3^q; \left|D_q^2 f(z)\right| \le d_1(\xi, q)\right\}$$

and

$$S^q_{\frac{M}{d_1(\xi,q)}} = \left\{ f \in S^q_M; \left| D_q f(z) \right| \le \frac{M}{d_1(\xi,q)} \right\}.$$

Proof. Let $f \in S^q_{3,d_1(\xi,q)}$. Since $f \in A\left(\overline{\mathbb{D}}\right)$, we know that $f(0) = a_0 = 0$, $D_q f(0) = a_1 = 1$. Also since $\mathcal{W}_\xi(f;q,z)$ is continuous from Theorem 7.5, we can take q-derivative of it. Thus, we have

$$\frac{1}{d_1(\xi,q)} \mathcal{W}_\xi(f;q,0) = 0, \quad \frac{1}{d_1(\xi,q)} D_q \mathcal{W}_\xi(f;q,0) = a_1 = 1.$$

Also, since

$$D_q^2 \mathcal{W}_\xi(f;q,z) = \frac{(q+1)}{\sqrt{\xi}\,\Gamma_{q^2}\left(\frac{1}{2}\right)} \int_0^{\frac{\sqrt{\xi}}{\sqrt{1-q^2}}} D_q^2 f\left(ze^{-it}\right) e^{-2it} E_{q^2}\left(-q^2\frac{t^2}{\xi}\right) d_q t,$$

and $\left| D_q^2 f(z) \right| \le \left| d_1(\xi,q) \right|$, it follows that

$$\left| \frac{1}{d_1(\xi,q)} D_q^2 \mathcal{W}_\xi(f;q,z) \right|$$

$$\le \frac{(q+1)}{\left| d_1(\xi,q) \right| \sqrt{\xi}\,\Gamma_{q^2}\left(\frac{1}{2}\right)} \int_0^{\frac{\sqrt{\xi}}{\sqrt{1-q^2}}} \left| D_q^2 f\left(ze^{-it}\right) \right| \left| e^{-2it} \right| E_{q^2}\left(-q^2\frac{t^2}{\xi}\right) d_q t$$

$$\le \frac{(q+1)}{\sqrt{\xi}\,\Gamma_{q^2}\left(\frac{1}{2}\right)} \int_0^{\frac{\sqrt{\xi}}{\sqrt{1-q^2}}} E_{q^2}\left(-q^2\frac{t^2}{\xi}\right) d_q t = 1,$$

that is, $\frac{1}{d_1(\xi,q)} \mathcal{W}_\xi(f) \in S^q_3$.

Now, let $f \in S^q_M$, that is, $\left| D_q f(z) \right| \le M$. It follows that

$$\left| \frac{1}{d_1(\xi,q)} D_q \mathcal{W}_\xi(f;q,z) \right|$$

$$\le \frac{(q+1)}{\left| d_1(\xi,q) \right| \sqrt{\xi}\,\Gamma_{q^2}\left(\frac{1}{2}\right)} \int_0^{\frac{\sqrt{\xi}}{\sqrt{1-q^2}}} \left| D_q f\left(ze^{-it}\right) \right| \left| e^{-it} \right| E_{q^2}\left(-q^2\frac{t^2}{\xi}\right) d_q t$$

$$\le \frac{M}{\left| d_1(\xi,q) \right|} \frac{(q+1)}{\sqrt{\xi}\,\Gamma_{q^2}\left(\frac{1}{2}\right)} \int_0^{\frac{\sqrt{\xi}}{\sqrt{1-q^2}}} E_{q^2}\left(-q^2\frac{t^2}{\xi}\right) d_q t = \frac{M}{\left| d_1(\xi,q) \right|},$$

which implies that $\frac{1}{d_1(\xi,q)} \mathcal{W}_\xi(f) \in S^q_{\frac{M}{d_1(\xi,q)}}$. ∎

7.2.5 *Exact Order of Approximation*

For exact order of approximation, we give a modification of the operator (7.3).

For $\xi > 0$, $q \in (0,1)$, the complex q-Gauss–Weierstrass integral of $f \in A\left(\overline{\mathbb{D}}\right)$ is defined as

$$W_{\xi}^{*}\left(f;q,z\right) := \frac{(q+1)}{2\sqrt{\xi}\,\Gamma_{q^2}\left(\frac{1}{2}\right)} \int_{0}^{\frac{\sqrt{\xi}}{1-q^2}} \left(f\left(ze^{-it}\right)+f\left(ze^{it}\right)\right) E_{q^2}\left(-q^2\frac{t^2}{\xi}\right) d_q t$$

for $z \in \overline{\mathbb{D}}$. The approximation properties of the $W_{\xi}^{*}\left(f;q,z\right)$ are expressed by the following theorem.

Theorem 7.8.   (*i*) *Let $f \in A\left(\overline{\mathbb{D}}\right)$. For all $\xi \in (0,1]$ and $z \in \overline{\mathbb{D}}$, it follows*

$$\left|W_{\xi}^{*}\left(f;q,z\right)-f\left(z\right)\right| \le C\omega_2\left(f;\sqrt{\xi}\right)_{\partial\mathbb{D}}$$

(*ii*) *Let us suppose that $f(z) = \sum\limits_{k=0}^{\infty} a_k z^k$ for all $z \in \mathbb{D}_R$, $R > 1$. If f is not constant for*
s $= 0$ *and not a polynomial of degree $\le s-1$ for $s \in \mathbb{N}$, then for all $1 \le r < r_1 < R$, $\xi \in (0;1]$, and $s \in \mathbb{N} \cup \{0\}$*

$$\left\|\left(W_{\xi}^{*}\right)^{(s)}(f)-f^{(s)}\right\|_r \sim \xi$$

where the constants in the equivalence depend only on f, q, p, r, r_1.

Proof. (*i*) We get

$$W_{\xi}^{*}\left(f;q,z\right)-f\left(z\right)$$

$$= \frac{(q+1)}{2\sqrt{\xi}\,\Gamma_{q^2}\left(\frac{1}{2}\right)} \int_{0}^{\frac{\sqrt{\xi}}{1-q^2}} \left(f\left(ze^{-it}\right)-2f\left(z\right)+f\left(ze^{it}\right)\right) E_{q^2}\left(-q^2\frac{t^2}{\xi}\right) d_q t.$$

For $|z| = 1$, we can write

$$\left|W_{\xi}^{*}\left(f;q,z\right)-f\left(z\right)\right|$$

$$\le \frac{(q+1)}{2\sqrt{\xi}\,\Gamma_{q^2}\left(\frac{1}{2}\right)} \int_{0}^{\frac{\sqrt{\xi}}{1-q^2}} \left|f\left(ze^{-it}\right)-2f\left(z\right)+f\left(ze^{it}\right)\right| E_{q^2}\left(-q^2\frac{t^2}{\xi}\right) d_q t$$

$$\le \frac{(q+1)}{2\sqrt{\xi}\,\Gamma_{q^2}\left(\frac{1}{2}\right)} \int_{0}^{\frac{\sqrt{\xi}}{1-q^2}} \omega_2\left(f;t\right)_{\partial\mathbb{D}} E_{q^2}\left(-q^2\frac{t^2}{\xi}\right) d_q t$$

$$\leq \omega_2 \left(f; \sqrt{\xi}\right)_{\partial \mathbb{D}} \frac{(q+1)}{2\sqrt{\xi}\Gamma_{q^2}\left(\frac{1}{2}\right)} \int_0^{\frac{\sqrt{\xi}}{1-q^2}} \left(\frac{t}{\sqrt{\xi}} + 1\right)^2 E_{q^2}\left(-q^2\frac{t^2}{\xi}\right) d_q t.$$

We can write the q-derivative of the equality $t = \sqrt{\xi}\sqrt{u}$ as

$$D_{q^2}(t) = \sqrt{\xi}\frac{\sqrt{u} - \sqrt{q^2 u}}{(1-q^2)u}$$

$$= \sqrt{\xi}\frac{1}{(q+1)\sqrt{u}}.$$

Also, using the change of variable formula for q-integral with $\beta = \frac{1}{2}$, we have

$$\frac{(q+1)}{2\xi\sqrt{\xi}\Gamma_{q^2}\left(\frac{1}{2}\right)} \int_0^{\frac{\sqrt{\xi}}{1-q^2}} t^2 E_{q^2}\left(-q^2\frac{t^2}{\xi}\right) d_q t = \frac{1}{2\Gamma_{q^2}\left(\frac{1}{2}\right)} \int_0^{\frac{1}{1-q^2}} u^{\frac{1}{2}} E_{q^2}\left(-q^2 u\right) d_{q^2} u$$

$$= \frac{\Gamma_{q^2}\left(\frac{3}{2}\right)}{2\Gamma_{q^2}\left(\frac{1}{2}\right)} < \infty$$

and

$$\frac{(q+1)}{\xi\Gamma_{q^2}\left(\frac{1}{2}\right)} \int_0^{\frac{\sqrt{\xi}}{1-q^2}} t E_{q^2}\left(-q^2\frac{t^2}{\xi}\right) d_q t = \frac{1}{\Gamma_{q^2}\left(\frac{1}{2}\right)} \int_0^{\frac{1}{1-q^2}} E_{q^2}\left(-q^2 u\right) d_{q^2} u$$

$$= \frac{1}{\Gamma_{q^2}\left(\frac{1}{2}\right)} < \infty.$$

Thus, we have desired result.

(ii) We follow here the ideas in the proof of [76, pp. 269–272]. We can easily see that for $r \geq 1$,

$$\omega_2\left(f; \sqrt{\xi}\right) \leq C_{r,q}(f)\xi,$$

where

$$\omega_2\left(f; \sqrt{\xi}\right)_{\partial \mathbb{D}_r} = \sup\left\{\Delta_u^2 f\left(re^{it}\right) : |u| < \sqrt{\xi}\right\}.$$

From (i) we have

$$\left\|\mathcal{W}_\xi^*(f) - f\right\|_r \leq C_{r,q}(f)\xi$$

for all $\xi \in (0,1]$ and $z \in \overline{\mathbb{D}_r}$ (see [76]).

Now, we find the upper estimate in (ii) by using the Cauchy's formulas. Let γ be a circle of radius $r_1 > 1$ and center 0. For $u \in \gamma$, we get

$$\left| f^{(s)}(z) - W_{\xi}^{*(s)}(f)(z) \right| = \frac{s!}{2\pi} \left| \int_{\gamma} \frac{f(u) - \left(W_{\xi}^{*} \right)(f)(u)}{(u-z)^{s+1}} du \right|.$$

This equality implies that

$$\left\| D_{q}^{(s)} f - D_{q}^{(s)} W_{\xi}^{*}(f) \right\|_{r} \leq \left\| f^{(s)} - W_{\xi}^{*(s)}(f) \right\|_{r}$$

$$\leq C_{r_1,q}(f) \, \xi \frac{s! r_1}{(r_1 - r)^{s+1}}.$$

For the lower estimate in (*ii*), firstly, let us show the W_{ξ}^{*} operator as series. Using (*i*), for the $W_{\xi}^{*}(f)$ operator, we get

$$W_{\xi}^{*}(f)(z) = \sum_{k=0}^{\infty} a_k d_k^{*}(\xi, q) z^{k},$$

where

$$d_k^{*}(\xi, q) = \frac{(q+1)}{\sqrt{\xi}\,\Gamma_{q^2}\left(\frac{1}{2}\right)} \int_{0}^{\frac{\sqrt{\xi}}{\sqrt{1-q^2}}} \cos(kt) E_{q^2}\left(-q^2 \frac{t^2}{\xi}\right) d_q t.$$

By the mean value theorem applied to $h(t) = \cos kt$ on $[0, t]$, we get

$$|d_k^{*}(\xi, q)| \leq \frac{(q+1)}{\sqrt{\xi}\,\Gamma_{q^2}\left(\frac{1}{2}\right)} \int_{0}^{\frac{\sqrt{\xi}}{\sqrt{1-q^2}}} |\cos kt| E_{q^2}\left(-q^2 \frac{t^2}{\xi}\right) d_q t$$

$$\leq \frac{(q+1)}{\sqrt{\xi}\,\Gamma_{q^2}\left(\frac{1}{2}\right)} \int_{0}^{\frac{\sqrt{\xi}}{\sqrt{1-q^2}}} (1+kt) E_{q^2}\left(-q^2 \frac{t^2}{\xi}\right) d_q t$$

$$= 1 + k \frac{\sqrt{\xi}}{\Gamma_{q^2}\left(\frac{1}{2}\right)}. \tag{7.7}$$

Using q-derivative and taking $z = re^{i\varphi}$, we have

$$\left[D_q^{(s)} f(z) - D_q^{(s)} \left(W_{\xi}^{*} \right)(f)(z) \right] e^{-ip\varphi}$$

$$= \sum_{k=s}^{\infty} a_k [k]_q [k-1]_q \cdots [k-s+1]_q \, r^{k-s} e^{i(k-s-p)\varphi} \left[1 - d_k^{*}(\xi, q) \right].$$

Integrating from $-\pi$ to π, we obtain

$$\frac{1}{2\pi}\int\limits_{-\pi}^{\pi}\left[D_q^{(s)}f(z)-D_q^{(s)}\left(\mathcal{W}_\xi^*\right)(f)(z)\right]e^{-ip\varphi}d\varphi$$

$$= a_{s+p}[s+p]_q[s+p-1]_q\cdots[p+1]_q r^p\left[1-d_{s+p}^*(\xi,q)\right].$$

Then, passing to absolute value and using (7.7), we easily obtain for $\xi\in(0,1]$

$$\left\|D_q^{(s)}f-D_q^{(s)}\left(\mathcal{W}_\xi^*\right)(f)\right\|_r$$

$$\geq\left|a_{s+p}\right|[s+p]_q[s+p-1]_q\cdots[p+1]_q r^p\left|1-d_{s+p}^*(\xi,q)\right|$$

$$\geq\left|a_{s+p}\right|[s+p]_q[s+p-1]_q\cdots[p+1]_q r^p\left|1-\left|d_{s+p}^*(\xi,q)\right|\right|$$

$$\geq\left|a_{s+p}\right|[s+p]_q[s+p-1]_q\cdots[p+1]_q r^p(s+p)\frac{\sqrt{\xi}}{\Gamma_{q^2}\left(\frac{1}{2}\right)}$$

$$\geq\left|a_{s+p}\right|[s+p]_q[s+p-1]_q\cdots[p+1]_q r^p(s+p)\frac{\xi}{\Gamma_{q^2}\left(\frac{1}{2}\right)}.$$

Using this inequality, we have for $p\geq1$ and $\xi\in(0,1]$

$$\left\|f-\mathcal{W}_\xi^*(f)\right\|_r\geq\left|a_p\right|r^p\frac{\xi}{\Gamma_{q^2}\left(\frac{1}{2}\right)}.$$

Thus, we can say that if there exists a subsequence $(\xi_k)_k$ in $(0,1]$ with $\lim_{k\to\infty}\xi_k=0$ and such that $\lim_{k\to\infty}\frac{\left\|f-\mathcal{W}_{\xi_k}^*(f)\right\|_r}{\xi_k}=0$, then $a_p=0$ for all $p\geq1$, that is, f is constant on $\overline{\mathbb{D}}_r$.

Therefore, if f is not constant, then for $\xi\in(0,1]$, there exists a constant $C_{r,q}(f)>0$ such that $\left\|f-\mathcal{W}_\xi^*(f)\right\|_r\geq\xi C_{r,q}(f)$.

Now, we consider $s\geq1$. We can write

$$\left\|D_q^{(s)}f-D_q^{(s)}\left(\mathcal{W}_\xi^*\right)(f)\right\|_r\geq\left|a_{s+p}\right|[s+p]_q[s+p-1]_q\cdots[p+1]_q r^p(s+p)\frac{\xi}{\Gamma_{q^2}\left(\frac{1}{2}\right)}$$

for $\xi\in(0,1]$ and for all $p\geq0$. Similarly, if there exists a subsequence $(\xi_k)_k$ in $(0,1]$ with $\lim_{k\to\infty}\xi_k=0$ and such that $\lim_{k\to\infty}\frac{\left\|D_q^{(s)}f-D_q^{(s)}\left(\mathcal{W}_{\xi_k}^*\right)(f)\right\|_r}{\xi_k}=0$, then $a_{s+p}=0$ for all $p\geq0$, that is, f is a polynomial degree $\leq s-1$ on $\overline{\mathbb{D}}_r$.

Therefore, if f is not a polynomial of degree $\leq s-1$, then for $\xi\in(0,1]$, there exists a constant $C_{r,q}(f)>0$ such that

$$\left\|D_q^{(s)}f-D_q^{(s)}\left(\mathcal{W}_\xi^*\right)(f)\right\|_r\geq\xi C_{r,q}(f).\qquad\blacksquare$$

References

1. U. Abel, V. Gupta, An estimate of the rate of convergence of a Bezier variant of the Baskakov–Kantorovich operators for bounded variation functions. Demons. Math. **36**(1), 123–136 (2003)
2. U. Abel, M. Ivan, Some identities for the operator of Bleimann, Butzer and Hahn involving divided differences. Calcolo **36**, 143–160 (1999)
3. U. Abel, V. Gupta, R.N. Mohapatra, Local approximation by a variant of Bernstein Durrmeyer operators. Nonlinear Anal. Ser. A: Theor. Meth. Appl. **68**(11), 3372–3381 (2008)
4. M. Abramowitz, I.A. Stegun (eds.), *Handbook of Mathematical Functions with Formulas, Graphs and Mathematical Tables*. National Bureau of Standards Applied Mathematics, Series 55, Issued June (Dover, New York, 1964)
5. R.P. Agarwal, V. Gupta, On q-analogue of a complex summation-integral type operators in compact disks. J. Inequal. Appl. **2012**, 111 (2012). doi:10.1186/1029-242X-2012-111
6. O. Agratini, Approximation properties of a generalization of Bleimann, Butzer and Hahn operators. Math. Pannon. **9**, 165–171 (1988)
7. O. Agratini, A class of Bleimann, Butzer and Hahn type operators. An. Univ. Timişoara Ser. Math. Inform. **34**, 173–180 (1996)
8. O. Agratini, Note on a class of operators on infinite interval. Demons. Math. **32**, 789–794 (1999)
9. O. Agratini, Linear operators that preserve some test functions. Int. J. Math. Math. Sci. **8**, 1–11 (2006) [Art ID 94136]
10. O. Agratini, G. Nowak, On a generalization of Bleimann, Butzer and Hahn operators based on q-integers. Math. Comput. Model. **53**(5–6), 699–706 (2011)
11. J.W. Alexander, Functions which map the interior of the unit circle upon simple region. Ann. Math. Sec. Ser. **17**, 12–22 (1915)
12. F. Altomare, R. Amiar, Asymptotic formula for positive linear operators. Math. Balkanica (N.S.) **16**(1–4), 283–304 (2002)
13. F. Altomare, M. Campiti, *Korovkin Type Approximation Theory and Its Application* (Walter de Gruyter Publications, Berlin, 1994)
14. R. Álvarez-Nodarse, M.K. Atakishiyeva, N.M. Atakishiyev, On q-extension of the Hermite polynomials $H_n(x)$ with the continuous orthogonality property on R. Bol. Soc. Mat. Mexicana (3) **8**, 127–139 (2002)
15. G.A. Anastassiou, Global smoothness preservation by singular integrals. Proyecciones **14**(2), 83–88 (1995)
16. G.A. Anastassiou, A. Aral, Generalized Picard singular integrals. Comput. Math. Appl. **57**(5), 821–830 (2009)
17. G.A. Anasstasiou, A. Aral, On Gauss–Weierstrass type integral operators. Demons. Math. **XLIII**(4), 853–861 (2010)

A. Aral et al., *Applications of q-Calculus in Operator Theory*,
DOI 10.1007/978-1-4614-6946-9, © Springer Science+Business Media New York 2013

18. G.A. Anastassiou, S.G. Gal, *Approximation Theory: Moduli of Continuity and Global Smoothness Preservation* (Birkhäuser, Boston, 2000)
19. G.A. Anastassiou, S.G. Gal, Convergence of generalized singular integrals to the unit, univariate case. Math. Inequal. Appl. **3**(4), 511–518 (2000)
20. G.E. Andrews, R. Askey, R. Roy, *Special Functions* (Cambridge University Press, Cambridge, 1999)
21. T.M. Apostal, *Mathematical Analysis* (Addison-Wesley, Reading, 1971)
22. A. Aral, On convergence of singular integrals with non-isotropic kernels. Comm. Fac. Sci. Univ. Ank. Ser. A1 **50**, 88–98 (2001)
23. A. Aral, On a generalized λ-Gauss–Weierstrass singular integrals. Fasc. Math. **35**, 23–33 (2005)
24. A. Aral, On the generalized Picard and Gauss Weierstrass singular integrals. J. Comput. Anal. Appl. **8**(3), 246–261 (2006)
25. A. Aral, A generalization of Szász–Mirakyan operators based on q-integers. Math. Comput. Model. **47**(9–10), 1052–1062 (2008)
26. A. Aral, Pointwise approximation by the generalization of Picard and Gauss–Weierstrass singular integrals. J. Concr. Appl. Math. **6**(4), 327–339 (2008)
27. A. Aral, O. Doğru, Bleimann Butzer and Hahn operators based on q-integers. J. Inequal. Appl. **2007**, 12 pp. (2007) [Article ID 79410]
28. A. Aral, S.G. Gal, q-Generalizations of the Picard and Gauss–Weierstrass singular integrals. Taiwan. J. Math. **12**(9), 2501–2515 (2008)
29. A. Aral, V. Gupta, q-Derivatives and applications to the q-Szász Mirakyan operators. Calcalo **43**(3), 151–170 (2006)
30. A. Aral, V. Gupta, On q-Baskakov type operators. Demons. Math. **42**(1), 109–122 (2009)
31. A. Aral, V. Gupta, On the Durrmeyer type modification of the q Baskakov type operators. Nonlinear Anal.: Theor. Meth. Appl. **72**(3–4), 1171–1180 (2010)
32. A. Aral, V. Gupta, Generalized q-Baskakov operators. Math. Slovaca **61**(4), 619–634 (2011)
33. A. Aral, V. Gupta, Generalized Szász Durrmeyer operators. Lobachevskii J. Math. **32**(1), 23–31 (2011)
34. N.M. Atakishiyev, M.K. Atakishiyeva, A q-analog of the Euler gamma integral. Theor. Math. Phys. **129**(1), 1325–1334 (2001)
35. A. Attalienti, M. Campiti, Bernstein-type operators on the half line. Czech. Math. J. **52**(4), 851–860 (2002)
36. D. Aydin, A. Aral, Some approximation properties of complex q-Gauss–Weierstrass type integral operators in the unit disk. Oradea Univ. Math. J. Tom **XX (1)**, 155–168 (2013)
37. V.A. Baskakov, An example of sequence of linear positive operators in the space of continuous functions. Dokl. Akad. Nauk. SSSR **113**, 249–251 (1957)
38. C. Berg, From discrete to absolutely continuous solution of indeterminate moment problems. Arap. J. Math. Sci. **4**(2), 67–75 (1988)
39. G. Bleimann, P.L. Butzer, L. Hahn, A Bernstein type operator approximating continuous function on the semi-axis. Math. Proc. A **83**, 255–262 (1980)
40. K. Bogalska, E. Gojtka, M. Gurdek, L. Rempulska, The Picard and the Gauss–Weierstrass singular integrals of functions of two variable. Le Mathematiche **LII**(Fasc 1), 71–85 (1997)
41. F. Cao, C. Ding, Z. Xu, On multivariate Baskakov operator. J. Math. Anal. Appl. **307**, 274–291 (2005)
42. E.W. Cheney, *Introduction to Approximation Theory* (McGraw-Hill, New York, 1966)
43. I. Chlodovsky, Sur le développement des fonctions d éfinies dans un interval infini en séries de polynômes de M. S. Bernstein. Compos. Math. **4**, 380–393 (1937)
44. W. Congxin, G. Zengtai, On Henstock integral of fuzzy-number-valued functions, I. Fuzzy Set Syst. **115**(3), 377–391 (2000)
45. O. Dalmanoglu, Approximation by Kantorovich type q-Bernstein operators, in *Proceedings of the 12th WSEAS International Conference on Applied Mathematics*, Cairo, Egypt (2007), pp. 113–117
46. P.J. Davis, *Interpolation and Approximation* (Dover, New York, 1976)

47. M.-M. Derriennic, Sur l'approximation de functions integrable sur $[0,1]$ par des polynomes de Bernstein modifies. J. Approx. Theor. **31**, 323–343 (1981)
48. M.-M. Derriennic, Modified Bernstein polynomials and Jacobi polynomials in q-calculus. Rend. Circ. Mat. Palermo, Ser. II **76**(Suppl.), 269–290 (2005)
49. A. De Sole, V.G. Kac, On integral representation of q-gamma and q-beta functions. AttiAccad. Naz. Lincei Cl. Sci. Fis. Mat. Natur. Rend. Lincei (9) Mat. Appl. **16**(1), 11–29 (2005)
50. R.A. DeVore, G.G. Lorentz, *Constructive Approximation* (Springer, Berlin, 1993)
51. Z. Ditzian, V. Totik, *Moduli of Smoothness* (Springer, New York, 1987)
52. O. Doğru, On Bleimann, Butzer and Hahn type generalization of Balázs operators, Dedicated to Professor D.D. Stancu on his 75th birthday. Studia Univ. Babeş-Bolyai Math. **47**, 37–45 (2002)
53. O. Doğru, O. Duman, Statistical approximation of Meyer–König and Zeller operators based on the q-integers. Publ. Math. Debrecen **68**, 190–214 (2006)
54. O. Dogru, V. Gupta, Monotonocity and the asymptotic estimate of Bleimann Butzer and Hahn operators on q integers. Georgian Math. J. **12**, 415–422 (2005)
55. O. Dogru, V. Gupta, Korovkin type approximation properties of bivariate q-Meyer König and Zeller operators. Calcolo **43**, 51–63 (2006)
56. D. Dubois, H. Prade, Fuzzy numbers: an overview, in *Analysis of Fuzzy Information, vol.1: Mathematics and Logic* (CRC Press, Boca Raton, 1987), pp. 3–39
57. O. Duman, C. Orhan, Statistical approximation by positive linear operators. Stud. Math. **161**, 187–197 (2006)
58. J.L. Durrmeyer, Une formule d'inversion de la Transformee de Laplace, Applications a la Theorie des Moments, These de 3e Cycle, Faculte des Sciences de l' Universite de Paris, 1967
59. T. Ernst, The history of q-calculus and a new method, U.U.D.M Report 2000, 16, ISSN 1101-3591, Department of Mathematics, Upsala University, 2000
60. S. Ersan, O. Doğru, Statistical approximation properties of q-Bleimann, Butzer and Hahn operators. Math. Comput. Model. **49**(7–8), 1595–1606 (2009)
61. Z. Finta, N.K. Govil, V. Gupta, Some results on modified Szász–Mirakyan operators. J. Math. Anal. Appl. **327**(2), 1284–1296 (2007)
62. Z. Finta, V. Gupta, Approximation by q Durrmeyer operators. J. Appl. Math. Comput. **29**(1–2), 401–415 (2009)
63. Z. Finta, V. Gupta, Approximation properties of q-Baskakov operators. Cent. Eur. J. Math. **8**(1), 199–211 (2009)
64. J.A. Friday, H.I. Miller, A matrix characterization of statistical convergence. Analysis **11**, 59–66 (1991)
65. A.D. Gadzhiev, The convergence problem for a sequence of positive linear operators on unbounded sets and theoems analogous to that of P.P. Korovkin. Dokl. Akad. Nauk. SSSR **218**(5), 1001–1004 (1974) [in Russian]; Sov. Math. Dokl. **15**(5), 1433–1436 (1974) [in English]
66. A.D. Gadjiev, On Korovkin type theorems. Math. Zametki **20**, 781–786 (1976) [in Russian]
67. A.D. Gadjiev, A. Aral, The weighted L_p-approximation with positive operators on unbounded sets, Appl. Math. Letter **20**(10), 1046–1051 (2007)
68. A.D. Gadjiev, Ö. Çakar, On uniform approximation by Bleimann, Butzer and Hahn operators on all positive semi-axis. Trans. Acad. Sci. Azerb. Ser. Phys. Tech. Math. Sci. **19**, 21–26 (1999)
69. A.D. Gadjiev, Orhan, Some approximation theorems via statistical convergence. Rocky Mt. J. Math. **32**(1), 129–138 (2002)
70. A.D. Gadjiev, R.O. Efendiev, E. Ibikli, Generalized Bernstein Chlodowsky polynomials. Rocky Mt. J. Math. **28**(4), 1267–1277 (1998)
71. A.D. Gadjiev, R.O. Efendiyev, E. İbikli, On Korovkin type theorem in the space of locally integrable functions. Czech. Math. J. **53**(128)(1), 45–53 (2003)
72. S.G. Gal, Degree of approximation of continuous function by some singular integrals. Rev. Anal. Numér. Théor. Approx. **XXVII**(2), 251–261 (1998)

73. S.G. Gal, Remark on degree of approximation of continuous function by some singular integrals. Math. Nachr. **164**, 197–199 (1998)

74. S.G. Gal, Remarks on the approximation of normed spaces valued functions by some linear operators, in *Mathematical Analysis and Approximation Theory* (Mediamira Science Publisher, Cluj-Napoca, 2005), pp. 99–109

75. S.G. Gal, *Shape-Preserving Approximation by Real and Complex Polynomials* (Birkhäuser, Boston, 2008)

76. S.G. Gal, *Approximation by Complex Bernstein and Convolution-Type Operators* (World Scientific, Singapore, 2009)

77. S.G. Gal, V. Gupta, Approximation of vector-valued functions by q-Durrmeyer operators with applications to random and fuzzy approximation. Oradea Univ. Math. J. **16**, 233–242 (2009)

78. S.G. Gal, V. Gupta, Approximation by a Durrmeyer-type operator in compact disk. Ann. Univ. Ferrara **57**, 261–274 (2011)

79. S.G. Gal, V. Gupta, Quantative estimates for a new complex Durrmeyer operator in compact disks. Appl. Math. Comput. **218**(6), 2944–2951 (2011)

80. S. Gal, V. Gupta, N.I. Mahmudov, Approximation by a complex q Durrmeyer type operators. Ann. Univ. Ferrara **58**(1), 65–87 (2012)

81. G. Gasper, M. Rahman, *Basic Hypergeometrik Series*. Encyclopedia of Mathematics and Its Applications, vol. 35 (Cambridge University Press, Cambridge, 1990)

82. T.N.T. Goodman, A. Sharma, A Bernstein type operators on the simplex. Math. Balkanica **5**, 129–145 (1991)

83. N.K. Govil, V. Gupta, Convergence rate for generalized Baskakov type operators. Nonlinear Anal. **69**(11), 3795–3801 (2008)

84. N.K. Govil, V. Gupta, Convergence of q-Meyer–König-Zeller–Durrmeyer operators. Adv. Stud. Contemp. Math. **19**, 97–108 (2009)

85. V. Gupta, Rate of convergence by the Bézier variant of Phillips operators for bounded variation functions. Taiwan. J. Math. **8**(2), 183–190 (2004)

86. V. Gupta, Some approximation properties on q-Durrmeyer operators. Appl. Math. Comput. **197**(1), 172–178 (2008)

87. V. Gupta, A. Aral, Convergence of the q analogue of Szász-Beta operators. Appl. Math. Comput. **216**, 374–380 (2010)

88. V. Gupta, A. Aral, Some approximation properties of q Baskakov Durrmeyer operators. Appl. Math. Comput. **218**(3), 783–788 (2011)

89. V. Gupta, Z. Finta, On certain q Durrmeyer operators. Appl. Math. Comput. **209**, 415–420 (2009)

90. V. Gupta, M.A. Noor, Convergence of derivatives for certain mixed Szász-Beta operators. J. Math. Anal. Appl. **321**, 1–9 (2006)

91. V. Gupta, C. Radu, Statistical approximation properties of q Baskakov Kantorovich operators. Cent. Eur. J. Math. **7**(4), 809–818 (2009)

92. V. Gupta, H. Sharma, Recurrence formula and better approximation for q Durrmeyer operators. Lobachevskii J. Math. **32**(2), 140–145 (2011)

93. V. Gupta, G.S. Srivastava, On the rate of convergence of Phillips operators for functions of bounded variation. Comment. Math. **XXXVI**, 123–130 (1996)

94. V. Gupta, H. Wang, The rate of convergence of q-Durrmeyer operators for $0 < q < 1$. Math. Meth. Appl. Sci. **31**(16), 1946–1955 (2008)

95. V. Gupta, T. Kim, J. Choi, Y.-H. Kim, Generating functions for q-Bernstein, q-Meyer–König–Zeller and q-Beta basis. Autom. Comput. Math. **19**(1), 7–11 (2010)

96. W. Heping, Korovkin-type theorem and application. J. Approx. Theor. **132**, 258–264 (2005)

97. W. Heping, Properties of convergence for the q-Meyer–König and Zeller operators. J. Math. Anal. Appl. **335**(2), 1360–1373 (2007)

98. W. Heping, Properties of convergence for ω, q Bernstein polynomials. J. Math. Anal. Appl. **340**(2), 1096–1108 (2008)

99. W. Heping, F. Meng, The rate of convergence of q-Bernstein polynomials for $0 < q < 1$. J. Approx. Theor. **136**, 151–158 (2005)

100. W. Heping, X. Wu, Saturation of convergence of q-Bernstein polynomials in the case $q \geq 1$. J. Math. Anal. Appl. **337**(1), 744–750 (2008)
101. V.P. Il'in, O.V. Besov, S.M. Nikolsky, *The Integral Representation of Functions and Embedding Theorems* (Nauka, Moscow, 1975) [in Russian]
102. A. Il'inski, S. Ostrovska, Convergence of generalized Bernstein polynomials. J. Approx. Theor. **116**, 100–112 (2002)
103. F.H. Jackson, On a q-definite integrals. Q. J. Pure Appl. Math. **41**, 193–203 (1910)
104. V. Kac, P. Cheung, *Quantum Calculus* (Springer, New York, 2002)
105. T. Kim, q-Generalized Euler numbers and polynomials. Russ. J. Math. Phys. **13**(3), 293–298 (2006)
106. T. Kim, Some identities on the q-integral representation of the product of several q-Bernstein-type polynomials. Abstr. Appl. Anal. **2011**, 11 pp. (2011). doi:10.1155/2011/634675 [Article ID 634675]
107. T.H. Koornwinder, q-Special functions, a tutorial, in *Deformation Theory and Quantum Groups with Applications to Mathematical Physics*, ed. by M. Gerstenhaber, J. Stasheff. Contemporary Mathematics, vol. 134 (American Mathematical Society, Providence, 1992)
108. S.L. Lee, G.M. Phillips, Polynomial interpolation at points of a geometric mesh on a triangle. Proc. R. Soc. Edinb. **108A**, 75–87 (1988)
109. B. Lenze, Bernstein–Baskakov–Kantorovich operators and Lipscitz type maximal functions, in *Approximation Theory* (Kecskemét, Hungary). Colloq. Math. Soc. János Bolyai, vol. 58 (1990), pp. 469–496
110. A. Lesniewicz, L. Rempulska, J. Wasiak, Approximation properties of the Picard singular integral in exponential weighted spaces. Publ. Mat. **40**, 233–242 (1996)
111. Y.-C. Li, S.-Y. Shaw, A proof of Hölder's inequality using the Cauchy–Schwarz inequality. J. Inequal. Pure Appl. Math. **7**(2), 1–3 (2006) [Article 62]
112. A.-J. López-Moreno, Weighted silmultaneous approximation with Baskakov type operators. Acta Math. Hung. **104**, 143–151 (2004)
113. G.G. Lorentz, *Bernstein Polynomials*. Math. Expo., vol. 8 (University of Toronto Press, Toronto, 1953)
114. G.G. Lorentz, *Approximation of Functions* (Holt, Rinehart and Wilson, New York, 1966)
115. L. Lupas, A property of S. N. Bernstein operators. Mathematica (Cluj) **9**(32), 299–301 (1967)
116. L. Lupas, On star shapedness preserving properties of a class of linear positive operators. Mathematica (Cluj) **12**(35), 105–109 (1970)
117. A. Lupas, A q-analogue of the Bernstein operator, in *Seminar on Numerical and Statistical Calculus* (Cluj-Napoca, 1987), pp. 85–92. Preprint, 87-9 Univ. Babes-Bolyai, Cluj. MR0956939 (90b:41026)
118. N. Mahmoodov, V. Gupta, H. Kaffaoglu, On certain q-Phillips operators. Rocky Mt. J. Math. **42**(4), 1291–1312 (2012)
119. N.I. Mahmudov, On q-parametric Szász–Mirakjan operators. Mediterr. J. Math. **7**(3), 297–311 (2010)
120. N.I. Mahmudov, P. Sabancıgil, q-Parametric Bleimann Butzer and Hahn operators. J. Inequal. Appl. **2008**, 15 pp. (2008) [Article ID 816367]
121. N.I. Mahmudov, P. Sabancigil, On genuine q-Bernstein–Durrmeyer operators. Publ. Math. Debrecen **76**(4) (2010)
122. G. Mastroianni, A class of positive linear operators. Rend. Accad. Sci. Fis. Mat. Napoli **48**, 217–235 (1980)
123. C.P. May, On Phillips operator. J. Approx. Theor. **20**(4), 315–332 (1977)
124. I. Muntean, *Course of Functional Analysis. Spaces of Linear and Continuous Mappings*, vol. II (Faculty of Mathematics, Babes-Bolyai' University Press, Cluj-Napoca, 1988) [in Romanian]
125. S. Ostrovska, q-Bernstein polynomials and their iterates. J. Approx. Theor. **123**, 232–255 (2003)
126. S. Ostrovska, On the improvement of analytic properties under the limit q-Bernstein operator. J. Approx. Theor. **138**(1), 37–53 (2006)

127. S. Ostrovska, On the Lupas q-analogue of the Bernstein operator. Rocky Mt. J. Math. **36**(5), 1615–1629 (2006)
128. S. Ostrovska, The first decade of the q-Bernstein polynomials: results and perspectives. J. Math. Anal. Approx. Theor. **2**, 35–51 (2007)
129. S. Ostrovska, The sharpness of convergence results for q Bernstein polynomials in the case $q > 1$. Czech. Math. J. **58**(133), 1195–1206 (2008)
130. S. Ostrovska, On the image of the limit q Bernstein operator. Math. Meth. Appl. Sci. **32**(15), 1964–1970 (2009)
131. S. Pethe, On the Baskakov operator. Indian J. Math. **26**(1–3), 43–48 (1984)
132. G.M. Phillips, On generalized Bernstein polynomials, in *Numerical Analysis*, ed. by D.F. Griffiths, G.A. Watson (World Scientific, Singapore, 1996), pp. 263–269
133. G.M. Phillips, Bernstein polynomials based on the q- integers, The heritage of P.L. Chebyshev: A Festschrift in honor of the 70th-birthday of Professor T. J. Rivlin. Ann. Numer. Math. **4**, 511–518 (1997)
134. G.M. Phillips, *Interpolation and Approximation by Polynomials* (Springer, Berlin, 2003)
135. R.S. Phillips, An inversion formula for Laplace transforms and semi-groups of linear operators. Ann. Math. Sec. Ser. **59**, 325–356 (1954)
136. C. Radu, On statistical approximation of a general class of positive linear operators extended in q-calculus. Appl. Math. Comput. **215**(6), 2317–2325 (2009)
137. P.M. Rajković, M.S. Stanković, S.D. Marinkovic, Mean value theorems in q-calculus. Math. Vesnic. **54**, 171–178 (2002)
138. T.J. Rivlin, *An Introduction to the Approximation of Functions* (Dover, New York, 1981)
139. A. Sahai, G. Prasad, On simultaneous approximation by modified Lupaş operators. J. Approx. Theor. **45**, 122–128 (1985)
140. I.J. Schoenberg, On polynomial interpolation at the points of a geometric progression. Proc. R. Soc. Edinb. **90A**, 195–207 (1981)
141. K. Seip, A note on sampling of bandlinited stochastic processes. IEEE Trans. Inform. Theor. **36**(5), 1186 (1990)
142. R.P. Sinha, P.N. Agrawal, V. Gupta, On simultaneous approximation by modified Baskakov operators. Bull. Soc. Math. Belg. Ser. B **43**(2), 217–231 (1991)
143. Z. Song, P. Wang, W. Xie, Approximation of second-order moment processes from local averages. J. Inequal. Appl. **2009**, 8 pp. (2009) [Article Id 154632]
144. H.M. Srivastava, V. Gupta, A certain family of summation-integral type operators. Math. Comput. Model. **37**, 1307–1315 (2003)
145. D.D. Stancu, Approximation of functions by a new class of linear polynomial operators. Rev. Roumanine Math. Pures Appl. **13**, 1173–1194 (1968)
146. E.M. Stein, G. Weiss, *Singular Integrals and Differentiability Properties of Functions* (Princeton University Press, Princeton, 1970)
147. E.M. Stein, G. Weiss, *Introduction to Fourier Analysis on Euclidean Spaces* (Princeton University Press, Princeton, 1971)
148. O. Szász, Generalization of S. Bernstein's polynomials to infinite interval. J. Research Nat. Bur. Stand. **45**, 239–245 (1959)
149. J. Thomae, Beitrage zur Theorie der durch die Heinsche Reihe. J. Reine. Angew. Math. **70**, 258–281 (1869)
150. T. Trif, Meyer, König and Zeller operators based on the q-integers. Rev. Anal. Numér. Théor. Approx. **29**, 221–229 (2002)
151. V.S. Videnskii, On some class of q-parametric positive operators, Operator Theory: Advances and Applications **158**, 213–222 (2005)
152. V.S. Videnskii, On q-Bernstein polynomials and related positive linear operators, in *Problems of Modern Mathematics and Mathematical Education* Hertzen readings (St.-Petersburg, 2004), pp. 118–126 [in Russian]
153. I. Yuksel, N. Ispir, Weighted approximation by a certain family of summation integral-type operators. Comput. Math. Appl. **52**(10–11), 1463–1470 (2006)
154. L.A. Zadeh, Fuzzy sets. Inform. Contr. **8**, 338–353 (1965)

Index

Printed in the United States
By Bookmasters